Carbon-Based Electrochemical Materials for Energy Storage

Carbon-Based Electrochemical Materials for Energy Storage

Guest Editor

Huicong Xia

MDPI

Basel • Beijing • Wuhan • Barcelona • Belgrade • Novi Sad • Cluj • Manchester

Guest Editor
Huicong Xia
College of Materials Science
and Engineering
Zhengzhou University
Zhengzhou
China

Editorial Office
MDPI AG
Grosspeteranlage 5
4052 Basel, Switzerland

This is a reprint of the Special Issue, published open access by the journal *Molecules* (ISSN 1420-3049), freely accessible at: https://www.mdpi.com/journal/molecules/special_issues/2GXMJ514VX.

For citation purposes, cite each article independently as indicated on the article page online and as indicated below:

Lastname, A.A.; Lastname, B.B. Article Title. *Journal Name* **Year**, *Volume Number*, Page Range.

ISBN 978-3-7258-6448-5 (Hbk)
ISBN 978-3-7258-6449-2 (PDF)
https://doi.org/10.3390/books978-3-7258-6449-2

Contents

Preface

In recent years, the demand for high-performance power sources has significantly shaped global research efforts aimed at developing advanced energy storage materials and devices. The ongoing need for efficient and sustainable energy solutions has prompted scientists and engineers to explore innovative technologies that can effectively address the challenges associated with energy storage. This Special Issue, titled "Carbon-Based Electrochemical Materials for Energy Storage," seeks to illuminate the latest advancements and trends in carbon-based electrode materials, which are critical to the development of next-generation batteries, supercapacitors, and fuel cells.

The aim of this Special Issue is to consolidate and disseminate state-of-the-art research on carbon-based materials, which are increasingly recognized for their potential to revolutionize energy storage applications. Carbon-based materials have garnered extensive attention due to their abundance, cost-effectiveness, non-toxic nature, and diverse array of electrochemical properties. These characteristics render them highly appealing as electrode materials in various energy storage systems.

Huicong Xia
Guest Editor

molecules

Article

Atypical Analysis of a Graphite-Based Anode Prepared Using Aqueous Processes

Jun-Yi Liao [1], Chia-Chin Chang [2,3], Yuh-Lang Lee [1,*] and Ten-Chin Wen [1,*]

1 Department of Chemical Engineering, National Cheng Kung University, Tainan 70101, Taiwan; morley23133463@gmail.com
2 Department of Greenergy, National University of Tainan, Tainan 70005, Taiwan; ccchang@mail.nutn.edu.tw
3 Graduate Institute of Energy and Sustainability Technology, National Taiwan University of Science and Technology, Taipei City 10607, Taiwan
* Correspondence: yllee@mail.ncku.edu.tw (Y.-L.L.); tcwen@mail.ncku.edu.tw (T.-C.W.)

Abstract

In order to form a solid electrolyte interphase (SEI) layer using aqueous processes, a graphite anode called MG-AQP was designed by wrapping and crosslinking graphite particles with aqueous composites (AQCs), which contained zwitterionic polymer, zwitterion molecules, and lithium salts. First, MG-AQP was used to fabricate a full lithium-ion battery (LIB) cell with $Li[Ni_{0.8}Mn_{0.1}Co_{0.1}]O_2$ (NMC811) as the cathode, denoted as LIB-MG-AQP//NMC811, to demonstrate its performance via a 0.5 C-rate break-in and 1 C-rate cycling. Accordingly, this showed that LIB-MG-AQP exhibits outstanding cyclic stability. To evaluate its electrochemical performance, MG-AQP and lithium metal were used to fabricate a half cell named LIBs-MG-AQP. According to the initial cyclic voltammetry curve, almost no surface reaction for forming an SEI layer exists in LIBs-MG-AQP, illustrating its high initial coulombic efficiency of 92% at a 0.5 C-rate break-in. These outstanding results are due to the fact that the AQC has fewer cracks, thus blocking solvent molecules from passing from the electrolyte into the graphite anode. This study provides new insights to optimize graphite anodes via 0.5 C-rate break-in rather than conventional SEI formation to save time and energy.

Keywords: atypical analysis; graphite anode; aqueous process; break-in

1. Introduction

Conventionally, graphite is used as an anode material for lithium-ion batteries (LIBs) due to its availability, stability, and good conductivity [1–6]. Carboxymethyl cellulose (CMC), commonly employed as a binder for graphite anodes, provides high viscosity and strong adhesion, enabling uniform slurry dispersion and improved particle cohesion [7–9]. However, there are still some drawbacks with the usage of CMC as a binder for graphite anodes [10].

For instance, under formation processes, many side products are formed, such as LiF, Li_2CO_3, LiOH, $ROCO_2Li$, and ROLi in solid electrolyte interface (SEI) layers, as well as the generation of C_2H_4, CO_2, H_2, and HF gases due to irreversible reactions between lithium ions, solvents, and anionic molecules with electron transfer [11–14]. Consequently, a significant number of lithium ions are consumed during the initial process. Moreover, the formation of weak SEI layers occurs in graphite anodes with the application of conventional electrolytes, as reported in some studies [15–18]. Alternatively, graphite anodes prepared with a CMC binder exhibit pronounced surface cracks, which can be attributed to the tendency of linear CMC polymers to aggregate into clusters during the drying process, leading to portions of the graphite surface becoming exposed and cracked. In this case, under high C-rates, lithium ions favor the low resistance pathways provided by the cracks instead of traveling through the intact electrode surfaces. Nevertheless, the weakness of the SEI layer becomes so severe that many lithium ions are consumed in the reforming of the SEI layer rather than contributing to reversible energy storage, resulting in poor performance and cyclic stability. Generally, in the conventional design of graphite anodes, it is necessary to form an SEI layer at 0.1 C-rate. Also demonstrated in previous studies is the fact that due to the crystallization properties of CMC at low temperatures, lithium ions with extremely high resistances travel from the electrolyte to the graphite surface to undergo energy storage. The above statements show the many drawbacks of CMC being used for graphite anodes.

To address the limitations of conventional binders in graphite anodes, in our previous studies, CMC was reacted with sulfobetaine methacrylate (SBMA) as a zwitterionic monomer, and zwitterion molecules were added to obtain a product denoted as CMC-SBMA[a], which was applied as a graphite binder [19,20]. It has been observed that zwitterion molecules can improve the formation of a strong SEI layer comprising rich LiF. Moreover, zwitterion molecules can enhance the transportation of lithium ions with high diffusivities and low resistances. The strong interactions between the zwitterion molecules can reduce the crystallization tendency of CMC-SBMA[a], enabling this binder to deliver excellent performance at low temperatures. Thus, the addition of zwitterion molecules provides superior performance and cyclic stability in graphite anodes.

However, it is inevitable that some cracks occur on graphite anodes produced using CMC-SBMA[a] as a binder due to the linear polymer and zwitterion molecules. At a high C-rate, solvent molecules carrying lithium ions can penetrate these cracks and reach the graphite surface, where they participate in surface reactions that produce a weak and highly resistive SEI [21,22]. In this study, we solve this problem by developing a novel aqueous fabrication method for the development of graphite anodes. First, aqueous composites containing the zwitterionic polymer CMC-SBMA[a], zwitterion molecules, and lithium salts with crosslinkers, which form the network structure to wrap the graphite particles, are used to develop anodes for lithium-ion batteries (LIBs). We hypothesize that the aqueous composites will prevent cracks from forming on the graphite surface and block solvent molecules from passing from the electrolyte to the graphite anode. Accordingly, this study illustrates how our aqueous processes save time in the SEI formation of graphite anodes. The electrochemical performance of the anode at 0.5 C as a high C-rate break-in is atypically analyzed and the results are described below.

2. Results and Discussion

2.1. *Demonstration of Full Cells*

In this study, MG-AQP and MG-CMC as the anode and $Li[Ni_{0.8}Mn_{0.1}Co_{0.1}]O_2$(NMC811) as the cathode were used to fabricate LIBs in full-coin cells for LIBs-MG-AQP//NMC811 and LIBs-MG-CMC//NMC811, respectively. Both LIBs-MG-AQP//NMC811 and LIBs-MG-CMC//NMC811 undergo a break-in process at a 0.5 C-rate under atypical analysis; their charging and discharging curves are shown in Figure 1a,b. The curve of NMC811 under a 0.5 C-rate break-in is presented in Figure S1. Accordingly, the initial curve of LIBs-MG-AQP//NMC811 shows a smoother slope than that of LIBs-MG-CMC//NMC811. LIBs-MG-AQP//NMC811 and LIBs-MG-CMC//NMC811 exhibit voltage plateaus at 3.84 V and 3.81 V at SOC 50% during the discharging process, representing the lower resistances of LIBs-MG-AQP//NMC811 than LIBs-MG-CMC//NMC811. The charging and discharging capacities of LIBs-MG-AQP//NMC811 are 168.7 mAhg^{-1} and 157 mAhg^{-1}, while those of LIBs-MG-CMC//NMC811 are 200 mAhg^{-1} and 125.04 mAhg^{-1}. The initial coulombic efficiencies of LIBs-MG-AQP//NMC811 and LIBs-MG-CMC//NMC811 are 92% and 63%, respectively. For the cyclic stabilities after 10 cycles at a 1 C-rate, as shown in Figure 1c,d, the coulombic efficiencies of LIBs-MG-AQP//NMC811 become stable at nearly 100%. Conversely, MG-CMC//NMC811 exhibits stable capacities and coulombic efficiencies after 70 cycles. For the final cycles, LIBs-MG-AQP//NMC811 and LIBs-MG-CMC//NMC811 remain at 83% and 25% at 280 cycles. It is observed that the numerous surface reactions required for forming an SEI take place in MG-CMC//NMC811 during the break-in process at a 0.5 C-rate, substantially reducing the amount of lithium ions for energy storage. Therefore, MG-CMC//NMC811 is necessary to form the SEI layer under a low C-rate, as shown in Figure S2. However, LIBs-MG-AQP//NMC811 exhibits excellent stability during the break-in process at a 0.5 C-rate, indicating its high and stable capacity during cycling analysis. Therefore, it seems that the new insights obtained from the development of aqueous graphite anodes can be used to demonstrate the energy and time saved by these devices. A detailed electrochemical analysis of graphite anodes is presented below.

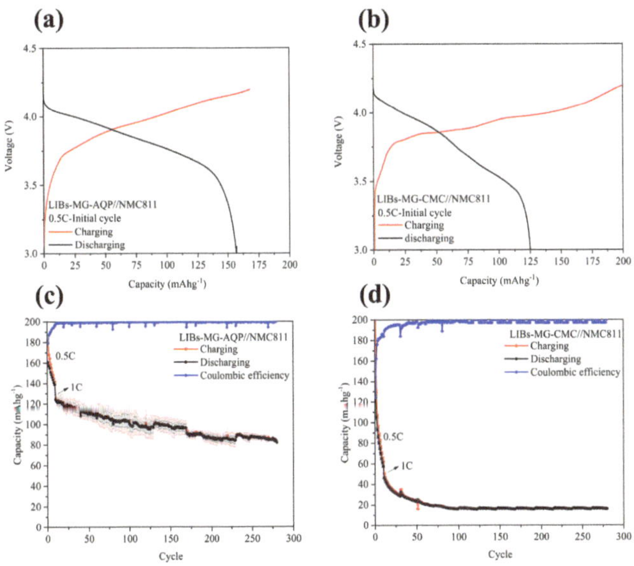

Figure 1. The first charging and discharging curves of (**a**) LIBs-MG-AQP//NMC811 and (**b**) LIBs-MG-CMC//NMC811 at 0.5 C-rate break-in. The cyclic stabilities of (**c**) LIBs-MG-AQP//NMC811 and (**d**) LIBs-MG-CMC//NMC811 for 280 cycles.

2.2. Electrochemical Performance of Half Cells

To present the electrochemical performance of the half cells fabricated in an anode containing the lithium metals LIBs-MG-AQP and LIBs-MG-CMC at a 0.5 C-rate break-in, charging and discharging curves are shown in Figure 2a,b. The initial coulombic efficiencies of LIBs-MG-AQP and LIBs-MG-CMC are 88.44% and 73.7%, which are nearly in line with the efficiency values for full cells of LIBs-MG-AQP//NMC811 and LIBs-MG-CMC//NMC811. Additionally, the voltage values at SOC 50% are 0.172 and 0.275 V vs. Li/Li^+ during the initial delithiation process, demonstrating that lower resistances exist in

LIBs-MG-AQP than in LIBs-MG-CMC. These results also correspond to the performance obtained from the CV analysis shown in Figure 2c,d. As can be observed from Figure 2c,d, initially, many surface reactions occur in LIBs-MG-CMC from 0.63 V vs. Li/Li^+ to 0.214 V vs. Li/Li^+ during the lithiation process, while LIBs-MG-AQP exhibits only few surface reactions from 0.48 V vs. Li/Li^+ to 0.215 V vs. Li/Li^+, indicating that it is efficient for suppressing and stagnating the surface reactions when forming an SEI from solvent molecules with lithium ions. During the delithiation process, the anodic peak for LIBs-MG-AQP is positioned more toward the left than that for LIBs-MG-CMC, indicating the lower energy consumption of LIBs-MG-AQP than LIBs-MG-CMC. The following figures show the discharging and charging curves at 1C-rate cycling as well as Nyquist plots of both LIBs-MG-AQP and LIBs-MG-CMC to demonstrate the behavior of resistance.

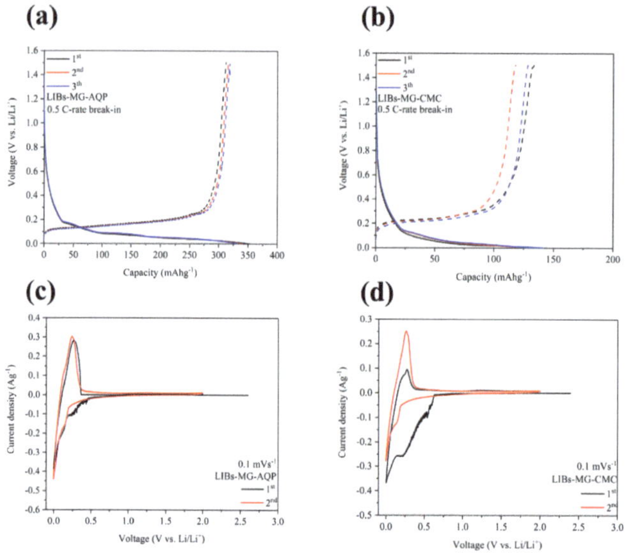

Figure 2. The first lithiation and delithiation curves of (**a**) LIBs-MG-AQP and (**b**) LIBs-MG-CMC at a 0.5 C-rate break-in. The initial cyclic voltammetry curves of (**c**) LIBs-MG-AQP and (**d**) LIBs-MG-CMC.

The discharging and charging curves of both LIBs-MG-AQP and LIBs-MG-CMC before cycling and at 1 C-rate cycling at 50, 150, and 250 cycles are presented in Figure 3a–d. LIBs-MG-CMC exhibits increasing voltage plateaus under 1 C-rate cycling during the delithiation process from 0.36 V vs. Li/Li^+ at 50 cycles to 0.43 vs. Li/Li^+ at 250 cycles. By contrast, there are almost no changes in the voltage plateau for LIBs-MG-AQP from 0.208 V vs. Li/Li^+ at 50 cycles to 0.213 V vs. Li/Li^+ at 250 cycles. Correspondingly, the Nyquist plots of LIBs-MG-AQP and LIBs-MG-CMC before cycling and under 1 C-rate cycling at 50, 150, and 250 cycles are presented in Figure 3e–h. To quantify the values of resistances, the data for the resistance of the SEI (R_{SEI}), interfacial charge transfer resistance (R_{CT}), and lithium ion diffusion coefficient (D_{Li^+}), calculated using Equation (1), during cycling are presented in Figure 3i–k.

$$D_{Li^+} = \left(\frac{RT}{\sqrt{2}F^2 AC_{Li^+} \times W} \right)^2 \tag{1}$$

where D_{Li^+} (cm^2s^{-1}) denotes the diffusivities of lithium in the anode; n is the charge transfer number; F (C mol^{-1}) is Faraday's constant of 96,485; A (cm^2) is the area of the circle-shaped anode; C_{Li^+} (mol cm^{-3}) is the concentration of lithium ion in bulk; R is the gas constant for 8.314 × 10^3 (kJ (mol K)$^{-1}$); T is the temperature (K); and W ($\Omega s^{-0.5}$) is the Walberg coefficient.

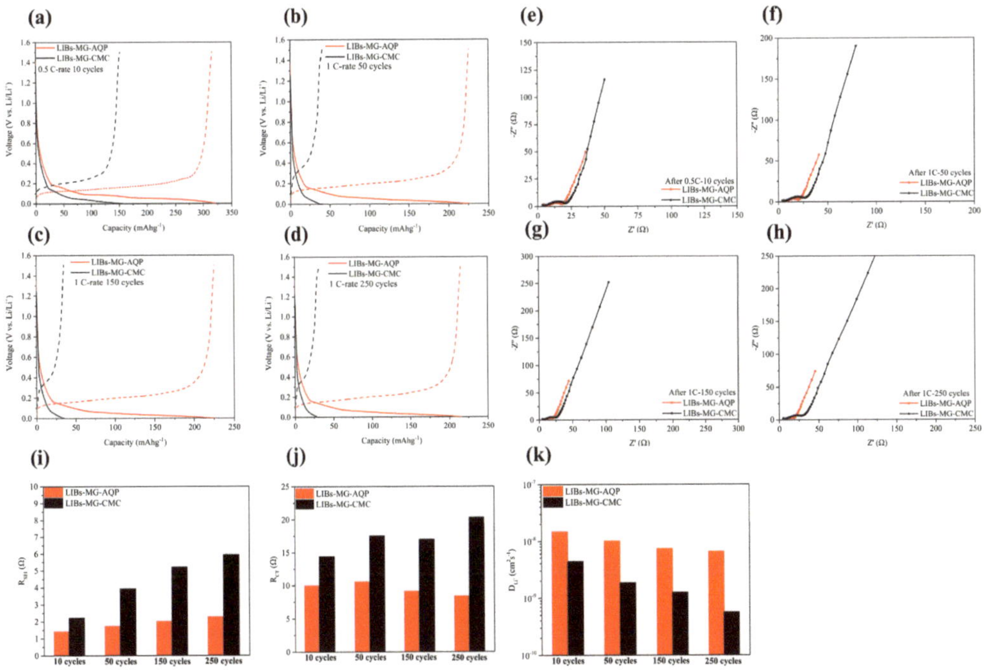

Figure 3. Curves showing the voltage versus capacity of LIBs-MG-AQP and LIBs-MG-CMC at (**a**) a 0.5 C-rate of 10 cycles, (**b**) a 1 C-rate of 50 cycles, (**c**) a 1 C-rate of 150 cycles, and (**d**) a 1 C-rate of 250 cycles. Nyquist plots of LIBs-MG-AQP and LIBs-MG-CMC at (**e**) a 0.5 C-rate of 10 cycles, (**f**) a 1 C-rate of 50 cycles, (**g**) a 1 C-rate of 150 cycles, and (**h**) a 1 C-rate of 250 cycles. The values of (**i**) R_{SEI}, (**j**) R_{SEI}, and (**k**) $D_{Li}{}^+$ for LIBs-MG-AQP and LIBs-MG-CMC at a 0.5 C-rate of 10 cycles, a 1 C-rate of 50 cycles, a 1 C-rate of 150 cycles, and a 1 C-rate of 250 cycles.

It is observed that the R_{SEI} and R_{CT} values increase in LIBs-MG-CMC after 250 cycles. However, for LIBs-MG-AQP, these values remain almost the same from 50 cycles to 250 cycles. Moreover, the $D_{Li}{}^+$ values for LIBs-MG-AQP maintain almost the same values, while those for LIBs-MG-CMC continue dropping up to 250 cycles. In this case, many zwitterion molecules and lithium salts contained within the AQC undergo mutual dissociation to form the free lithium ions and unassembled zwitterion molecules that provide an adequate pathway to improve mobility and diffusivity. The AQC is also designed for application in graphite anodes, preventing solvent molecules from passing from the electrolyte to the graphite anodes, which can improve the coulombic efficiency and stability.

2.3. SEM Images of Graphite Anodes

SEM images of MG-AQP and MG-CMC before and after cycling are presented in Figure 4a–d, respectively. Figure 4a,b show that the morphologies of graphite particles in MG-AQP is smoother than in MG-CMC before cycling. Additionally, the more uniform dispersions for graphite and binder present in MG-AQP and MG-CMC are also visible. Figure 4c,d, representing MG-AQP and MG-CMC after cycling, show that MG-CMC exhibits a considerably greater number of large cracks than MG-AQP. It is obvious that the many cracks on the graphite in MG-CMC contribute to the persistently substantial consumption of lithium ions, which form the SEI layer, which corresponds to the low coulombic efficiency of LIBs-MG-CMC//NMC811. Conversely, due to the network structure and lack of cracks in the AQC, the lithium ions in MG-AQC are mostly used for energy storage rather than the formation of the SEI. Nevertheless, MG-AQP still exhibits a few cracks before cycling, as the sharp edges of the graphite cannot be completely wrapped by the AQC. In this case, at a high current rate, there is still a risk that lithium ions accompanied by solvent molecules preferentially penetrate through these cracks into the inner graphite, resulting in extensive surface reactions with the solvent and reduced cycling stability. Accordingly, this has driven us to utilize more spheroidal graphite in anodes and ensure that it is wrapped more uniformly in future research in order to achieve more superior performance.

Figure 4. SEM images of (**a**) MG-AQP and (**b**) MG-CMC anodes before cycling, and (**c**) MG-AQP and (**d**) MG-CMC after 280 cycles.

2.4. XPS Analysis of Graphite Anodes

The XPS curves of the C, O, F, and P elements from the MG-AQP and MG-CMC anodes before and after cycling are presented in Figure 5a,b in subfigures (i), (ii), (iii), and (iV), respectively. In Figure 5a, the curves of the C and O element of MG-AQP nearly overlap before and after cycling, suggesting that the graphite particles are not covered by a thick SEI layer and no significant O-rich SEI species have formed. This indicates that solvent molecules cannot be transported through our designed AQC. As shown in Figure 5a (iii) and (iV), some F- and P-based species have formed around the graphite particles, showing that a few PF_6^- species participate in the strong formation of LiF and Li_3PO_4 at the SEI. By contrast, according to Figure 5b, the C and O spectra exhibit suppressed peak intensities and significant peak shifts, indicating the formation of thick SEI layers. In addition, F- and P-based species are more prominently detected on the particle surfaces. These XPS results clearly demonstrate that the AQC composites effectively suppress excessive solvent decomposition and facilitate the formation of thin, stable, and inorganic-rich SEI layers (LiF and Li_3PO_4). In contrast, the conventional CMC binder induces the accumulation of thick, unstable SEI films. The differences between the XPS results for the MG-AQP and MG-CMC anodes are consistent with the above electrochemical performance, indicating that the MG-AQP anodes exhibit high initial coulombic efficiency and undergo minimal surface reactions with solvent molecules for forming the SEI. The findings also suggest that almost no cracks are present on the MG-AQP surfaces, as further corroborated by the following analyses.

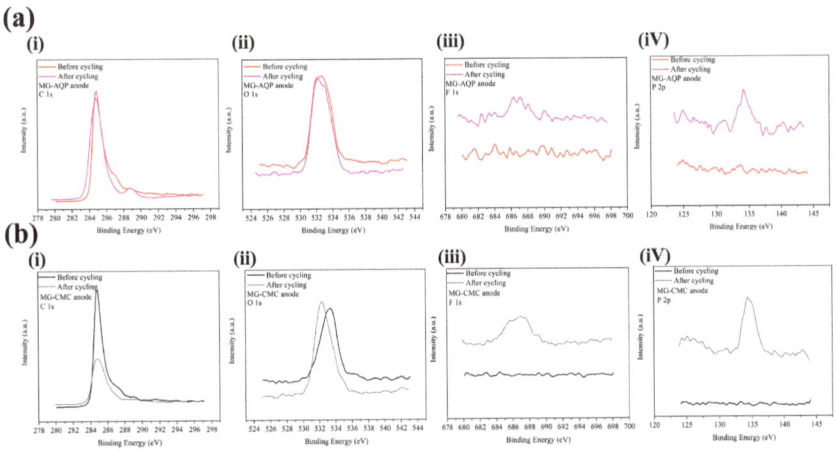

Figure 5. The XPS analysis for (**a**) MG-AQP and (**b**) MG-CMC of (i) C, (ii) O, (iii) F, and (iV) F elements before and after cycling.

2.5. Property Differences Between AQC and CMC

Finally, the properties of both the AQC and CMC were found to coincide with the results from the SEM images of MG-AQP and MG-CMC. First, the SEM images of the AQC and CMC are presented in Figure 6a,b. It can be observed that AQC possesses few cracks on the surface, while there are significant large cracks on the CMC, indicating that many solvent molecules could penetrate through the cracks. The swelling ratio (SW) versus time for both the AQC and CMC with DMC solvent are presented in Figure 6c. After three days, the SWs of the AQC and CMC are 1.47% and 80.39%, indicating that the solvent molecules are not more permeable through the AQP than the CMC. These results provide convincing evidence that applying the AQC to graphite anodes at a break-in of 0.5 C-rate can replace that of 0.1 C-rate SEI formation.

In terms of the thermal stabilities of the AQC and CMC, Figure 7a,b show that the temperatures of the dehydration process used for the decomposition of AQC and CMC range from 250 °C to 350 °C and from 255 °C to 320 °C. It seems that the AQC presents lower temperatures for thermal decomposition than CMC. However, as can be seen from the differential TGA (dTGA) curves in Figure 7b, the AQC exhibits a broader curve and higher temperature on the final day of the dehydration process than CMC, indicating a slower rate of heat release. It is possible that the use of AQCs in graphite prevents the thermal runaway of LIBs.

The mechanism of the AQCs and CMC used to fabricate the graphite anode is presented in Figure 8. The large numbers of dissociated zwitterion molecules and lithium ions in the AQC provide many uniform pathways for ion transportation, improving diffusivity and lowering resistance. Compared to conventional CMC binders, graphite anodes with AQCs can exhibit lower energy consumption during charging or lithiation but higher energy during discharging or delithiation. Additionally, in our previous studies, zwitterion molecules were found to improve the dispersion of slurries, thus saving more power when fabricating graphite anodes compared to CMC-based graphite anodes. Moreover, due to the fewer cracks in the AQC, solvent molecules do not travel from the electrolyte to the graphite anode. Therefore, it is not necessary to form an SEI layer conventionally at a 0.1 C-rate. Instead, the 0.5 C-rate break-in used in this study is treated as an atypical analysis to show that LIBs-MG-AQP//NMC811 exhibits excellent initial coulombic efficiency and cyclic stability for saving energy as well as time. Furthermore, this work highlights the need for in situ spectroscopy such as Raman spectra, XRD analysis, and molecular dynamics simulations, which will be presented in future studies.

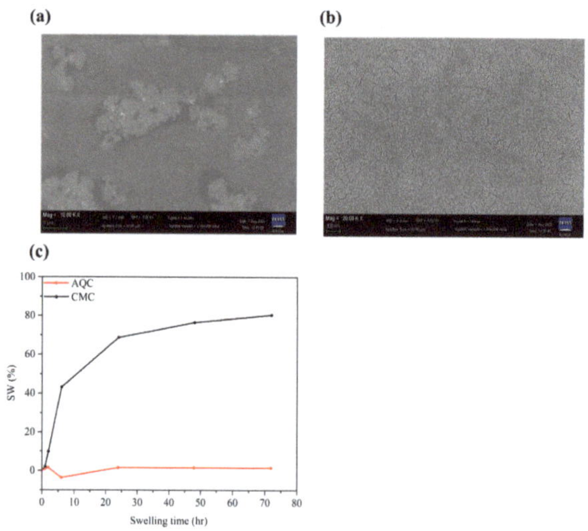

Figure 6. SEM images of (**a**) AQC and (**b**) CMC, as well as (**c**) SW ratio versus time for AQC and CMC.

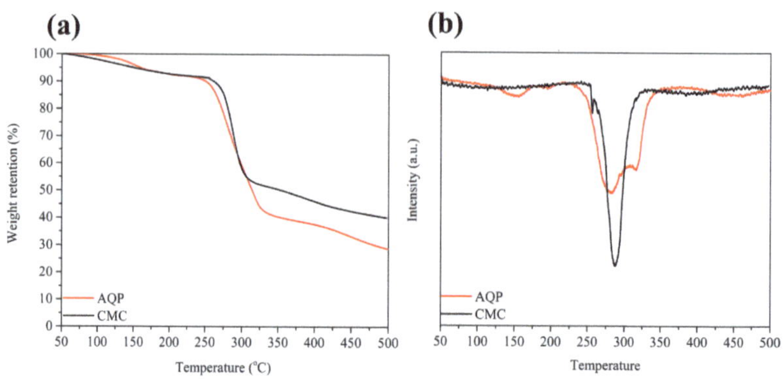

Figure 7. (**a**) The TGA curves and (**b**) dTGA curves of the AQC and CMC from 50 °C to 500 °C.

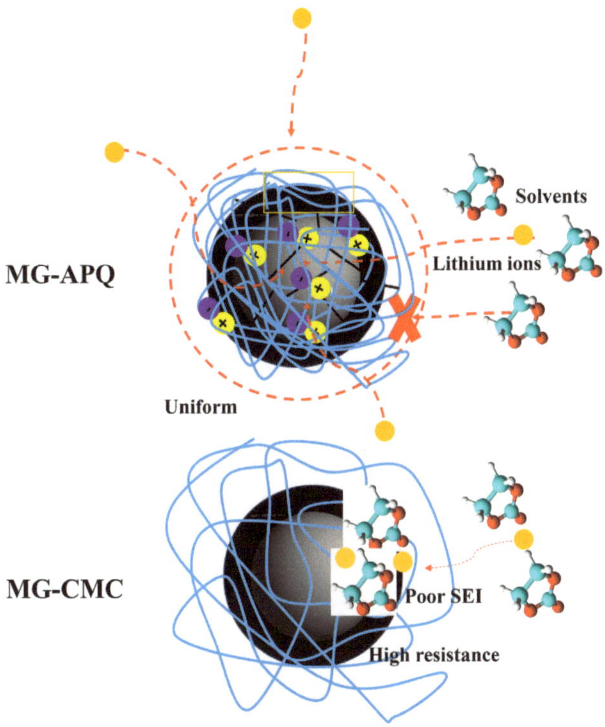

Figure 8. Schematic diagram of the MG-AQP and MG-CMC mechanisms as an atypical analysis.

3. Materials and Methods

3.1. Materials

Carboxymethyl cellulose (CMC, BVH8) was purchased from Ashland Inc. (authorized distributor of Ashland Inc., Taipei, Taiwan). Sulfobetaine methacrylamide (SBMA), lithium trifluoromethanesulfonate ($LiCF_3SO_3$) (LiOTf), and glutaraldehyde (GA, 25 wt%) were purchased from Sigma-Aldrich (via Merck Ltd. Taiwan, Taipei, Taiwan). Lithium hydroxide hydrate ($LiOH{\cdot}H_2O$, 98 wt%) and betaine (99 wt%) were purchased from Thermo Scientific (via DKSH Taiwan, Taipei, Taiwan). Dimethyl carbonate (DMC, 99 wt%) was purchased from Alfa Aesar (via DKSH Taiwan, Taipei, Taiwan). Graphite (MG30) was provided from China Steel Chemical Corporation, Kaohsiung, Taiwan. Super P was provided by Timcal (via Taiwan Maxwave Co., Ltd., Taipei, Taiwan). $Li[Ni_{0.8}Mn_{0.1}Co_{0.1}]O_2$ (NMC811) (~95.5%) was purchased from Ningbo Ronbay New Energy, Ningbo, China. Polyvinylidene fluoride (PVdF-5130) powder was purchased from Solvay S.A. (via a subsidiary of Solvay S.A, Taipei, Taiwan). Methylpyrrolidone (NMP) was purchased from the uniregion bio-tech company, New Taipei City, Taiwan.

3.2. Preparation of Aqueous Composites

In accordance with our previous study, first, LiOH and SBMA were added to 2 wt% of CMC aqueous solution equimolarly. CMC was reacted with SBMA at 50 °C for 16 h through oxa-Michael addition to obtain the CMC-SBMA[a] product. Subsequently, aqueous composites (AQCs) were prepared via mixing CMC-SBMA[a] with betaine and LiOTf equimolarly based on the amount of SBMA in the AQC.

3.3. Preparation of Graphite Anode

MG30, as one type of graphite, was wrapped using aqueous composites. MG30 particles were mixed in the aqueous composites evenly at a mass ratio of 128.24:1 (MG30 particles/AQC), and then wrapping the particles tightly, the aqueous composites were crosslinked into a network structure, being formed at 50 °C by adding glutaraldehyde (GA, 1 wt% relative to CMC-SBMA[a]) as a crosslinker. After the crosslinking reaction, 1wt% Super P, 0.75 wt% aqueous composites and 1.32 wt% styrene–butadiene rubber (SBR) were added to the solution with 55 wt% of solid content to prepare the anode slurry. Subsequently, a copper clad laminate was coated with graphite slurry with a weight film thickness of 200 μm. Finally, this graphite anode, called MG-AQP, was obtained for use in fabricating LIBs. The copper-clad laminate with graphite slurry was dried at 100 °C, vacuum-dried overnight, and pressed using a rolling machine.

For comparison, the conventionally fabricated graphite slurry comprised 96.18 wt% MG30 with 1.5 wt% CMC, 1wt% Super P, and 1.32 wt% styrene–butadiene rubber (SBR), with 55 wt% solid content in 45 wt% of water as a solvent. Similar to the above preparation, the graphite anode, denoted as MG-CMC, was obtained.

3.4. Preparation of NMC811 Cathode Slurry

The NMC811 cathode comprised 95.5 wt% of NMC811 particles, 2.5 wt% of PVdF, and 2 wt% of Super P. Similar to the graphite anode slurries, 55 wt% solid content in 45 wt% of NMP was used as a solvent. The obtained cathode was used to fabricate LIBs after the aluminum-clad laminate covered with NMC811 slurry was dried at 100 °C, vacuum-dried overnight, and pressed using a rolling machine [23,24].

3.5. Fabrication of the LIBs

MG-AQP and MG-CMC, in circle shapes with a diameter of 13 mm, were used to fabricate LIBs with lithium metal in the coin cells, called LIBs-MG-AQP and LIBs-MG-CMC, respectively. Here, 1M $LiPF_6$ in EC/DMC (1:1) as a liquid electrolyte, infiltrating fiber-reinforced plastic as a separator, was sandwiched between the graphite anodes and lithium metal. Information regarding LIBs-MG-AQP and LIBs-MG-CMC is provided in Table S1.

Furthermore, in order to demonstrate the actual performance of the LIBs, both MG-AQP and MG-CMC as anodes with $Li[Ni_{0.8}Mn_{0.1}Co_{0.1}]O_2$(NMC811) as a cathode were used to fabricate the LIBs for LIBs-MG-AQP//NMC811 and LIBs-MG-CMC//NMC811, whose information is provided in Table S2. Similarly, 1M $LiPF_6$ in EC/DMC (1:1) as an electrolyte, infiltrating fiber-reinforced plastic as a separator, was sandwiched between the graphite anodes and NMC811 cathodes.

3.6. Characterization

3.6.1. Analysis of Aqueous Composites and CMC

Both AQC and CMC were analyzed by a thermo-gravimetric analyzer (TGA, Perkin Elmer TGA4000; Waltham, MA, USA) from 50 to 500 °C with 2 °C min^{-1} to obtain their thermal properties. Images of the aqueous composites and CMC were also obtained by a scanning electron microscope (SEM) with Pt sputter coating to observe the morphologies at 10 kV.

Swelling analyses of the AQC and CMC were performed to determine whether solvent molecules diffuse into the materials. In this study, the aqueous composite and CMC were dipped into a DMC solvent. The weights of the aqueous composite and CMC were measured at various times to determine the swelling ratios, as per Equation (2).

$$SW\ (t) = \frac{w\ (t) - w_0}{w_0} \times 100 \quad (2)$$

where $SW\ (t)$ represents the swelling ratio, w represents the weight of the membranes, and w_0 represents the weight of the dry membranes.

3.6.2. Analysis of Half-Coin Cells

Both LIBs-MG-AQP and LIBs-MG-CMC were broken in with a 0.5 C-rate from 1.5 V to 0 V using a battery charge–discharge tester (BAT-750B) for ten cycles as an atypical analysis. In addition, LIBs-MG-AQP and LIBs-MG-CMC were analyzed by cyclic voltammetry (CV) with a scan rate of 0.1 mVs^{-1} from 2 V to 0 V to observe their electrochemical behavior under the break-in process. Then, they were analyzed at a 1 C-rate for cyclic stability. Here, we chose 10, 50, 150, and 250 cycles for LIBs-MG-AQP and LIBs-MG-CMC, which were analyzed by electrochemical impedance spectroscopy (EIS) with a lithium-ion state of charge of 0 (SOC = 0). After cycling, the graphite anodes from the LIBs-MG-AQP and LIBs-MG-CMC were analyzed by SEM to observe their morphologies. X-ray photoelectron spectroscopy (XPS) was employed to analyze the graphite anodes of the LIBs-MG-AQP and LIBs-MG-CMC before and after cycling, in order to clarify the formation of the SEI.

3.6.3. Analysis of Full-Coin Cells

Similarly, the 0.5 C-rate as the break-in process for 10 cycles was applied for both LIBs-MG-AQP//NMC811 and LIBs-MG-CMC//NMC811. After 10 cycles, the stability of the LIBs-MG-AQP//NMC811 and LIBs-MG-CMC//NMC811 was analyzed at a 1 C-rate for 250 cycles.

4. Conclusions

In this study, we designed aqueous composites, called AQCs, which contain zwitterionic polymers, zwitterion molecules, and lithium salts. Graphite, wrapped using an AQC with a crosslinker, was used to fabricate the anode via aqueous processes, denoted as MG-AQP. The atypical analysis of its electrochemical performance demonstrates that a 0.5 C-rate break-in achieves an excellent initial coulombic efficiency of 92% and long-term cycle stability of 83% retention after 280 cycles for LIBs-MG-AQP//NMC811. For half cells, LIBs-MG-AQP shows fewer surface reactions according to CV analysis, as well as low and unchangeable resistance after cycling. Structural and physicochemical analyses coincided with the results of electrochemical performance, indicating that the AQC exhibits minimal cracking, thus limiting solvent penetration. These observations indicate that the AQC both physically blocks solvent access to graphite, reducing SEI-forming reactions, and chemically facilitates ion transportation via the dissociation of lithium ions and zwitterion molecules. This study highlights a novel approach for saving time and energy when producing excellent graphite anodes. Moreover, this strategy offers a practical pathway toward the time- and energy-efficient activation of LIBs, suggesting its potential for sustainable and scalable graphite anode fabrication for LIBs in the future.

Supplementary Materials: The following supporting information can be downloaded at: https://www.mdpi.com/article/10.3390/molecules30193947/s1. Table S1: Information on coin cells for LIBs-MG-AQP and LIBs-MG-CMC; Table S2: Information on coin cells for LIBs-MG-AQP//NMC811 and LIBs-MG-CMC//NMC811; Figure S1: The initial delithiation and lithiation curves of NMC811 at a 0.5 C-rate; Figure S2: (a) The initial charging and discharging curves and (b) cyclic stability of LIBs-MG-CMC//NMC811 at a 1 C-rate.

Author Contributions: K.-Y.L.: Validation, formal analysis, investigation, writing—original draft, data curation, visualization. C.-C.C.: Methodology, investigation, data curation, equipment provision. Y.-L.L.: Methodology, supervision, project administration. T.-C.W.: Conceptualization, methodology, supervision, writing—review and editing, visualization, project administration. All authors have read and agreed to the published version of the manuscript.

Funding: The authors gratefully acknowledge the financial support from the projects of the National Science and Technology Council with BBL Technology Corporation (114-2622-E-006-008-).

Data Availability Statement: The data presented in this study are available on request.

Acknowledgments: The National Science and Technology Council and BBL Technology Corporation are acknowledged for their financial support from the Industrial/Academic Cooperation Projects. The authors gratefully acknowledge the financial support provided by projects from the National Science and Technology Council with BBL Technology Corporation (114-2622-E-006-008-). Equipment was supplied by Chia-Chin Chang's laboratory, Department of Greenergy, National University of Tainan. The authors gratefully acknowledge the usage of an EDS High-Resolution Transmission Electron Microscope, JEOL TEM-2100F/EDS, X-ray diffractometer (XRD, Rigaku/SmartLab), analytical field emission scanning electron microscope (AFE-SEM, AURIGA), and X-ray photoelectron spectroscopy (XPS) equipment, which belonged to the Core Facility Center of National Cheng Kung University. Kuan-Yi Liao gratefully acknowledges the scholarship support from the Graduate Research Fellowship of the National Science and Technology Council and Google LLC.

Conflicts of Interest: The authors declare no conflicts of interest.

References

1. Zhao, W.; Zhao, C.; Wu, H.; Li, L.; Zhang, C. Progress, challenge and perspective of graphite-based anode materials for lit batteries: A review. *J. Energy Storage* **2024**, *81*, 110409. [CrossRef]
2. Zhao, L.; Ding, B.; Qin, X.; Wang, Z.; Lv, W.; He, Y.; Kang, F. Revisiting the roles of natural graphite in ongoing lithiur batteries. *Adv. Mater.* **2022**, *34*, 2106704. [CrossRef] [PubMed]
3. Liu, Y.; Shi, H.; Wu, Z. Recent status, key strategies and challenging perspectives of fast-charging graphite anodes for lithiun batteries. *Energy Environ. Sci.* **2023**, *16*, 4834–4871. [CrossRef]
4. He, S.; Liu, L.; Wang, J.; Zhang, W.; Wang, J.; Li, H. Considering Critical Factors of Silicon/Graphite Anode Materials for Pra High-Energy Lithium-Ion Battery Applications. *Energy Fuels* **2020**, *35*, 944–964. [CrossRef]
5. Wu, J.; Cao, Y.; Zhao, H.; Mao, J.; Guo, Z. The critical role of carbon in marrying silicon and graphite anodes for high-en lithium-ion batteries. *Carbon Energy* **2019**, *1*, 57–76. [CrossRef]
6. Gao, C.; Cui, X.; Wang, C.; Wang, M.; Wu, S.; Quan, Y.; Wang, P.; Zhao, D.; Li, S. 3D-printed hierarchical porous and multidi sional conductive network based on conducting polymer/graphene oxide. *J. Mater.* **2024**, *10*, 234–244. [CrossRef]
7. Park, J.H.; Kim, S.H.; Ahn, K.H. Role of carboxymethyl cellulose binder and its effect on the preparation process of anode slu for Li-ion batteries. *Colloids Surf. A Physicochem. Eng. Asp.* **2023**, *664*, 131130. [CrossRef]
8. Ishii, M.; Makino, S.; Nakamura, H. The role of carboxymethyl cellulose on the rheology of anode slurries in lithium-ion batt *Curr. Opin. Colloid Interface Sci.* **2024**, *74*, 101858. [CrossRef]
9. Kim, K.J.; Ahn, K.H. Effect of carboxymethyl cellulose on silicon dispersion and the performance of graphite/Si-based electr for lithium-ion batteries. *Powder Technol.* **2025**, *452*, 120452. [CrossRef]
10. Liao, K.Y.; Chang, C.C.; Lee, Y.L.; Wen, T.C. Carboxymethyl cellulose grafted with lithium acrylate as anode binder for impro performance of lithium-ion batteries at low temperatures. *J. Taiwan Inst. Chem. Eng.* **2025**, *174*, 106188. [CrossRef]
11. Gong, T.; Duan, X.; Shan, Y.; Huang, L. Gas generation in lithium-ion batteries: Mechanisms, failure pathways, and thermal s implications. *Batteries* **2025**, *11*, 152. [CrossRef]
12. Lin, J. Strategies for constructing stable solid electrolyte interface on the surface of lithium metal negative electrodes. *J. Phys. Ser.* **2024**, *2798*, 012001. [CrossRef]
13. Ren, F. Contribution of Raman Micro-Spectroscopy to the Study of the Electrolyte for Li-Metal Batteries. Ph.D. Thesis, Unive de Montpellier, Montpellier, France, 2022.

Verma, P. Electrochemical and Chemical Surface Modifications of Carbons for Li-Ion Batteries. Ph.D. Thesis, ETH Zurich, Zurich, Switzerland, 2011.
Agubra, V.A.; Fergus, J.W. The formation and stability of the solid electrolyte interface on the graphite anode. *J. Power Sources* **2014**, *268*, 153–162. [CrossRef]
Bommier, C.; Ji, X. Electrolytes, SEI formation, and binders: A review of nonelectrode factors for sodium-ion battery anodes. *Small* **2018**, *14*, 1703576. [CrossRef]
Beheshti, S.H.; Javanbakht, M.; Omidvar, H.; Hosen, M.S.; Hubin, A.; Van Mierlo, J.; Berecibar, M. Development, retainment, and assessment of the graphite–electrolyte interphase in Li-ion batteries regarding the functionality of SEI-forming additives. *iScience* **2022**, *25*, 103862. [CrossRef]
Lv, X.; Liu, J.; Li, C.; Yu, F. Probing a solid electrolyte interphase layer with sub-nanometer pores using redox mediators. *eScience* **2024**, *6*, 100351. [CrossRef]
Liao, K.Y.; Li, W.C.; Wen, T.C. Constructing supercapacitors with biopolymer bearing zwitterion as hydrogel electrolyte and binder for superior performance at −40 °C. *J. Power Sources* **2024**, *598*, 234191. [CrossRef]
Liao, K.Y.; Chang, C.C.; Lee, Y.L.; Wen, T.C. Applying Carboxymethyl cellulose-based aqueous binder with zwitterion molecules in graphite anode for lithium-ion batteries. *Appl. Surf. Sci.* **2025**, *711*, 164099. [CrossRef]
Li, J.; Fan, S.; Xiu, H.; Wu, H.; Huang, S.; Wang, S.; Yin, D.; Deng, Z.; Xiong, C. TiO_2-Coated Silicon Nanoparticle Core-Shell Structure for High-Capacity Lithium-Ion Battery Anode Materials. *Nanomaterials* **2023**, *13*, 1144. [CrossRef]
Zhang, K.; Cao, W.Z.; Wang, J.; Zhao, Z.N.; Yin, W.A.; Lv, Z.; Zhang, J.F.; Wang, R.; Wu, F.; Tan, G.Q. High-capacity and low-expansion $MnCO_3$@cyclized-PAN composite anodes for high-performance lithium-ion batteries. *Rare Met.* **2025**, *44*, 3575–3581. [CrossRef]
Chang, C.C.; Liang, P.C.; Chen, W.H.; Luo, D.; Balasubramanian, D. Comparative analysis of Li-ion batteries with carbonate-based liquid and PVdF-based gel polymer electrolytes: Performance, temperature sensitivity, aging rate, and life cycle assessment. *ACS Appl. Energy Mater.* **2024**, *7*, 8035–8053. [CrossRef]
Wang, J.; Batara, B.; Xu, K.; Zhang, K.; Hua, W.; Peng, Y.; Liu, W.; Putri, A.H.I.; Xu, Y.; Sun, A.X.; et al. Co-precipitation of Ni-rich $Me(OH)_2$ Precursors for High-Performance $LiNixMnyCo_{1\text{-}x\text{-}}yO_2$ Cathodes: A Review. *Energy Environ. Mater.* **2025**, *8*, e70078. [CrossRef]

molecules

Article

Investigating the Mars–van Krevelen Mechanism for CO Capture on the Surface of Carbides

Naveed Ashraf * and Younes Abghoui *

Science Institute of the University of Iceland, 102 Reykjavik, Iceland
* Correspondence: naa8@hi.is (N.A.); younes@hi.is (Y.A.)

Abstract

Electrochemical reduction processes enable the CO to be converted into a useful chemical fuel. Our study employs density functional theory calculations to analyze the (110) facets of the transition metal carbide surfaces for CO capture, incorporating the Mars–van Krevelen (MvK) mechanism. All the possible adsorption sites on the surface, including carbon, metal, and bridge sites, were fully investigated. The findings indicate that the carbon site is more active relative to the other adsorption sites examined. The CO hydrogenation paths have been comprehensively investigated on all the surfaces, and the free energy diagrams have been constructed towards the product. The results conclude that the TiC is the most promising candidate for the formation of methane, exhibiting an onset potential of −0.44 V. The predicted onset potential for CrC, MoC, NbC, VC, WC, ZrC, and HfC are −0.86, −0.61, −0.61, −0.93, −0.87, −0.61, and −0.81 V, respectively. Our calculated results demonstrate that MvK is selectively relevant to methane synthesis. Additionally, we investigated the stability of these surfaces against decomposition and conversion to pure metals concerning thermodynamics and kinetics. It was found that these carbides could remain stable under ambient conditions. The exergonic adsorption of hydrogen on carbon sites, requiring smaller potential values for product formation, and stability against decomposition indicate that these surfaces are highly suitable for CO reduction reactions using the MvK mechanism.

Keywords: carbides; electrocatalysis; CO reduction; DFT calculations; Mars–van Krevelen mechanism

1. Introduction

The increasing quantity of carbon dioxide (CO_2) in the atmosphere poses a significant threat to human existence and global ecosystems. Despite the ongoing efforts of researchers and scientists, the current pace of progress remains insufficient, highlighting the urgent need for effective measures to mitigate this climate crisis. Human activities, specifically the combustion of fossil fuels, industrial processes, and vehicular emissions, are the primary sources of CO_2 emissions. With the expansion of the world population, the need for energy concurrently increases [1]. To address this issue, there is an immediate need to reduce CO_2 formation and transform it into valuable energy sources. CO_2 reduction via the electrochemical method offers a promising route to transform CO_2 into useful fuels and chemicals, thereby contributing to the closure of the global carbon loop. However, the electrochemical reduction of CO_2 presents significant scientific challenges. One major hurdle is the difficulty in steering the reaction toward a single desired product, as multiple reduction pathways can occur simultaneously. In aqueous systems, the hydrogen evolution reaction (HER) often competes with the CO_2 reduction reaction (CO_2RR), reducing the overall selectivity of the process [2]. To overcome these limitations, the development of efficient and selective catalysts is crucial. Among various candidates, transition metal carbides (TMCs) are considered cost-effective and promising materials for heterogeneous catalysis [3]. These materials are valued for their remarkable chemical, thermal, and electrical properties, making them suitable for use in electrochemical CO_2 conversion technologies [4].

These catalysts exhibit exceptional chemical, thermal, and structural stability, together with outstanding electrical and thermal conductivity [3,5,6]. TMCs display complex morphologies defined by an altered, adjusted electrical conductivity owing to the robust interplay between the carbon and metal atoms. As a result, these materials display unique properties concerning their transition metals and exhibit catalytic performance [7–12]. These TMC materials have transition metal, ionic crystal, and covalent solid properties [13] that can furnish catalytic characteristics for many applications, including the electrochemical formation of CO from CO_2 and other compounds [14], hydrogen production [15], biogas formation [16], and benzene formation [17]. Previous research indicated that TMC prompted interest in several types of applications, such as desulfurization [18], dry reforming of methane [19], formation of methane into gas [12,20,21], and oxidation of CO [22]. The features of TMCs demonstrate potential for catalytic activities [23,24]. The conversion of CO_2 into CO requires a 2-electron proton step, which can be easily achieved, but converting CO to a further useful product requires several electron-proton steps and high potentials. The primary objective of this work is to clarify the reaction pathway for the transformation of carbon monoxide (CO) into methane (CH_4) via the MvK mechanism. This study aims to elucidate the thermodynamics of the CORR, free energy landscapes, and active surface sites that regulate CO hydrogenation to methane.

Prior studies suggest that TMC provides a substantial redox environment conducive to the reduction of CO into useful compounds, highlighting the capability of these catalysts [25,26]. This study is the inaugural comprehensive analysis of (110) facets of TMC to evaluate the phenomenon of CORR through the MvK mechanism. These carbides are typically in polycrystalline shapes, with the (100) and (110) facets being the predominant orientations. We have already examined the (100) facets, and here we concentrate on the (110) facets.

The concept of the MvK mechanism, originally proposed to describe oxidation catalysis over metal oxides, is characterized by the participation of lattice oxygen as the active oxidizing species. The oxidation of a substrate in this redox process results in the elimination of lattice oxygen, creating transitory oxygen vacancies on the catalyst surface. These vacancies enable the binding and activation of a gas-phase oxidant, usually molecular oxygen, which then replenishes the oxygen vacancies, thus reinstating the original oxidation state of the metal oxides and concluding the catalytic process.

Recent advancements have considerably expanded the MvK mechanism's application beyond metal oxides, illustrating its relevance to many non-oxide systems, including metal sulfides, chlorides, nitrides, carbides, and hydrides. Specifically, metal carbides such as molybdenum carbide and iron carbide demonstrate MvK-type behavior in significant syngas-related transformations [27]. In chemical cyclic procedures, the homologation of methane, its transformation into other hydrocarbons, has been successfully performed by a two-step redox approach that includes metal carbide formation followed by reduction. Methane initially functions as the carbon supply and reducing agent, interacting with transition metals to produce metal carbides. Once created, these carbides undergo reduction with hydrogen, which promotes the liberation of hydrocarbon compounds [28]. A recently published study offers an in-depth look at the generation and regulations of surface vacancies in several catalytic materials, including oxides, carbides, nitrides, sulfides, and hydrides, relevant to the MvK mechanism [29–31]. The formation of intermediates, such as carbonates and hydroxyls, has been observed to participate in Mars–van Krevelen mechanisms. In the previous studies, the nitrogen reduction reaction has also been studied through the MvK mechanism theoretically [32–35] and experimentally for the nitrides [36], and this mechanism has been experimentally proven for Co_3Mo_3N catalysts [37]. These examples highlight the relevance of the Mars–van Krevelen (MvK) mechanism as a viable pathway in various catalytic reactions, particularly those involving transition metal carbides (TMCs), which are the focus of this study. This research aims to explore the stepwise protonation of the (110) surfaces of TMCs leading to methane formation, following the generation of surface carbon vacancies. The inherent carbon atoms in the carbide structure play a crucial role in facilitating methane evolution through interaction with four protons, which results in the creation of a carbon vacancy [38]. This vacancy is subsequently filled by the carbon atom from an adsorbed CO molecule, while the oxygen atom combines with protons to form water. This process effectively regenerates and cleans the catalytic surface, enabling continuous operation.

2. Methodology

Density functional theory (DFT) calculations were performed using version 6.3.2 of the Vienna Ab initio Simulation Package (VASP) [39], employing the Revised Perdew–Burke–Ernzerhof (RPBE) functional. The (110) surface orientation of transition metal carbides (TMCs) in the rocksalt (RS) crystal structure was selected for modeling. The RPBE functional offers a well-balanced trade-off between computational cost and accuracy, making it suitable for simulating surface reaction phenomena. Its reliability has been validated in prior studies focused on CO_2RR, where predicted trends were found consistent with experimental observations [40–42].

A $4 \times 4 \times 1$ Monkhorst–Pack k-point was used for Brillouin zone sampling, with the cutoff energy of 400 eV. The surface slab consisted of five atomic layers and a total of 40 atoms, equally divided between metal and carbon atoms. Each layer contained four metal and four carbon atoms. A vacuum gap of 20 Å was introduced along the z-direction to avoid periodic interactions, while the x and y boundaries were kept periodic. As indicated in Figure 1, the upper three layers, along with adsorbates, were permitted to undergo complete relaxation during structural optimization, whereas the bottom two layers were constrained.

Reaction mechanisms for the CO reduction reaction (CORR) on TMC surfaces were explored using a thermochemical model (TCM), a well-established approach in the literature. The free energy diagrams for surface hydrogenation via the MvK mechanism along various pathways were computed with respect to the computational hydrogen electrode (CHE) introduced by Nørskov et al. [43]. The influence of applied electrochemical potential was incorporated through the (CHE) method. The Climbing Image Nudged Elastic Band (CI-NEB) method was used to examine carbon diffusion from the subsurface to the surface [44]. All catalyst structures were optimized until the forces converged below 0.03 eV/Å.

The formula employed to compute the adsorption energy of various species on the TMC surface is as follows [44]:

$$\Delta E_{ads} = E_{AB} - E_A - E_B \quad (1)$$

E_{AB} represents the total energy of a system with fully adsorbed intermediates and species on carbides, E_A defines the energy of the carbide catalyst in the absence of adsorbate, and E_B indicates the total energy of the adsorbate. To assess the thermodynamic feasibility of each reaction step, we calculated Gibbs free energy changes using standard DFT-based thermochemistry corrections, including zero-point energy, entropy, and pH adjustments.

$$\Delta G = \Delta E_{DFT} + \Delta E_{ZPE} - T\Delta S + \Delta G_{pH} + \Delta H_{0K \to T} \quad (2)$$

ΔE_{DFT} denotes the overall energy of the system as computed by DFT calculation, ΔE_{ZPE} signifies the calculated zero-point energy of the adsorbates. $T\Delta S$ indicates the entropy shift between the beginning and final structures of each reaction. ΔG_{pH} denotes the energy correction of the H^+ and is calculated as follows:

$$\Delta G_{pH} = \ln 10 \times k_B T \times pH \quad (3)$$

The $\Delta H_{0K \to T}$ term denotes changes in the internal energy of the system due to the effect of temperature, which can be calculated by this formula $\int_0^T C_p\,(T')\,dT'$, where $C_p\,(T')$ is the quantity of heat at constant pressure, and the integral of this equation suggests that at ambient temperature, the heat contribution is minimal. The values of the gas phase of the molecules were taken from the tables in the textbooks [45]. In our previous study, we already calculated the electronic properties of carbides [32]. The theoretical onset potential (OP) required to drive the reaction was calculated using the computational hydrogen electrode model by using Equation (4). The step with the highest free energy change (ΔG_{max}) determines the limiting potential.

$$OP = -\Delta G_{max}/e \quad (4)$$

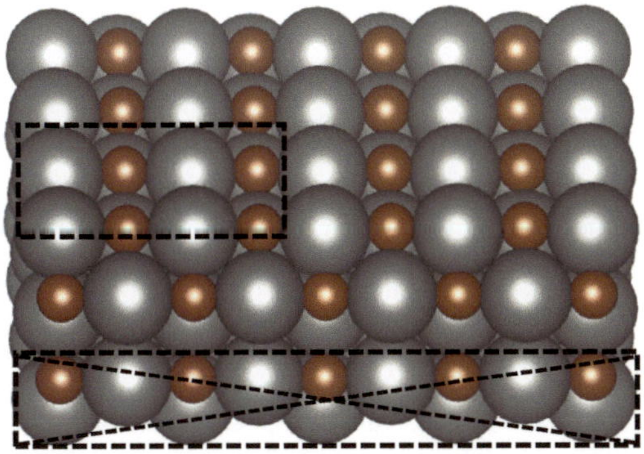

Figure 1. Illustration of the transition metal carbide (TMC) surface with a (110) orientation. Carbon atoms are represented in brown, while metal atoms are shown in gray. The dotted box at the top highlights the unit cell used for the study. The cross-hatched dotted region indicates that the bottom two layers were constrained during structural relaxation.

3. Results and Discussion

3.1. Adsorption and MvK Mechanism

In the initial step of the study, all reactant species were adsorbed onto the pristine surfaces of the selected catalysts, evaluating adsorption at various potential active sites, namely the metal, bridge, and carbon sites across transition metal carbides (TMCs) such as CrC, MoC, NbC, VC, TiC, ZrC, HfC, and WC with the (110) surface orientation. The findings indicate that the carbon sites of the carbides show the highest catalytic progress, indicating more favorable exergonic adsorption energies as compared to the bridge and metal sites of the carbides. During the electrochemical process, surface poisoning may affect electrochemical performance, so investigation of the likelihood of O, H, and OH species binding the active sites on the surface is important. It was identified that with further protonation, O and OH convert into $H_2O_{(l)}$ via the following equations:

$$*O + H^+ + e^- \rightarrow *OH \quad (5)$$

$$*OH + H^+ + e^- \rightarrow H_2O_{(l)} \quad (6)$$

This means that if O and OH adsorb on the surface of the carbides, it is of no concern, as further protonation steps convert them into H_2O, which has weak adsorption on the surface as compared to other key species (shown in Figure 2). Hence, the likelihood of O and OH becoming poisonous is low.

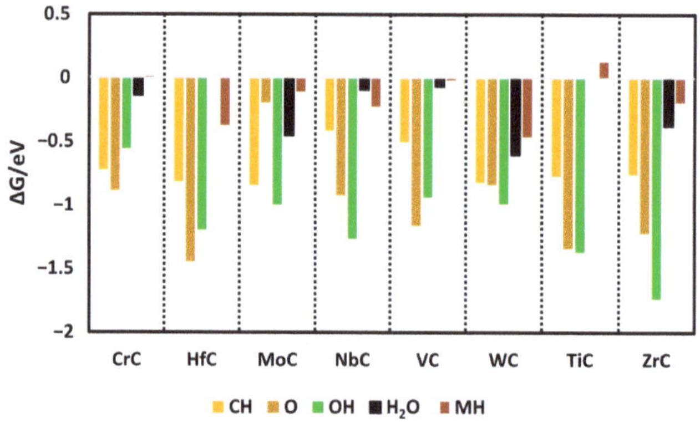

Figure 2. Calculated Gibbs free energies of all species on the surface of carbides. In the case of HfC and TiC, water does not bind.

During the experimental synthesis of TMC, there is a possibility that a defect in the form of carbon can be formed, or there is a possibility that, during the electrochemical process, due to the continuous protonation, four protons attach to the C of carbides and release methane by leaving the carbon vacancy. Thus, the release of methane generates the carbon vacancy on the carbide's surface. This carbon vacancy of carbides' surface can be occupied by carbon from ambient CO, and one oxygen atom of CO is converted into a water molecule after finishing the electrochemical step with a clean and non-defective surface. This ongoing electrochemical cyclic process generates a product like methane, as illustrated in Figure 3.

The reactions associated with the MvK mechanism for the formation of the vacancy are specified in Equations (7) and (8), accompanied by the Kröger–Vink notation [46].

$$* + 4(H^+ + e^-) \rightarrow CH_{4(g)} + {}^{C}Vac \quad (7)$$

$${}^{C}Vac + CO + 2(H^+ + e^-) \rightarrow H_2O + * \quad (8)$$

Here, the asterisk (*) represents the clean surface of the carbide catalysts, while "^{C}Vac" denotes the carbon vacancy formed following methane (CH_4) desorption.

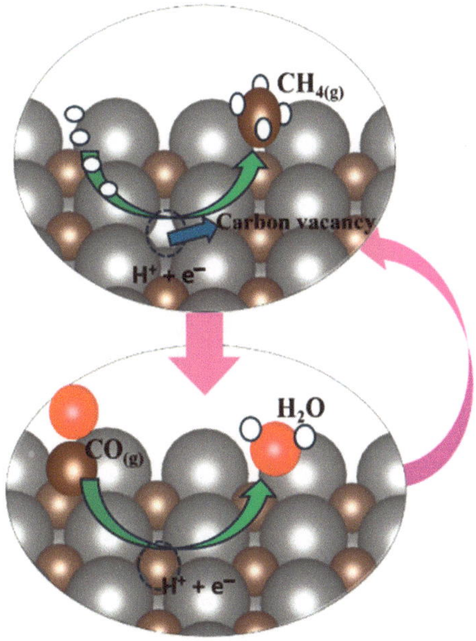

Figure 3. Representation of the generation of carbon vacancies via the MvK mechanism. The vacancy accommodates the C of atmospheric CO to fill the vacancy, and the O of CO is released in the form of water, and as a result, the surface will be regenerated and clean.

3.2. Electrochemical Performance

Among the TMC catalysts investigated, TiC stands out as the most suitable catalyst for CH_4 production, demonstrating the least OP value of −0.44 V. Both CO adsorption and water formation proceed spontaneously, as depicted in the free energy profile presented in Figure 4. The initial protonation step occurs at the carbon site of TiC, leading to the exergonic formation of a CH intermediate with a free energy of −0.76 eV, thereby promoting CH_4 generation. The most energy-demanding step, called the potential-determining step (PDS), is observed among the CH and CH_2 intermediates (second carbon on the TiC surface), with an associated energy barrier of 0.44 eV.

In this investigation, we also evaluated the adsorption free energies of hydrogen atoms on various sites across the carbide surfaces. The results show that the proton adsorption on the metal site of TiC is endothermic with an adsorption free energy of 0.12 eV, as illustrated in the free energy profile in Figure 4 with the red path. While the adsorption of free energy of hydrogen was computed at each step, the carbon site consistently emerged as the most energetically favorable. We categorize the performance of materials on the basis of electrochemical performance, and our studied carbides are expected to produce methane at onset potentials below −1.0 V. The smaller (closer to zero) the onset potential, the better for CO reduction. The order of activity for the carbides studied here is as follows: TiC (−0.44 V) > MoC, NbC, and ZrC (−0.61 V) > CrC (−0.86 V) > WC (−0.87 V) > VC (−0.93 V), as shown in Figure 5. All of these surfaces are promising for CORR with onset potential below −1.0 V; however, TiC is the most promising surface not only due to the smaller onset potential needed but also due to the minimum number of proton-electron transfer steps to make methane.

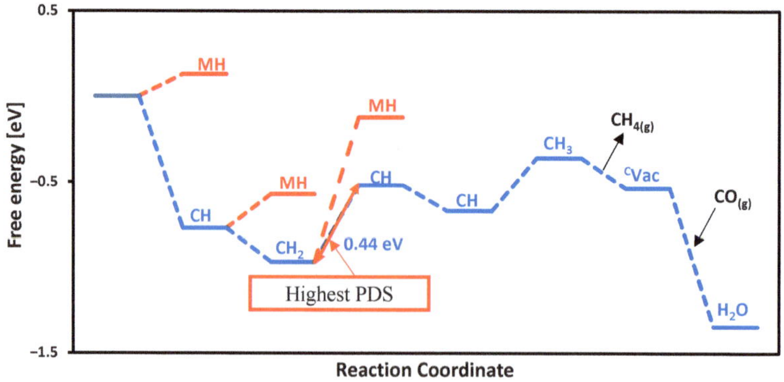

Figure 4. Calculated free energy diagram for CH_4 production on the TiC surface via the MvK mechanism. CH is the proton on the carbon site, and MH is the proton on the metal site of TiC. The highest PDS is between CH_2 and CH.

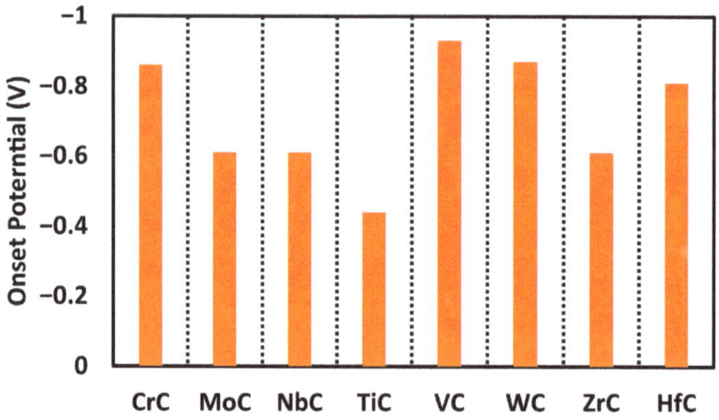

Figure 5. Comparison of the obtained onset potential for each catalyst for the production of methane.

At each protonation stage, all adsorption sites of carbides, such as carbon, metal, and bridge, were systematically explored. However, only the most thermodynamically favorable reaction pathways are reported. The MvK mechanism involves four proton–electron transfer steps to release methane. Depending on the specific catalyst, additional proton adsorption may occur on metal (MH), bridge (BrH), or different carbon sites (CH), as illustrated for MoC in Figure 6. In the first protonation step, a proton adsorbs on the first carbon atom. In the second step, a proton adsorbs on the second carbon atom, and this continues until all 4 carbon atoms of the surface are occupied by protons. In the fifth protonation step, the proton absorbs on the first bridge site, followed by adsorption on the second bridge site in the sixth step. The seventh protonation is more exergonic, as the first carbon forms a CH_2 intermediate, which subsequently leads to the formation of CH_4, as shown in Figure 6. For most surfaces studied, methane generation followed by carbon vacancy refilling via atmospheric CO is a facile process, effectively completing the catalytic cycle.

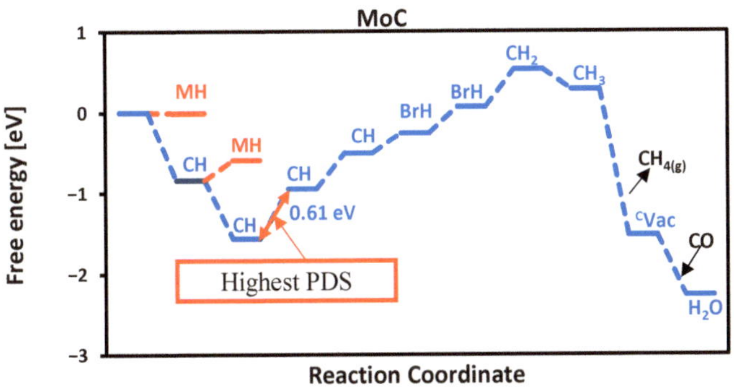

Figure 6. Free energy diagram of MoC. The highest PDS is between CH and CH.

3.3. Vacancy Stability and Carbon Migration

Within the MvK framework, surface vacancies can extend into subsurface layers, potentially enabling the upward diffusion of the carbon atoms from the interior to occupy surface defects. If sustained, this process could eventually result in the depletion of bulk carbon atoms in the form of methane, effectively reducing the transition of metal carbide (TMC) to its pure metallic form. This study explores the thermodynamic feasibility and kinetic barriers associated with carbon migration from subsurface layers to the surface. To analyze the energy profile of this transformation, the CI-NEB method was performed to identify the minimum energy path and corresponding energy barriers [47]. Specifically, we calculated the activation energy (Ea) required for a carbon atom to migrate from the second to the topmost layer. Among the studied materials, TiC emerged as a noteworthy candidate, exhibiting a diffusion barrier of 2.94 eV. This high barrier indicates that carbon migration under ambient conditions is energetically demanding, confirming the structural robustness of TiC. Thus, both thermodynamic and kinetic considerations suggest that TiC is resistant to decomposition, as shown in Figure 7.

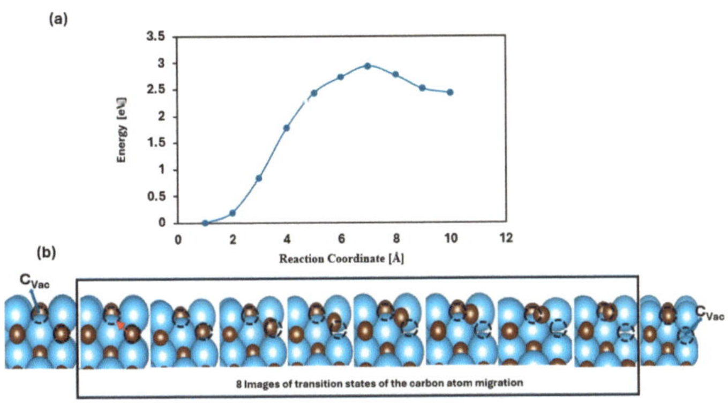

Figure 7. Schematic illustration depicting methane release and carbon migration: (**a**) shows the calculated energy barrier associated with the diffusion of a carbon atom from the subsurface layer to the top surface layer; (**b**) the first image illustrates the carbon vacancy (C_{Vac}), the next eight images represent the transition states corresponding to the movement of carbon atoms of carbide from the 2nd layer toward the surface, and the last image represents the vacancy at the 2nd layer. The red arrow in the second image represents the direction of motion of the carbon atom from the sublayer towards the surface carbon vacancy.

The energy difference between vacancies located at the surface and in the sublayer was determined using the equation $\Delta E = E_{vac,2} - E_{vac,1}$. The term $E_{vac,2}$ corresponds to the vacancy energy of the sublayer of the carbides, and $E_{vac,1}$ represents the vacancy energy on the surface. The results, summarized in Table 1, indicate that the catalysts demonstrate both thermodynamic and kinetic stability. Although the thermodynamic value for WC suggests it is thermodynamically unstable, further kinetic study shows that it is stable, as shown in Table 1. All of the kinetic values for the migration of carbon from the sublayer to the first layer in our studies are greater than 1.0 eV, which indicates that the rate of the migration reaction will be slow, and this indicates more stability.

Table 1. Calculated thermodynamics and kinetics values of migration of a carbon atom from the sublayer to the first layer.

Catalyst	CrC	HfC	MoC	NbC	TiC	VC	WC	ZrC
E_a	2.77	3.41	2.21	2.46	2.94	1.46	1.92	3.28
ΔE	1.37	2.62	0.1	1.35	2.43	1.37	−0.68	2.62

3.4. Electronic Structure Analysis

Furthermore, we investigated the electronic properties of the materials by calculating the density of states (DOS) and performing charge analysis. Figure 8 illustrates the DOS for TiC, identified as the most promising catalyst. The results reveal an overlap between the carbon p orbitals and metal d orbitals, with states extending across the Fermi level—indicating the metallic nature of these transition metal carbides.

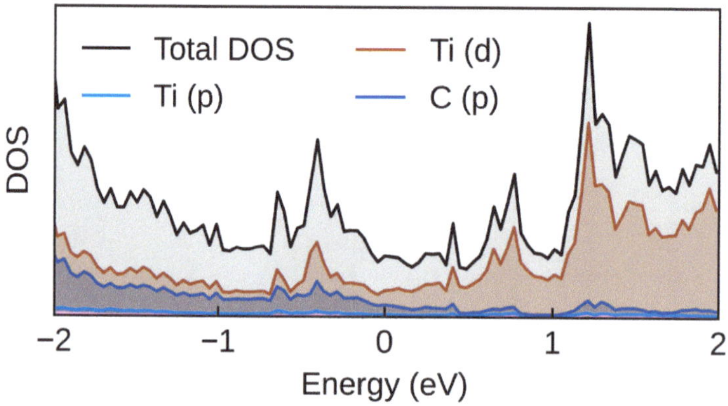

Figure 8. Calculated DOS of TiC.

The computed Bader charges for hydrogen adsorbed on each carbide are as follows: VC (0.10), CrC (0.08), MoC (0.08), NbC (0.05), WC (0.02), HfC (0.02), TiC (0.01), and ZrC (0.02). A positive Bader charge reflects electron transfer from the catalyst surface to the adsorbed species, while a negative value indicates electron donation from the gas molecule to the surface. To calculate the charge density of H adsorption on our most promising candidate, TiC, we used Equation (9) [48].

$$\Delta\rho = \rho_{AB} - \rho_A - \rho_B \quad (9)$$

where $\Delta\rho$ indicates the total change in charge density, ρ_{AB} represents the charge density of adsorbed H on the surface of TiC, ρ_A is the charge density of TiC, and ρ_B is the charge density of H. Yellow and cyan colors are used to show charge accumulation and depletion during H adsorption on TiC as depicted in Figure 9a. Furthermore, the electron localization function (ELF) has been computed to assess the electron density behavior of hydrogen on the TiC surface, as illustrated in Figure 9b, which facilitates the differentiation between confined regions and more dispersed areas. The ELF varies from 0 to 1, indicating total delocalization (as observed in metallic systems) as well as strong localization (as seen in covalent interactions) [49]. Our findings elucidate the collective electron interactions, emphasizing the attributes of covalent bonding.

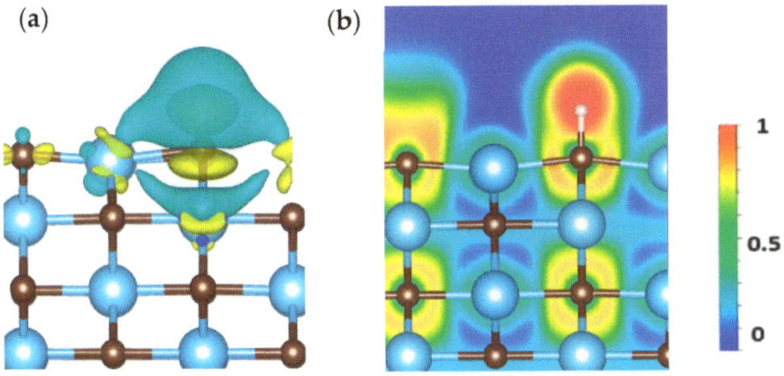

Figure 9. (**a**) represents the calculated charge density difference, and (**b**) represents the electron localization function (ELF) of H adsorption on the surface of the TiC.

3.5. Experimental Validation

In the previous studies, theoretical works demonstrated how DFT calculations are helpful to provide key guidelines to experimentalists. For example, in a previous study, Ti_2C_3 and Mo_2C were investigated theoretically and experimentally for CO_2 reduction. The DFT calculations indicated that intermediates were chemisorbed on the surface of catalysts, which is challenging for reduction towards further intermediates and products [50]. In addition, experimental studies for Ti_2CT_x (T_x = F and O) and Mo_2CT_x were carried out for formic acid formation at −1.0 V, and then computational studies via DFT calculations explained how the termination group (-F) helped to tune the adsorption strength of CO_2 and its intermediates [51]. This study also gives a detailed analysis for experimenters on how carbon on the surface of the carbides causes the formation of CH_4, and the carbon vacancy accommodates the atmospheric CO to fill the vacancy to regenerate the surface.

4. Conclusions

Transition metal carbides (TMCs) with (110) facets have been widely explored as catalysts for carbon monoxide reduction reactions. To identify the most reactive surface sites, we calculated the adsorption free energies of various species on all potential adsorption sites, including carbon, bridge, and metal positions. Our findings reveal that carbon sites exhibit superior activity compared to other examined locations. To further evaluate catalytic performance, we investigated the MvK mechanism for CO reduction on TMC surfaces. The results highlight the relevance of the MvK mechanism for methane production, with TiC emerging as a particularly effective catalyst, showing an onset potential of −0.44 V. The onset potential was determined to be −0.93 V for VC, −0.86 V for CrC, −0.87 V for WC, −0.81 V for HfC, and −0.61 V for MoC, NbC, and ZrC, respectively. Additionally, we evaluated the structural stability of these carbides and the persistence of surface vacancies by analyzing both the thermodynamic and kinetic aspects of carbon atom migration from sublayers to the surface. The calculated diffusion barriers indicate that this migration is energetically demanding, confirming the robust stability of these catalysts. The favorable hydrogen adsorption on carbon sites, low onset potentials, and substantial diffusion barriers collectively suggest that TMC (110) surfaces are highly suitable for methane production via the MvK pathway.

Author Contributions: N.A.: writing—original draft, methodology, investigation, and formal analysis; Y.A.: visualization, supervision, software, resources, project administration, conceptualization, and writing—review and editing. All authors have read and agreed to the published version of the manuscript.

Funding: The financial support was provided by the Research Fund of the University of Iceland and the Technology Development Fund of Rannis via grant No. 2215496-0611.

Data Availability Statement: The original contributions presented in this study are included in the article. Further inquiries can be directed to the corresponding authors.

Conflicts of Interest: The authors declare no conflict of interest.

References

1. Ahmad, T.; Zhang, D. A critical review of comparative global historical energy consumption and future demand: The story so far. *Energy Rep.* **2020**, *6*, 1973–1991. [CrossRef]
2. Möller, T.; Ju, W.; Bagger, A.; Wang, X.; Luo, F.; Thanh, T.N.; Varela, A.S.; Rossmeisl, J.; Strasser, P. Efficient CO_2 to CO electro on solid Ni–N–C catalysts at industrial current densities. *Energy Environ. Sci.* **2019**, *12*, 640–647. [CrossRef]
3. Prats, H.; Stamatakis, M. Atomistic and electronic structure of metal clusters supported on transition metal carbides: Implica for catalysis. *J. Mater. Chem. A* **2022**, *10*, 1522–1534. [CrossRef]
4. Awais, M.; Ashraf, N.; Abghoui, Y. Mechanistic roadmap for CO_2 to methane conversion on tailored carbonitride surfaces. . *Surf. Sci.* **2025**, *710*, 163815. [CrossRef]
5. Reddy, K.P.; Dama, S.; Mhamane, N.B.; Ghosalya, M.K.; Raja, T.; Satyanarayana, C.V.; Gopinath, C.S. Molybdenum ca catalyst for the reduction of CO_2 to CO: Surface science aspects by NAPPES and catalysis studies. *Dalton Trans.* **201** 12199–12209. [CrossRef] [PubMed]
6. Hafeez, J.; Islam, M.U.; Ali, S.M.; Khalid, S.; Ashraf, N. Computational exploring the potential of pure and Ag-decorated V for detecting volatile organic compounds (VOCs). *Mater. Sci. Semicond. Process.* **2024**, *182*, 108710. [CrossRef]
7. Deng, Y.; Ge, Y.; Xu, M.; Yu, Q.; Xiao, D.; Yao, S.; Ma, D. Molybdenum carbide: Controlling the geometric and electronic stru of noble metals for the activation of O–H and C–H bonds. *Acc. Chem. Res.* **2019**, *52*, 3372–3383. [CrossRef] [PubMed]
8. Porosoff, M.D.; Yang, X.; Boscoboinik, J.A.; Chen, J.G. Molybdenum carbide as alternative catalysts to precious metals for hi selective reduction of CO_2 to CO. *Angew. Chem. Int. Ed.* **2014**, *53*, 6705–6709. [CrossRef]
9. Ashraf, N.; Iqbal, A.; Abghoui, Y. Exploring reaction mechanisms for CO_2 reduction on Carbides. *J. Mater. Chem. A* **202** 30340–30350. [CrossRef]
10. Awais, M.; Ashraf, N.; Abghoui, Y. Engineering innovative catalysts for efficient CO_2 reduction toward carbon neutrali *Environ. Chem. Eng.* **2025**, *13*, 116621. [CrossRef]
11. Ashraf, N.; Abghoui, Y. Innovative catalysis for CO reduction: Paving the way towards Greener future. *Int. J. Hydrogen En* **2025**, *136*, 383–391. [CrossRef]
12. Mubaraka, F.; Rafique, H.; Najeeb, J.; Akram, S.; Munir, H.; Naeem, S.; Kausar, N.; Ashraf, N. Synthesis of amino a functionalized iron oxide nanoparticles for response surface methodology-based statistical optimization of photocata degradation of methylene blue. *Int. J. Environ. Sci. Technol.* **2024**, *21*, 2489–2504. [CrossRef]
13. Viñes, F.; Sousa, C.; Liu, P.; Rodriguez, J.; Illas, F. A systematic density functional theory study of the electronic structure of and (001) surface of transition-metals carbides. *J. Chem. Phys.* **2005**, *122*, 174709. [CrossRef]
14. Posada-Pérez, S.; Vines, F.; Ramirez, P.J.; Vidal, A.B.; Rodriguez, J.A.; Illas, F. The bending machine: CO_2 activation hydrogenation on δ-MoC (001) and β-Mo_2C (001) surfaces. *Phys. Chem. Chem. Phys.* **2014**, *16*, 14912–14921. [CrossRef]
15. Ferri, T.; Gozzi, D.; Latini, A. Hydrogen evolution reaction (HER) at thin film and bulk TiC electrodes. *Int. J. Hydrogen En* **2007**, *32*, 4692–4701. [CrossRef]
16. Kunkel, C.; Vines, F.; Illas, F. Biogas upgrading by transition metal carbides. *ACS Appl. Energy Mater.* **2017**, *1*, 43–47. [Cross
17. Lee, J.S.; Yeom, M.H.; Lee, D.-S. Catalysis by Molybdenum Carbide in Activation of CC, CO and CH bonds. *J. Mol. Catal.* **199** L45–L51. [CrossRef]

Rodriguez, J.; Liu, P.; Dvorak, J.; Jirsak, T.; Gomes, J.; Takahashi, Y.; Nakamura, K. Adsorption and decomposition of SO_2 on TiC (001): An experimental and theoretical study. *Surf. Sci.* **2003**, *543*, L675–L682. [CrossRef]

Brungs, A.J.; York, A.P.; Green, M.L. Comparison of the group V and VI transition metal carbides for methane dry reforming and thermodynamic prediction of their relative stabilities. *Catal. Lett.* **1999**, *57*, 65–69. [CrossRef]

Claridge, J.B.; York, A.P.; Brungs, A.J.; Marquez-Alvarez, C.; Sloan, J.; Tsang, S.C.; Green, M.L. New catalysts for the conversion of methane to synthesis gas: Molybdenum and tungsten carbide. *J. Catal.* **1998**, *180*, 85–100. [CrossRef]

Viñes, F.; Rodriguez, J.A.; Liu, P.; Illas, F. Catalyst size matters: Tuning the molecular mechanism of the water–gas shift reaction on titanium carbide based compounds. *J. Catal.* **2008**, *260*, 103–112. [CrossRef]

Ono, L.K.; Roldan-Cuenya, B. Effect of interparticle interaction on the low temperature oxidation of CO over size-selected Au nanocatalysts supported on ultrathin TiC films. *Catal. Lett.* **2007**, *113*, 86–94. [CrossRef]

Sullivan, M.M.; Chen, C.-J.; Bhan, A. Catalytic deoxygenation on transition metal carbide catalysts. *Catal. Sci. Technol.* **2016**, *6*, 602–616. [CrossRef]

Ashraf, N.; Abghoui, Y. Dynamics of C_1 and C_2 Products Formation on (110) Facets of Carbides. *Surf. Interfaces* **2025**, *70*, 106793. [CrossRef]

Hiragond, C.B.; Kim, H.; Lee, J.; Sorcar, S.; Erkey, C.; In, S.-I. Electrochemical CO_2 reduction to CO catalyzed by 2D nanostructures. *Catalysts* **2020**, *10*, 98. [CrossRef]

Powar, N.S.; Hiragond, C.B.; Bae, D.; In, S.-I. Two-dimensional metal carbides for electro-and photocatalytic CO_2 reduction. *J. CO_2 Util.* **2022**, *55*, 101814. [CrossRef]

Xiao, T.-c.; Hanif, A.; York, A.P.; Nishizaka, Y.; Green, M.L. Study on the mechanism of partial oxidation of methane to synthesis gas over molybdenum carbide catalyst. *Phys. Chem. Chem. Phys.* **2002**, *4*, 4549–4554. [CrossRef]

Koerts, T.; Deelen, M.J.; Van Santen, R.A. Hydrocarbon formation from methane by a low-temperature two-step reaction sequence. *J. Catal.* **1992**, *138*, 101–114. [CrossRef]

Mine, S.; Toyao, T.; Hinuma, Y.; Shimizu, K.-I. Understanding and controlling the formation of surface anion vacancies for catalytic applications. *Catal. Sci. Technol.* **2022**, *12*, 2398–2410. [CrossRef]

Ahmed, I.; Jhung, S.H. Catalytic oxidation reactions for environmental remediation with transition metal nitride nanoparticles. *J. Environ. Chem. Eng.* **2024**, *12*, 112907. [CrossRef]

Daisley, A.; Hargreaves, J. Metal nitrides, the Mars-van Krevelen mechanism and heterogeneously catalysed ammonia synthesis. *Catal. Today* **2023**, *423*, 113874. [CrossRef]

Ellingsson, V.; Iqbal, A.; Skúlason, E.; Abghoui, Y. Nitrogen reduction reaction to ammonia on transition metal carbide catalysts. *ChemSusChem* **2023**, *16*, e202300947. [CrossRef]

Iqbal, A.; Skúlason, E.; Abghoui, Y. Catalytic Nitrogen Reduction on the Transition Metal Carbonitride (110) Facet: DFT Predictions and Mechanistic Insights. *J. Phys. Chem. C* **2024**, *128*, 10300–10307. [CrossRef]

Sinev, M.Y. Oxygen Activation and Pathways in High-Temperature Reactions of Light Alkane Oxidation: A Seeming Simplicity of Kinetic Description. *Kinet. Catal.* **2019**, *60*, 420–431. [CrossRef]

Zeinalipour-Yazdi, C.D.; Hargreaves, J.S.; Catlow, C.R.A. Low-T mechanisms of ammonia synthesis on Co_3Mo_3N. *J. Phys. Chem. C* **2018**, *122*, 6078–6082. [CrossRef]

Hanifpour, F.; Canales, C.P.; Fridriksson, E.G.; Sveinbjörnsson, A.; Tryggvason, T.K.; Yang, J.; Arthur, C.; Jónsdóttir, S.; Garden, A.L.; Ólafsson, S. Operando quantification of ammonia produced from computationally-derived transition metal nitride electrocatalysts. *J. Catal.* **2022**, *413*, 956–967. [CrossRef]

Zeinalipour-Yazdi, C.D.; Hargreaves, J.S.; Laassiri, S.; Catlow, C.R.A. The integration of experiment and computational modelling in heterogeneously catalysed ammonia synthesis over metal nitrides. *Phys. Chem. Chem. Phys.* **2018**, *20*, 21803–21808. [CrossRef] [PubMed]

Ashraf, N.; Abghoui, Y. Electrochemical synthesis of methane on (110) facets of carbides via MvK mechanism. *Electrochim. Acta* **2025**, *525*, 146069. [CrossRef]

Kresse, G.; Furthmüller, J. Software VASP, vienna (1999). *Phys. Rev. B* **1996**, *54*, 169.

Hussain, J.; Jónsson, H.; Skúlason, E. Calculations of product selectivity in electrochemical CO_2 reduction. *ACS Catal.* **2018**, *8*, 5240–5249. [CrossRef]

Jovanov, Z.P.; Hansen, H.A.; Varela, A.S.; Malacrida, P.; Peterson, A.A.; Nørskov, J.K.; Stephens, I.E.; Chorkendorff, I. Opportunities and challenges in the electrocatalysis of CO_2 and CO reduction using bifunctional surfaces: A theoretical and experimental study of Au–Cd alloys. *J. Catal.* **2016**, *343*, 215–231. [CrossRef]

Ashraf, N.; Betolaza, D.B.; Gunnarsson, H.I.; Khatibi, M.I.; Iqbal, A.; Abghoui, Y. How can phosphides catalyze CO_2 reduction reaction? *Electrochim. Acta* **2025**, *517*, 145755. [CrossRef]

43. Nørskov, J.K.; Rossmeisl, J.; Logadottir, A.; Lindqvist, L.; Kitchin, J.R.; Bligaard, T.; Jonsson, H. Origin of the overpotenti oxygen reduction at a fuel-cell cathode. *J. Phys. Chem. B* **2004**, *108*, 17886–17892. [CrossRef] [PubMed]
44. Li, G.; Han, J.; Wang, H.; Zhu, X.; Ge, Q. Role of dissociation of phenol in its selective hydrogenation on Pt (111) and Pd *ACS Catal.* **2015**, *5*, 2009–2016. [CrossRef]
45. Atkins, P.W.; De Paula, J.; Keeler, J. *Atkins' Physical Chemistry*; Oxford University Press: Oxford, UK, 2023.
46. Kröger, F.; Vink, H. Relations between the concentrations of imperfections in crystalline solids. In *Solid State Physics*; Els Amsterdam, The Netherlands, 1956; Volume 3, pp. 307–435.
47. Jónsson, H.; Mills, G.; Jacobsen, K.W. Nudged Elastic Band Method for Finding Minimum Energy Paths of Transitions. In *Cl and Quantum Dynamics in Condensed Phase Simulations*; World Scientific: Singapore, 1998; pp. 385–404.
48. Iqbal, A.; Skulason, E.; Abghoui, Y. Electrochemical Nitrogen Reduction to Ammonia at Ambient Condition on the (111) Fac Transition Metal Carbonitrides. *ChemPhysChem* **2024**, *25*, e202300991. [CrossRef]
49. Fan, K.; Ying, Y.; Li, X.; Luo, X.; Huang, H. Theoretical investigation of V_3C_2 MXene as prospective high-capacity anode ma for metal-ion (Li, Na, K, and Ca) batteries. *J. Phys. Chem. C* **2019**, *123*, 18207–18214. [CrossRef]
50. Attanayake, N.H.; Banjade, H.R.; Thenuwara, A.C.; Anasori, B.; Yan, Q.; Strongin, D.R. Electrocatalytic CO_2 reduction on abundant 2D Mo_2C and Ti_3C_2 MXenes. *Chem. Commun.* **2021**, *57*, 1675–1678. [CrossRef]
51. Handoko, A.D.; Chen, H.; Lum, Y.; Zhang, Q.; Anasori, B.; Seh, Z.W. Two-dimensional titanium and molybdenum ca MXenes as electrocatalysts for CO_2 reduction. *iScience* **2020**, *23*, 101181. [CrossRef]

molecules

MDPI

:le

ultiple Functions of Carbon Additives in NASICON-Type ectrodes for Stabilizing the Sodium Storage Performance

:he Tushev [1], Sonya Harizanova [1], Maria Shipochka [1], Radostina Stoyanova [1,2] and Violeta Koleva [1,*]

[1] Institute of General and Inorganic Chemistry, Bulgarian Academy of Sciences, 1113 Sofia, Bulgaria; tushev@svr.igic.bas.bg (T.T.); sonya@svr.igic.bas.bg (S.H.); shipochka@svr.igic.bas.bg (M.S.); radstoy@svr.igic.bas.bg (R.S.)

[2] National Centre of Excellence Mechatronics and Clean Technologies, 1113 Sofia, Bulgaria

* Correspondence: vkoleva@svr.igic.bas.bg

Abstract

Recently, there has been increased interest in NASICON-type electrodes for sodium-ion batteries due to their unique combination of intercalation properties, low cost, and safety. However, their commercialization is hindered by the low electrical conductivity. One strategy to overcome this issue is to integrate NASICON materials with carbon additives. This study shows that carbon additives improve the sodium storage performance of a NASICON-type electrode in various ways, depending on the additives' functional groups, texture, and conductivity properties. The proof-of-concept is based on a multi-electron phospho-sulphate electrode, $NaFeVPO_4(SO_4)_2$ (NFVPS) mixed with carbon black (C) and reduced graphene oxide (rGO). Carbon-coated samples are obtained via a simple ball milling procedure followed by thermal treatment in an argon flow. Sodium storage in the composites occurs through capacitive and Faradaic reactions. The Faradaic reaction is facilitated at the carbon black composite, while the capacitive reaction dominates for the rGO composite. NFVPS operates through two-electron reactions at 20 °C, while the increased temperatures favor the three-electron reaction. The rGO composite outperforms the carbon black composite in terms of cycling stability and rate capability at 20 and 40 °C. The role of the rGO and carbon black in electrochemical performance is discussed based on the different reactivity of hydroxyl/epoxide and carbonyl functional groups with the electrolyte salt, $NaPF_6$, and the solvent, polypropylene carbonate.

Keywords: sodium-ion batteries; mixed phosphate-sulphate electrodes; $NaFeVPO_4(SO_4)_2$; rGO-based composite; carbon black-based composite; electrode stability; cycling stability; rate capability

1. Introduction

Sodium-ion batteries (NIBs) are emerging as competitive alternatives to lithium-ion systems, combining low cost and inherent safety with a comparable intercalation chemistry [1]. In this context, NASICON-derived materials with compositions $A_xM_y(XO_4)_n$ (A = Na, Li, K, M = Fe, Mn, Co, etc. and X = P, S, Si, Mo, etc.) stand out as especially attractive electrode candidates [2]. The NASICON structure allows for the rapid diffusion of the Na^+ ions, enabling fast and reversible sodium intercalation without significant structural degradation [3]. Although NASICONs deliver excellent cycling stability and high rate capability, their intrinsically low electronic conductivity demands tailored electrode architectures to unlock optimal sodium-storage performance [4–6]. Three principal strategies address these limitations: structural engineering, morphology design and surface modification [2,4–6]. Structural engineering tunes the electronic structure via selective single-ion or multi-ion doping [7–9], while morphology design manipulates the nucleation and growth of NASICON particles to expose the crystallographic planes or axes where sodium intercalation occurs preferentially [10,11]. In comparison with these two strategies, surface modification enhances the overall conductivity by integrating conductive, porous carbon additives onto the phosphate framework [12–15]. Surface modification is more effective because it affects electronic conductivity, electrode-electrolyte interactions, electrolyte wetting of the electrode, corrosion resistance, thermal stability, and leaching of transition metal ions from NASICON phase.

Due to variety in the texture and conductivity properties, low cost, and easy synthesis procedure, the use of carbonaceous materials for the surface modification of NASICON-type electrodes has become the main research topic [4–6]. Through hydrothermal assisted sol-gel methods, electrospinning technique and high-boiling organic solvent-assisted colloidal synthesis, the core-shell nanocomposite, $Na_3V_2(PO_4)_3$@C, having excellent cycling stability, has been synthesized [16–20]. Mechanical ball milling followed by thermal reduction at various temperatures has also proven effective in preparing $Na_3V_2(PO_4)_3$@C with enhanced performance [21]. Another part of the study involves carbon doping with nitrogen and/or boron to accelerate the Na^+ transport in the carbon layer, thereby improving the rate capability and cycling stability of NASICON-type electrodes [22]. Pyrolysis of organic precursors is a common method for producing amorphous carbon coating [23]. This coating is then treated (namely by chemical vapour deposition, etc.) to form highly conductive, graphene-like coated carbon (for example, carbon-coated $Na_3V_2(PO_4)_3$ with 0.5% amorphous embedded in rGO with about 3%) [24]. The integration of the NASICON-material into flexible carbon network (such as carbon cloth, carbon nanofibers/nanotubes, graphene paper, etc.) has also been reported as effective method for improving the total conductivity of the electrode [20,25–28]. In summary, much research has been focused on the electrochemical performance of carbon-modified NASICON-type electrodes, while little research has addressed their total thermal stability [29]. Despite the good thermal stability of NASICON-type materials [30] the carbon additives in modified NASICON electrodes usually have lower thermal stability, resulting in lower total thermal stability. This is an important issue for the practical application of NASICON-type electrodes.

The present study is focused on the tracking the effect of carbon additive on both the thermal stability and sodium storage performance of multi-electron NASICON-type electrodes. The multi-electron phospho-sulphate, $NaFeVPO_4(SO_4)_2$, is selected as the NASICON-type electrode because it is able to reversibly intercalate from two to three moles of Na^+ ions between 1.5 and 4.2 volts [31,32]. This process is accompanied by limited variation in the lattice volume (up to 5.2%) [31]. In general, the sulphate salts are less thermally stable than the phosphate salts, but from electrochemical point of view the presence of SO_4^{2-} groups in the electrode material is valuable since SO_4^{2-} group increases the potential of the redox reactions, thus increasing the power density of the cells [31,32]. The carbon additives comprise the carbon black and reduced graphene oxide (rGO). These additives have been shown to be effective for improving the electrochemical performance of iron phospho-sulphate, $NaFe_2PO_4(SO_4)_2$@rGO, and mixed iron-vanadium and phospho-sulphate, $NaFeVPO_4(SO_4)_2$ [31,33]. When CNT is used, the composite $NaFe_{1.6}V_{0.4}(PO_4)(SO_4)_2$@CNT operates at a working voltage of approximately 3 V due to the Fe^{3+}/Fe^{2+} couple [34], which is contrary to the previous findings [31–33].

The carbon coated $NaFeVPO_4(SO_4)_2$ samples are obtained via a simple ball milling procedure followed by thermal treatment at 400 °C in an argon flow. The storage performance was analysed in model sodium half-cells at elevated temperatures. Ex-situ XRD and XPS analyses allow us to distinguish the roles of carbon black and rGO in the electrochemical performance of $NaFeVPO_4(SO_4)_2$ (hereinafter referred to as NFVPS).

2. Results

2.1. Characterization of NFVPS/rGO and NFVPS/C Composites

The active phase $NaFeVPO_4(SO_4)_2$ (NFVPS) adopts NASICON-type structure in trigonal space group *R*-3 (Figure S1a) where corner shared $(Fe,V)O_6$ octahedra and $(P,S)O_4$ tetrahedra form three-dimensional channels for the sodium ions movements [31]. The lattice parameters are a = 8.4691 (1) Å and c = 22.0162 (1) Å (Table S1). After ball milling of NFVPS with 15 wt.% carbon additives, carbon black and rGO (denoted as NFVPS/C and NFVPS/rGO, respectively), the NASICON crystal structure remains intact, and the lattice parameters approach those of untreated NFVPS (see Figure S1 and Table S1). The crystallinity of the carbon-modified samples is also preserved. The crystallite size varies from 53 to 48 nm.

The presence of carbon additives in the composites is evident from the HR-TEM images (Figure 1).

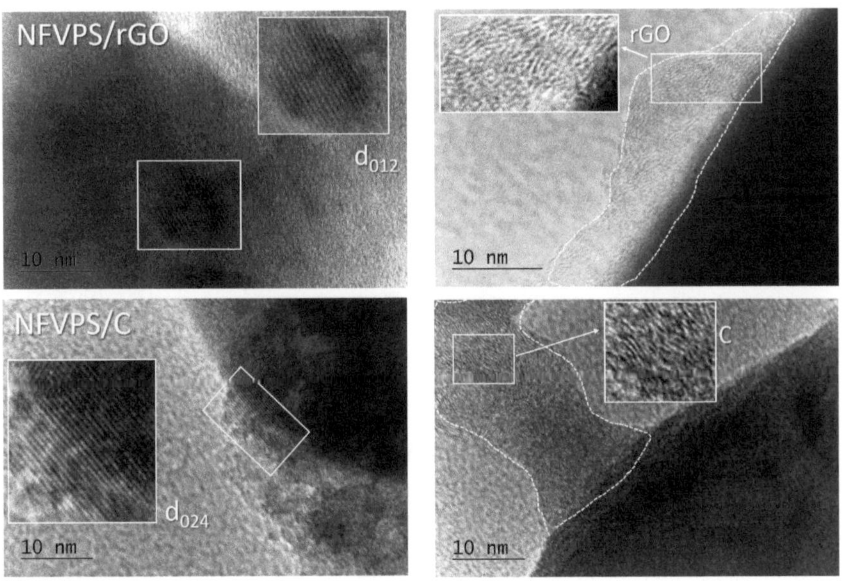

Figure 1. HR-TEM images of NFVPS/rGO (**top**) and NFVPS/C (**bottom**).

The carbon additives cover the phosphate-sulphate particles, but the carbon layers are inhomogeneous with thickness between 5 and 10 nm for NFVPS/rGO, and between 5 and 25 nm, for NFVPS/C. For the latter composite, the boundary between the NFVPS phase and carbon black cannot be clearly distinguished. It is of importance that both rGO and carbon black are in good interphase contact with the phospho-sulphate phase. Additionally, HR-TEM image of NFVPS/rGO reveals distinct fringes, with a calculated interplanar spacing of 0.60 nm that corresponds to the (012) plane of the trigonal NFVPS phase. This indicates that both the crystal structure and crystallinity of the phosphate-sulphate phase in the composite with rGO are intact. The same applies to the NFVPS/C composite, where the HR-TEM image shows lattice fringes with spacing of 0.30 nm corresponding to the (024) plane of the NFVPS lattice (Figure 1).

To determine the thermal stability of the carbon-modified NFVPS samples, the simultaneous thermogravimetry/differential thermal analysis/mass spectrometry analyses (TG/DTA/MS) were undertaken up to 1000 °C (Figure 2). The thermal decomposition of NFVPS begins at 440 °C with release of SO_2 gas and this process is accompanied by multiple endothermic effects on the DTA curve extending up to 830 °C (Figure 2a). The registration of CO_2 gas evolved between 430 and 650 °C reveals that the NFVPS sample contains residual in-situ-carbon generated from the citric acid presented in the precursor composition. The total mass loss for NFVPS is 26.5 mass %. These data reveal that the decomposition of NFVPS is a multistage process involving at least three stages, as determined by thermogravimetric analysis. Between 450 and 600 °C, NFVPS is probably decayed to $Fe_2(SO_4)_3$ and $Na_3V_2(PO_4)_3$, releasing SO_2 (mass loss of 9 mass % versus experimental value of 10%). Then, the in situ formed $Fe_2(SO_4)_3$ decomposes to iron oxide polymorphs, releasing SO_2, culminating in α-Fe_2O_3 at 700 °C [35]. Within this temperature range, V^{3+} ions can also be oxidized to V^{4+} ions [36], thus contributing to the decomposition of in-situ formed $Na_3V_2(PO_4)_3$. Above 700 °C, the final stage of decomposition takes place.

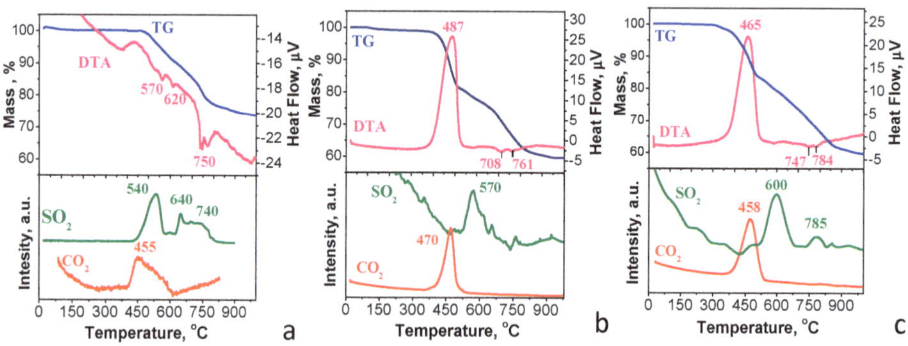

Figure 2. TG/DTA/MS curves of NFVPS (**a**), NFVPS/C (**b**) and NFVPS/rGO (**c**).

In comparison with carbon-free NFVPS sample, the decomposition of carbon-modified samples NFVPS/C and NFVPS/rGO starts at lower temperature (i.e., about 380 °C). This is related with the lower thermal stability of the carbon additives (Figure 2b,c). Moreover, the TG curves for NFVPS/C and NFVPS/rGO show two-step mass loss processes. These processes are accompanied by the initial release of CO_2 up to 500 °C, followed by the release of SO_2 up to 830 °C. The total mass loss reaches 40 mass %. Notably, SO_2 is released at a higher temperature in the composites than in the carbon-free NFVPS: 570 °C and 600 °C for NFVPS/C and NFVPS/rGO vs. 540 °C for NFVPS. This indicates that the presence of carbon additives on the NFVPS surface impacts the initial stage of NFVPS decomposition. This can be explained by the ability of carbonaceous materials to absorb SO_2 [37]. This retards the release of SO_2 during the NFVPS decomposition. It is worth mentioning that SO_2 is a toxic gas. Therefore, the enhanced temperature of SO_2 release in the composite supports its possible practical application. Once the carbon additive burns completely, the decomposition of the NFVPS becomes uncontrolled. Another difference in the thermal behavior of the carbon-free and carbon-modified NFVPS can be seen in the DTA curves (Figure 2). Due to the small amount of in-situ generated carbon in NFVPS, the endothermic process of the NFVPS decomposition dominates over the carbon oxidation and the overall thermal reaction registered on DTA curve is endothermic one (Figure 2a). The opposite is observed for the two composites. Due to the large amount of the carbon additive in the composites (15%) the exothermic process of the carbon oxidation to CO_2 dominates over the endothermic sulphate decomposition to SO_2 (Figure 2b,c).

The difference in the thermal properties of the composites can be related with the functional groups stabilized in the carbon additives. Figure 3 shows the XPS spectra of NFVPS/C and NFVPS/rGO in the region of the C1s binding energy.

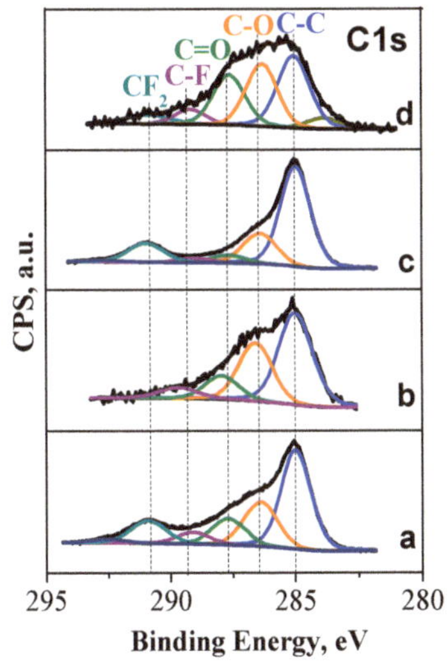

Figure 3. XPS spectra in the region of C1s binding energies of pristine NFVPS/rGO electrode (**a**) and after 200 cycles in Na-half cell (100 cycles at 20 °C and subsequent 100 cycles at 40 °C with a rate C/2) (**b**); pristine NFVPS/C electrode (**c**) and after 200 cycles in Na-half cell (100 cycles at 20 °C and subsequent 100 cycles at 40 °C with a rate C/2) (**d**).

The C1s spectra of NFVPS/C and NFVPS/rGO consist of a broad, asymmetric band that can be deconvoluted into peaks centered at 285.0, 286.5, 287.6, 289.3, and 290.8 eV. These peaks can be assigned to C C bonds in the carbon network as well as to single and double-bonded carbon-oxygen bonds: 286.5 eV for C–O and 287.6 eV for C=O. (It is worth mentioning that the peaks appearing above 289 eV are due to the C–F and CF_2 bonds in the PVDF used as a binder). The relative amounts of the hydroxyl/epoxide and carbonyl groups are listed on Table 1. The element contents determined from XPS spectra are given in Table S2.

Table 1. Ratio between the areas of the corresponding peaks from the deconvoluted spectra in the C1s region (Figure 3).

№	Electrodes	$\frac{C–O}{C–C}$	$\frac{C=O}{C–C}$
1	pristine NFVPS/rGO electrode	0.49	0.29
2	NFVPS/rGO electrode stopped at 1.5 V after 200 cycles: 100 cycles at 20 °C and next 100 cycles at 40 °C	0.61	0.25
3	pristine NFVPS/C electrode	0.33	0.08
4	NFVPS/C electrode stopped at 1.5 V after 200 cycles: 100 cycles at 20 °C and next 100 cycles at 40 °C	0.83	0.69

The comparison evidences that the epoxy and hydroxyl groups dominate the composite surface with carbon black. Meanwhile, both hydroxyl/epoxide groups and carbonyl groups contribute to the functionality of the composite with rGO. In comparison with C1s spectra, the assignments of O1s peaks to O species in the literature are not the straightforward procedure as in the case of C1s spectra [38,39]. However, the main conclusions from C1s spectra are supported by O1s spectra (Figure S2): the bands at around 532 and 533.0 eV that can be assigned to double-bonded O atoms in esters, carbonates, and acids, as well as to single-bonded O atoms in ketones, ethers, and alcohols [38,39]. Because the sample NFVPS/rGO has a higher concentration of the oxygen functional groups, it undergoes pyrolysis at a slightly lower temperature than that of NFVPS/C: 458 °C vs. 470 °C, respectively (Figure 2).

Figures 4 and S3 compare both the morphologies and porous characteristics of NFVPS and the carbon composites.

Figure 4. SEM images of NFVPS (**a**), NFVPS/rGO (**b**) and NFVPS/C (**c**); Pore size distribution curves (**d**).

For pristine NFVPS the SEM image shows large micrometers aggregates (above 5–10 μm) having big holes on their surface, the latter being formed from the release of gases during the thermal decomposition of the organic precursor (Figure 4). After ball milling of NFVPS with carbon additives, the aggregates break apart and become smaller. The sizes of the aggregates vary widely when rGO is added, ranging from 0.2 to 2–3 μm. In contrast, carbon black yields more regular aggregates with a close size distribution around 0.5 μm. This indicates that carbon additives stick to the NFVPS during ball milling (Figure 4). Furthermore, the nitrogen adsorption isotherms reveal a significant difference in the porosity of the three samples resulting from the presence of the carbon additives (Figure S3). The individual rGO and carbon black materials have very different texture properties as previously reported [40] and this reflects on the composites porosity. Because of high specific surface areas of the carbon additives (363 cm^2/g for rGO and 49 cm^2/g for carbon black [40]), the composites exhibit a two- to threefold increase in specific surface area compared to NFVPS: 31 cm^2/g for NFVPS/rGO and 17 cm^2/g for NFVPS/C versus 9 cm^2/g for NFVPS [31]. In the same order, the total pore volumes substantially increase: 0.12 cm^3/g for NFVPS/rGO and 0.10 cm^3/g for NFVPS/C vs. 0.03 cm^3/g for the uncoated sample.

Regarding pore size distribution, the three samples exhibit a large part of the mesopores with diameters in the range of 3–9 nm, as for NFVPS these pores are the predominant ones (Figure 4d). In contrast, for the composites there is additional contribution of larger mesopores: with diameters between 9 and 50 nm for NFVPS/C and between 15 and 50 nm for NFVPS/rGO, which are associated with the carbon additives [40]. From nitrogen adsorption studies it can be concluded that the high porosity of the composites could facilitate the electrolyte penetration in the electrode materials, thus enhancing the kinetics of sodium ion diffusion during the electrochemical cycling. On the other hand, the pore structure of NFVPS/rGO is more developed than that of NFVPS/C which is a prerequisite for its better electrochemical performance.

2.2. Electrochemical Studies

Figures 5 and S4 compare the CV curves of NFVPS/rGO and NFVPS/C composites at different scanning rate.

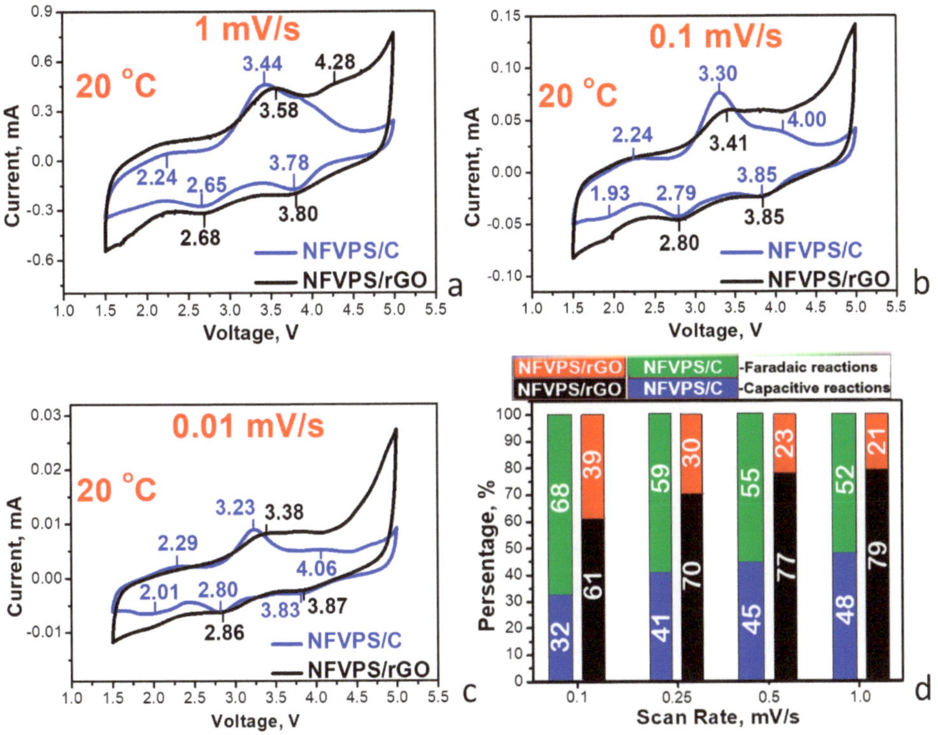

Figure 5. CV curves with scanning rates of 1 mV/s (**a**), 0.1 mV/s (**b**) and 0.01 mV/s (**c**) between 1.5 and 5.0 V for NFVPS/C and NFVPS/rGO in sodium-half cells at 20 °C. Contribution of Faradaic and Capacitive reactions (**d**).

Between 1.5 and 5.0 V, the CV curves display a series of redox peaks, corresponding to the $V^{3+}/V^{2+/4+}$ and Fe^{3+}/Fe^{2+} couples [31]. From the one hand, this evidences that electrochemical reaction proceeds in the same way for the composites. From the other hand, the redox peaks are more pronounced in the NFVPS/C composite, thus implying that capacitive reactions mainly occurs for the composite with rGO. Taking into account the correlation between the current and scanning rate [41], the contribution of the Faradaic and capacitive reactions towards sodium storage is evaluated (Figure 5d). The comparison shows that the Faradaic reaction dominates the electrochemical reaction at scanning rates lower than 1 mV/s, while the capacitive reaction becomes dominant at rates above 1 mV/s. Among the composites, the Faradaic reaction is facilitated by the composite with carbon black, while capacitive reaction dominates for the composite with rGO [42–44]. This can also be related with well-developed pore structure of NFVPS/rGO.

To compare the sodium storage performance of NFVPS/rGO and NFVPS/C, a cycling stability test was carried out at 20 °C and 40 °C at a C/2 rate. The testing protocol comprises the first 100 cycles at 20 °C and the next 100 cycles at 40 °C. The corresponding charge/discharge curves at a given cycle are presented on Figure 6.

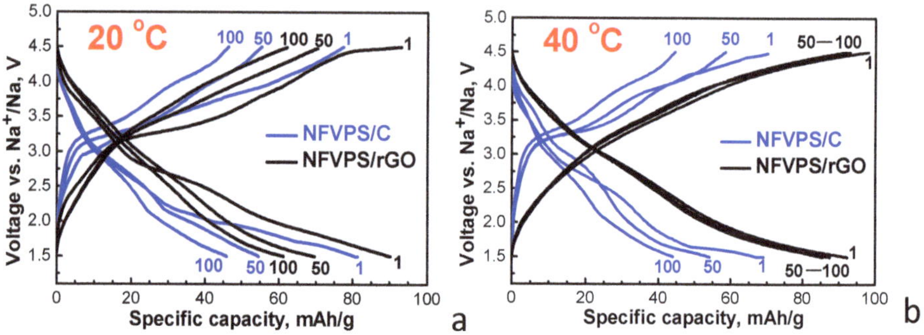

Figure 6. Charge-discharge curves (1st, 50th and 100th cycle) with a C/2 rate between 1.5 and 4.5 V of NFVPS/rGO and NFVPS/C at 20 °C (**a**) and 40 °C (**b**). The electrolyte is $NaPF_6$ in propylene carbonate (PC).

The charge/discharge curves of the two composites display similar profiles between 1.5 and 4.5 V, once again demonstrating that the sodium storage mechanism is the same as that observed in CV experiments (Figure 5).

At 20 °C, the first discharge capacity of both NFVPS/C and NFVPS/rGO is comparable: 80 mAh/g and 90 mAh/g, respectively (Figure 6). The discharge capacity of NFVPS/C decreases monotonously upon cycling, while the capacity of NFVPS/rGO quickly decreases up to 50 cycles and remains almost the same between 50 and 100 cycles (Figure 7). The cycling stability of NFVPS/C and NFVPS/rGO are around 55 and 70%, respectively (Figure 7). The Coulombic efficiency of NFVPS/C approaches 100%, whereas the efficiency of NFVPS/rGO slightly deviates from 100% (Figure 7).

After increasing the temperature of the cells from 20 to 40 °C, the discharge capacity for both NFVPS/C and NFVPS/rGO increases reaching a magnitude of 70 and 90 mAh/g.

As in the case of the 20 °C experiment, the discharge capacity of NFVPS/C continues to decrease upon cycling, while the discharge capacity of NFVPS/rGO slightly increases from 90 to 98 mAh/g after 100 cycles. This is related with an already formed surface layer at 20 °C, which appears to prevent the further interaction of the electrode with the electrolyte at 40 °C [31]. It is important to note that the cycling stability of both samples is better at 40 °C than at 20 °C: 70% for NFVPS/C and 95% for NFVPS/rGO, respectively.

Once again, the Coulombic efficiency is better for NFVPS/C. Thus, the electrochemical protocol discloses a significant difference in storage performance between NFVPS/C and NFVPS/rGO composites.

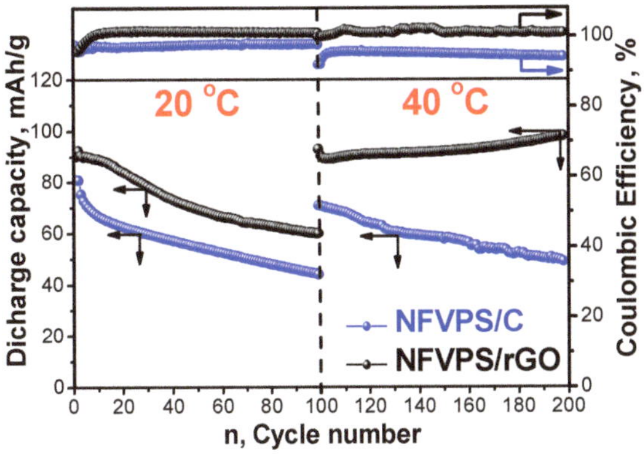

Figure 7. Cycling stability and Coulombic efficiency of NFVPS/rGO and NFVPS/C in Na half-cells at 20 °C and 40 °C at a C/2 rate. The cells operate firstly at 20 °C (cycles from 1 to 100) and then at 40 °C (cycles from 101 to 200).

The next electrochemical protocol is designed to determine the rate capability at elevated temperatures (Figure 8).

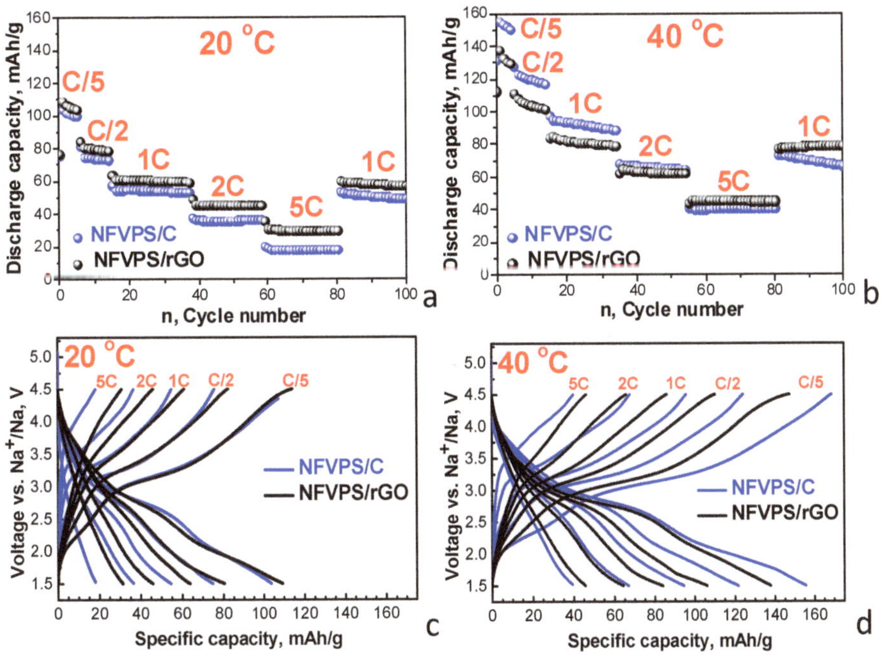

Figure 8. Rate capability in Na half-cells of NFVPS/rGO and NFVPS/C at rates ranging from C/10 to 5C and then to 1C at 20 °C (**a**) and 40 °C (**b**). Coulombic efficiencies are also shown (**a**,**b**); Second charge-discharge curves of NFVPS/rGO and NFVPS/C recorded at rates from C/10 to 5C at 20 °C (**c**) and 40 °C (**d**).

At 20 °C, NFVPS/C and NFVPS/rGO deliver almost the same discharge capacity at the slowest rate (i.e., varying between 110 and 100 mAh/g at a C/5 rate). This provides evidence for a two-electron reaction involving Fe^{3+}/Fe^{2+} and V^{3+}/V^{4+} couples occurring at NFVPS, which has a theoretical capacity of 128 mAh/g. At the fastest rate, NFVPS/rGO outperforms NFVPS/C: the discharge capacity is of 30 mAh/g versus 15 mAh/g at a 5C rate. When the rate returns from 5C to 1C, the capacity of NFVPS/rGO is restored to approximately 60 mAh/g and this capacity remains nearly unchanged after next 20 cycles. For NFVPS/C, capacity is nearly restored but decreases continuously with cycling. Furthermore, the capacity increases as temperature rises from 20 to 40 °C. At a C/5 rate, the initial capacities reach to 160 and 140 mAh/g for NFVPS/C and NFVPS/rGO, respectively. This suggests that the increased temperature favours the occurrence of a three-electron reaction involving Fe^{3+}/Fe^{2+} and $V^{3+}/V^{4+}/V^{5+}$ couples in NFVPS, which implies kinetic limitations for the three-electron reaction compared to the two-electron one. By increasing the rate from C/5 to 5C, the capacities of NFVPS/C and NFVPS/rGO become similar, around 40 mAh/g at 5C. This is another indication that the rate capability of NFVPS/rGO at 40 °C is better than that of NFVPS/C. Additionally, when the charging rate is returned from 5C to C/1, the capacities of NFVPS/rGO and NFVPS/C become close (77 mAh/g for NFVPS/rGO and 73 mAh/g for NFVPS/C). However, the capacity stability of NFVPS/rGO is much better than that of NFVPS/C after 20 cycles: 100% versus 92%, respectively. Therefore, regardless of the testing protocol, NFVPS/rGO outperforms NFVPS/C in terms of cycling stability and rate capability.

2.3. Ex-Situ Studies

The electrochemical performance of carbon-containing composites is due to the structural stability of NFVPS phase during the prolonged sodium extraction/insertion (Figure 9). For ex-situ XRD experiments, we analyzed the electrodes after 100 cycles at 20 °C, followed by 100 cycles at 40 °C, and then switch them off at 1.5 V (i.e., in the discharged state).

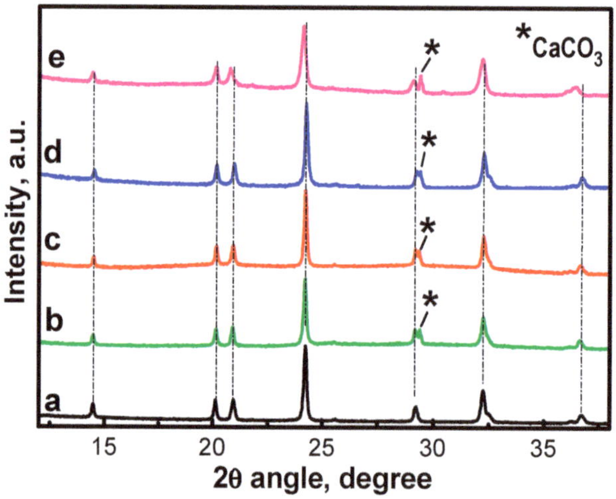

Figure 9. XRD pattern of powder NFVPS (**a**); Ex-situ XRD patterns of pristine NFVPS/rGO electrode (**b**) and after 200 cycles in Na-half cell (100 cycles at 20 °C and subsequent 100 cycles at 40 °C with a rate C/2) (**c**); pristine NFVPS/C electrode (**d**) and after 200 cycles in Na-half cell (100 cycles at 20 °C and subsequent 100 cycles at 40 °C with a rate C/2) (**e**). * $CaCO_3$ from plastic holder.

The ex-situ XRD patterns show that the NASICON-type structure of NFVPS is preserved with lattice parameters close to that for the pristine electrodes (Table S1). The structure stability of NFVPS is its important advantage from practical point of view. Apart from the structural stability, the electrode roughness also seems unchanged after prolonged cycling even at 40 °C as seen from the optical images (Figure S5).

The same electrodes are subjected to XPS analyses (Figure 9). This technique allows us to probe the surface of the cycled electrodes.

For the pristine electrodes, the spectra in the binding energy range of $V2p_{3/2}$ and Fe2p consist of peaks centered at around 517 eV and 713 eV, corresponding to V^{3+} and Fe^{3+} ions. Additionally, the S2p and P2p spectra show peaks due to SO_4 and PO_4 groups. Within the Na1s binding energy range, the spectra are dominated by a single peak corresponding to sodium atoms bound to PO_4 groups. After electrode cycling, the peaks due Na^+ in PO_4 environment, SO_4 and PO_4 groups retain their positions. This provides evidence for the stability of the SO_4 and PO_4 groups during the cycling. In the pristine electrodes Fe^{3+} ions are visible, but after cycling, both Fe^{3+} and Fe^{2+} present on the electrodes surfaces (discharged states). The spectra are barely visible in the $V2p_{3/2}$ range, suggesting that thicker surface layers are formed on the cycled electrodes.

To analyze the electrochemically formed surface layers, the XPS spectra in the range of F1s and C1s are of interest (Figure 10). The F1s spectra of the pristine electrodes exhibit a band at 688.1 eV, which is typical for F atoms in PVDF (polyvinylidene fluoride) binder [45]. After cycling, this band shows significant downshift to 686.6 eV for NFVPS/rGO and to 685.3 eV for NFVPS/C (Figure 9). The lowest energy band at 685.3 eV observed for NFVPS/C can be assigned to the formation of fluorides (i.e., F bonded primarily to Na and NaP_xF_y) and/or oxidized fluoride species, $P_xF_yO_z$ [46]. For NFVPS/rGO, the band position at 686.6 eV suggests for the deposition of oxidized fluoride species, $P_xF_yO_z$, rather than fluorides [47,48]. The fluorine products detected on the electrode surface can be related with the reactivity of the electrolyte salt $NaPF_6$ towards the carbon additives: the interaction of $NaPF_6$ with carbon black in NFVPS/C yields mainly fluorides, while rGO in NFVPS/rGO facilitates the deposition of oxofluorides.

The C1s spectra of the electrodes are also changed after the electrode cycling (Figure 3, Table 1). This finding suggests that the electrolyte solvent propylene carbonate (PC) reacts with the electrode surface. Furthermore, the extent of the interaction with PC depends on the origin of the carbon additives. As shown in Table 1, the electrode surface of the composite with carbon black is enriched with hydroxyl/epoxide functional groups. In the presence of rGO, the electrode surface contains both hydroxyl/epoxide and carbonyl functional groups, with a slight prevalence of hydroxyl/epoxide groups. This can be interpreted in terms of the reactivity of the electrolyte solvent, PC. There are two options for the decomposition of PC [49]. The energetically favorable pathway is through ring opening of the PC molecule and formation of alkylcarbonates, which further degrade to carbonates and alkenes. The second option involves ring opening, which yields alkoxyalkyl compounds. This is followed by the formation of epoxides and CO_2. According to the XPS study, it appears that PC degradation via the formation of alkoxyalkyl compounds proceeds more easily at the carbon black composite, while both PC degradation pathways occur on the rGO composite surface. The different reactivity of PC towards composite surfaces determines the composition of electrochemical surface layers, which in turn contribute to the cycling stability and rate capability. This is the first experimental observation for the effect of carbon additives on the degradation pathways of PC, which requires further experimental and theoretical studies.

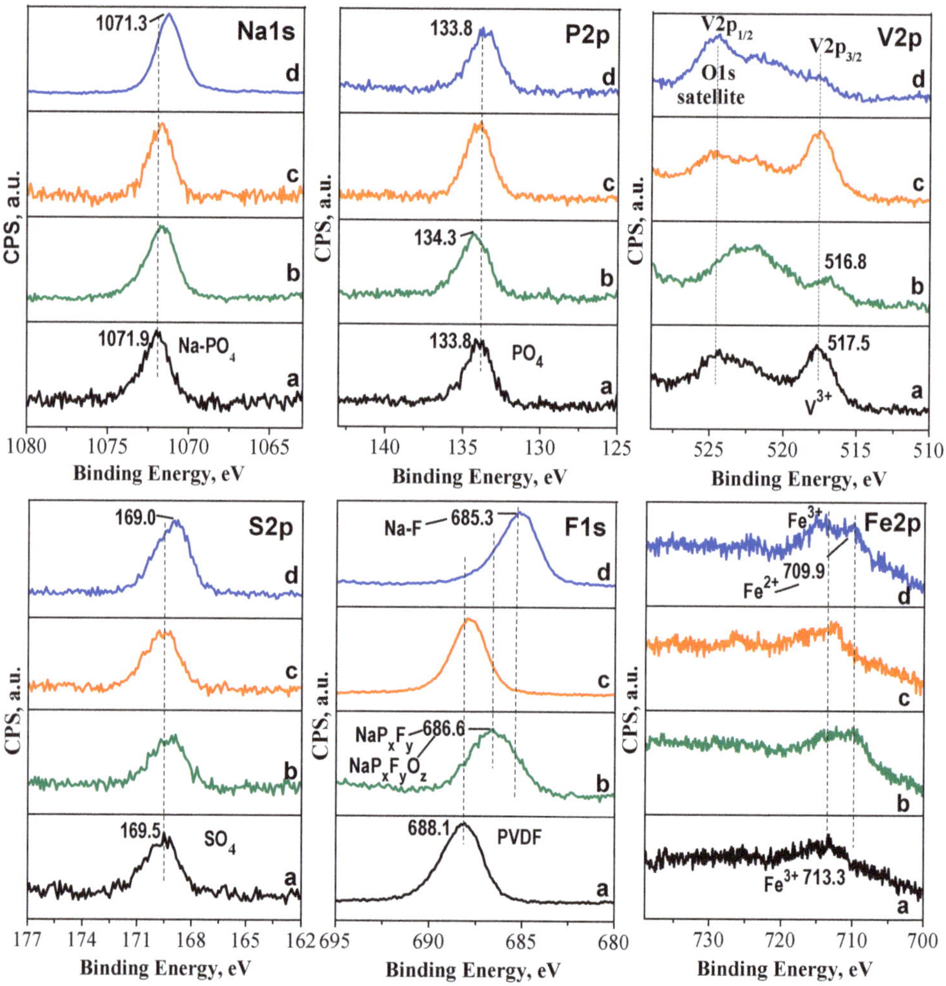

Figure 10. XPS spectra of pristine NFVPS/rGO electrode (**a**) and after 200 cycles in Na-half cell (100 cycles at 20 °C and subsequent 100 cycles at 40 °C with a rate C/2) (**b**); pristine NFVPS/C electrode (**c**) and after 200 cycles in Na-half cell (100 cycles at 20 °C and subsequent 100 cycles at 40 °C with a rate C/2) (**d**).

3. Materials and Methods

3.1. Synthesis

The powder $NaFeVPO_4(SO_4)_2$ (NFVPS) has been synthesized by a precursor method, following by annealing at 400 °C in Ar flow. The details are reported elsewhere [31]. Two types of carbon composites have been prepared, the one with traditional carbon black (Super C65, TIMCAL Ltd., Bodio, Switzerland) and another with rGO (Graphit Kropfmühl GmbH, Hauzenberg, Germany). A simple mechanical grinding method has been applied to obtain composites between NFVPS and 15 wt.% carbon additives, which are labelled as NFVPS/C and NFVPS/rGO. The ball-milling of the materials has been carried out for 4 h at a speed of 300 rpm using agate balls with ϕ of 10 mm, balls-to-powder ratio was 10:1. The grinding procedure has been performed by means of planetary mill "Pulverisette

6″ (Fritsch GmbH, Idar-Oberstein, Germany). After the milling the composites have been annealed at 400 °C for 3 h under Ar flow.

3.2. Characterization Methods

Powder X-ray diffraction technique (XRD) has been used to check the structure, phase composition and purity of pristine NFVPS, composites and electrodes after the electrochemical cycling (Bruker Advance D8 diffractometer with CuKα radiation, Karlsruhe, Germany). The lattice parameters have been calculated by WinPLOTR program. The thermal stability of NFVPS and composites has been studied by simultaneous thermogravimetry and differential thermal analysis combined with mass spectrometry for gas analysis (TG/DTA/MS) at a heating rate of 10 °C/min under Ar atmosphere (LABSYSTM Evo apparatus, SETARAM, Caluireet-Cuire, France). The specific surface area and total pores volume of the samples have been calculated by BET method based on the nitrogen adsorption-desorption isotherms (77.4 K) obtained by NOVA 1200e device (Quantachrome, Boynton Beach, FL, USA). The morphology of the powder samples and electrodes has been observed by scanning electron microscopy (SEM) using JSM 6390 microscope (JEOL, Tokyo, Japan). HR-TEM analysis has been performed with a JEOL 2100 microscope (Tokyo, Japan) with a GATAN Orius 832 SC1000 camera (Plesantan, CA, USA). The DigitalMicrograph software version 2 (Gatan, Inc., Pleasanton, CA, USA) to calculate the HR-TEM images was used. To probe the chemical and electronic state of the elements on the electrode surfaces X-ray photoelectron spectroscopy (XPS) analysis was carried out (VG Escalab II system with Al K_α radiation, Thermo Fisher Scientific, Waltham, MA, USA). The C1s line of adventitious carbon at 285.0 eV was used as an internal standard to calibrate the binding energies. The photoelectron spectra were corrected by subtracting a Shirley-type background and they were quantified using the peak area and Scofield's photo-ionization cross-section. The accuracy of the binding energy measured was ±0.1 eV. The optical images of the electrodes have been collected using ZEISS Stemi 508 stereo microscope (Carl Zeiss, Jena, Germany).

The electrochemical characterization has been performed in Swagelok type of sodium half-cells by means of 32 channel Biologic VMP-3e battery cycler (BioLogic, Seyssinet-Pariset, France). The potentiostatic and galvanostatic cycling tests have been carried out at 20 and 40 °C (KB-53 incubator, Binder GmbH, Tuttlingen, Germany).

The positive electrode contains 80% composite (NFVPS/C or NFVPS/rGO), 10% Super C65 carbon and 10% polyvinylidene fluoride (PVDF) which have been homogenized in N-methyl-2-pyrrolidone using planetary centrifugal mixer ARE-250 CE (THINKY, Tokyo, Japan). Then the slurry has been cast onto the carbon-coated aluminium folio (Goodfellow, Cambridge Ltd., Cambridge, UK) with a doctor blade film coater (ZAA 2600.A, Proceq SA, Schwerzenbach, Switzerland) and vacuum dried at 80 °C overnight. The electrode film with a thickness of 200 µm is then cut into 10 mm discs, pressed and vacuum dried at 120 °C for 10 h. The electrode loading is around 3 mg/cm^2. The negative electrode consists of sodium metallic disc. The electrolyte solution is 1 M $NaPF_6$ in PC (propylene carbonate) soaked in a glass fibre separator GF/D Whatman (Whatman International Ltd., Maidstone, UK). The sodium test cells have been assembled in an argon-filled glove box MB-Unilab Pro SP (1500/780) under controlled traces of O_2 and H_2O below 0.1 ppm (MBraun, Garching, Germany). The cyclic voltammograms (CV) have been scanned over the range 1.5–5.0 V (vs. Na^+/Na) with different rates between 10 and 0.01 mV/s. The contributions of the capacitive and Faradaic reactions are determined using the relationship between total current (i in A) at a specific potential (V) and scanning rate (v in V/s) according to Equation (1) [41,50]:

$$i(V) = k_1 v + k_2 v^{1/2} \quad (1)$$

where, k_1 and k_2 are adjustable parameters. In Equation (1), $k_1 v$ and $k_2 v^{1/2}$ correspond to the current attributed to capacitive reaction and diffusion-controlled reaction, respectively [41,50].

The galvanostatic tests in a voltage window of 1.5–4.5 V (vs. Na^+/Na) have been carried out at several C/n rates, where *n* is the number of hours needed for the insertion of two sodium per formula unit at the applied current intensity (1 C = 128 mA/g). The specific capacity has been calculated based on the mass of the active phase NFVPS in the positive electrodes. The cycling stability has been determined with a rate of C/2, while the rate performance has been studied at rates ranging from C/10 to 5C and then to 1C at 20 °C and 40 °C.

The chemicals PVDF, Na-electrolyte and N-methyl-2-pyrrolidone have been purchased from Sigma-Aldrich (St. Louis, MO, USA).

4. Conclusions

Adding carbon black (C) and reduced graphene oxide (rGO) to the NASICON-type multi-electron electrode, NFVPS, is crucial for its electrochemical performance. This complex phenomenon is a consequence of the oxygen-containing functional groups and porosity of the carbon additives. The carbon black contains predominately hydroxyl/epoxide functional groups, while rGO has a diversity of functional groups, namely hydroxyl/epoxide and carbonyl functional groups. Thus, both carbon black and rGO adhere well to the NFVPS particles, improving the thermal stability of NFVPS in the composites. Among carbon additives, the pore structure of NFVPS/rGO is more developed than that of NFVPS/C.

Sodium storage in carbon-modified NFVPS occurs through capacitive and Faradaic reactions. The Faradaic reaction is facilitated by the carbon black composite, while the capacitive reaction dominates in the rGO composite. NFVPS operates through two-electron reactions at 20 °C, while the increased temperatures favor the three-electron reaction. At 20 °C, the rGO composite outperforms the carbon composite in terms of cycling stability and rate capability. The discharge capacity remains at around 60 mAh/g at a 1C rate for 100 cycles. Increased temperature from 20 to 40 °C improves the performance of the rGO composite: the discharge capacity is around 95 mAh/g at a C/2 rate, and it maintains 95% cycling stability after 100 cycles. At 40 °C, the capacity of NFVPS/C at lower charging rate (i.e., between C/5 and 1C) is higher than that of the NFVPS/rGO composite, but at 5C the capacities of the two composites are similar. This shows that rate capability of NFVPS/rGO is better even at elevated temperatures.

The structural stability of NFVPS during cycling is one factor that contributes to the cycling stability of the composites. Additionally, the reactivity of the functional groups of the carbon additives toward electrolyte salt and solvent determines the composition of electrochemical surface layers. The hydroxyl/epoxide rich surface of the carbon black-composite is the most probably responsible for the deposition of fluorides, as well as facilitates the degradation of the propylene carbonate via the formation of alkoxyalkyl compounds. For the rGO-composite, the carbonyl groups, in addition to the hydroxyl/epoxide groups, cause the deposition of oxofluorides. Propylene carbonate degrades on NFVPS/rGO, forming alkylcarbonate and alkyloxyalkyl compounds.

This study demonstrates the multiple role of the carbon additives in the improving the sodium storage performance of the NASICON-type electrodes.

Supplementary Materials: The following supporting information can be downloaded at: https://www.mdpi.com/article/10.3390/molecules30173547/s1, Figure S1: Crystal structure of NFVPS (a); Rietveld refinements plots of NFVPS (b), NFVPS/rGO (c) and NFVPS/C (d); Table S1: Lattice parameters of powders NFVPS, NFVPS/rGO and NFVPS/C and of electrodes cycled in Na half-cells between 1.5 and 4.5 V in $NaPF_6$/PC electrolyte with C/2 rate; Figure S2: XPS spectra in the regions of O1s binding energies of pristine NFVPS/rGO electrode (a) and after 200 cycles in Na-half cell (100 cycles at 20 °C and subsequent 100 cycles at 40 °C with a rate C/2) (b); pristine NFVPS/C electrode (c) and after 200 cycles in Na-half cell (100 cycles at 20 °C and subsequent 100 cycles at 40 °C with a rate C/2) (d); Figure S3: Nitrogen adsorption/desorption isotherms (full/open symbols) with pore size distributions (insets) of NFVPS, NFVPS/rGO and NFVPS/C; Figure S4: CV curve of NFVPS/C and NFVPS/rGO between 1.5 and 5.0 V with a scanning rate of 10 mV/s at 20 °C in Na half-cell; Figure S5: Optical images of initial NFVPS/C and NFVPS/rGO electrodes and after cycling in Na half-cells at C/2 rate for 200 cycles (100 cycles at 20 °C and subsequent 100 cycles at 40 °C); Table S2: Element content determined from XPS spectra of the electrodes cycled between 1.5 and 4.5 V in Na half-cells with $NaPF_6$/PC electrolyte.

Author Contributions: Conceptualization, V.K. and R.S.; methodology, V.K., R.S.; formal analysis and investigation, T.T., M.S., S.H. and V.K.; writing—review and editing, V.K. and R.S.; visualization, T.T., V.K. and M.S.; funding acquisition, R.S.; project administration, R.S. All authors have read and agreed to the published version of the manuscript.

Funding: The research was funded by project "Master" (КП-06-ДО02/3 dated 18 May 2023).

Institutional Review Board Statement: Not applicable.

Informed Consent Statement: Not applicable.

Data Availability Statement: Data available on request.

Acknowledgments: We acknowledge the infrastructure support of the contract "National Center of Excellence Mechatronics and Clean Technologies" (BG16RFPR002-1.014-0006) under "Research, Innovation and Digitization for Smart Transformation" program 2021–2027. The authors are sincerely grateful to Robert Feher from Graphit Kropfmühl GmbH (Hauzenberg, Germany) for providing rGO additive and to Sergio Lavela for the discussion on the electrochemical performance of phosphate electrode materials.

Conflicts of Interest: The authors declare no conflicts of interest.

ferences

Zhao, L.; Zhang, T.; Li, W.; Li, T.; Zhang, L.; Zhang, X.; Wang, Z. Engineering of sodium-ion batteries: Opportunities and challenges. *Engineering* **2023**, *24*, 172–183. [CrossRef]

Fang, Y.; Zhang, J.; Xiao, L.; Ai, X.; Cao, Y. Phosphate Framework Electrode Materials for Sodium Ion Batteries. *Adv. Sci.* **2017**, *4*, 1600392–1600413. [CrossRef]

Mao, J.; Luo, C.; Gao, T.; Fan, X.; Wang, C. Scalable synthesis of $Na_3V_2(PO_4)_3$/C porous hollow spheres as a cathode for Na-ion batteries. *J. Mater. Chem. A* **2015**, *3*, 10378–10385. [CrossRef]

Yuan, Y.; Wei, Q.; Yang, S.; Zhang, X.; Jia, M.; Yuan, J.; Yan, X. Towards high-performance phosphate-based polyanion-type materials for sodium-ion batteries. *Energy Storage Mater.* **2022**, *50*, 760–782. [CrossRef]

Song, Z.; Liu, R.; Liu, W.-D.; Chen, Y.; Hu, W. Low-cost polyanion-type cathode materials for sodium-ion battery. *Adv. Energy Sustain. Res.* **2023**, *4*, 2300102–2300124. [CrossRef]

Liao, H.; Zhang, Z.; Zheng, Y.; Gao, Y. $NaFePO_4$ for sodium-ion batteries: Mechanism, synthesis and optimization strategies toward commercialization. *Energy Storage Mater.* **2024**, *65*, 103157–103177. [CrossRef]

Chen, G.; Huang, Q.; Wu, T.; Lu, L. Polyanion sodium vanadium phosphate for next generation of sodium-ion batteries—A review. *Adv. Funct. Mater.* **2020**, *30*, 2001289–2001312. [CrossRef]

8. Singh, B.; Wang, Z.; Park, S.; Sai Gautam, G.; Chotard, J.-N.; Croguennec, L.; Carlier, D.; Cheetham, A.K.; Masquelier, C.; Ca P. A chemical map of NASICON electrode materials for Sodium-ion batteries. *J. Mater. Chem. A* **2021**, *9*, 281–292. [CrossRe
9. Wang, L.; Wang, J.; Chen, H.; Dong, H.; Wang, H.; Wang, Y.; Xiao, Y.; Wang, J.; Chen, S. Fast Screening Suitable Doping Transition M to $Na_3V_2(PO_4)_2F_3$ for Sodium-Ion Batteries with High Energy Density in Wide-Temperature Range. *Adv. Mater.* **2025**, *37*, 25([CrossRef]
10. Liu, Y.; Wang, Y.; Wang, H.; Dou, S.; Tian, H.; Gan, W.; Yuan, Q. Dynamic template directed construction of three-dimens porous bismuth aerogels for high-rate Na-ion storage. *J. Mater. Chem. A* **2023**, *11*, 5945–5955. [CrossRef]
11. Sun, S.; Chen, Y.; Bai, Q.; Tian, Z.; Huang, Q.; Liu, C.; He, S.; Yang, Y.; Wang, Y.; Guo, L. Unravelling the regulation mechanis nanoflower shaped $Na_3V_2(PO_4)_3$ in methanol–water system for high performance sodium ion batteries. *Chem. Eng. J.* **2023** 138780–138791. [CrossRef]
12. Kate, R.S.; Jadhav, H.S.; Chothe, U.P.; Bhattacharjee, K.; Kulkarni, M.V.; Deokate, R.J.; Kale, B.B.; Kalubarme, R.S. Critical re of the recent progress and challenges of polyanion $Na_3V_2(PO_4)_3$ cathode materials in rechargeable sodium-ion batteries. *J. M Chem. A* **2024**, *12*, 7418–7451. [CrossRef]
13. Thirupathi, R.; Kumari, V.; Chakrabarty, S.; Omar, S. Recent progress and prospects of NASICON framework electrode Na-ion batteries. *Prog. Mater. Sci.* **2023**, *137*, 101128–101181. [CrossRef]
14. Tiwari, V.K.; Singh, R.K. Nanostructured coating strategies of cathode for improved sodium ion battery performance. *Chem. J.* **2023**, *471*, 144592–144615. [CrossRef]
15. Li, W.; Yao, Z.; Zhou, C.; Wang, X.; Xia, X.; Gu, C.; Tu, J. Boosting High-Rate Sodium Storage Performance of N-D Carbon-Encapsulated $Na_3V_2(PO_4)_3$ Nanoparticles Anchoring on Carbon Cloth. *Small* **2019**, *15*, 1902432–1902442. [CrossRe
16. A Jiang, Y.; Zeng, L.; Wang, J.; Li, W.; Pan, F.; Yu, Y. A carbon coated NASICON structure material embedded in porous ca enabling superior sodium storage performance: $NaTi_2(PO_4)_3$ as an example. *Nanoscale* **2015**, *7*, 14723–14729. [CrossRef]
17. Zhan, L.; Zhang, L.; Li, Y.; Cai, H.; Wu, Y. Performance and Stability Enhancement of Hole-Transporting Materials in Inv Perovskite Solar Cells. *ACS Appl. Energy Mater.* **2025**, *8*, 4355–4361. [CrossRef]
18. Duan, W.; Zhu, Z.; Li, H.; Hu, Z.; Zhang, K.; Cheng, F.; Chen, J. $Na_3V_2(PO_4)_3$@C core–shell nanocomposites for recharge sodium-ion batteries. *J. Mater. Chem. A* **2014**, *2*, 8668–8675. [CrossRef]
19. Li, H.; Bai, Y.; Wu, F.; Ni, Q.; Wu, C. $Na_3V_2(PO_4)_3$/C nanorods as advanced cathode material for sodium ion batteries. *Solid Ion.* **2015**, *278*, 281–286. [CrossRef]
20. Liu, R.; Li, S.; Wang, Z.; Xu, H.; Wang, W.; Jia, Y.; Zhang, L.; Xie, Z.; Wang, L. Carbon decorated $Na_3V_2(PO_4)_3$ nanoparticles high-rate-capability cathode for fast chargeable sodium-ion batteries. *Electrochim. Acta* **2025**, *552*, 145945–145953. [CrossRe
21. Li, G.; Jiang, D.; Wang, H.; Lan, X.; Zhong, H.; Jiang, Y. Glucose-assisted synthesis of $Na_3V_2(PO_4)_3$/C composite as an elec material for high-performance sodium-ion batteries. *J. Power Sources* **2014**, *265*, 325–334. [CrossRef]
22. Shen, W.; Li, H.; Wang, C.; Li, Z.; Xu, Q.; Liu, H.; Wang, Y. Improved electrochemical performance of the $Na_3V_2(PO_4)_3$ catho B-doping of the carbon coating layer for sodium-ion batteries. *J. Mater. Chem. A* **2015**, *3*, 15190–15201. [CrossRef]
23. Xu, Y.; Wei, Q.; Xu, C.; Li, Q.; An, Q.; Zhang, P.; Sheng, J.; Zhou, L.; Mai, L. Layer-by-Layer $Na_3V_2(PO_4)_3$ Embedded in Red Graphene Oxide as Superior Rate and Ultralong-Life Sodium-Ion Battery Cathode. *Adv. Energy Mater.* **2016**, *6*, 1600389–160 [CrossRef]
24. Zhu, C.; Song, K.; van Aken, P.A.; Maier, J.; Yu, Y. Carbon-Coated $Na_3V_2(PO_4)_3$ Embedded in Porous Carbon Matrix: An Ultr Na-Storage Cathode with the Potential of Outperforming Li Cathodes. *Nano Lett.* **2014**, *14*, 2175–2180. [CrossRef]
25. Ma, X.; Xia, J.; Wu, X.; Pan, Z.; Shen, P.K. Remarkable enhancement in the electrochemical activity of maricite NaFePC high-surface-area carbon cloth for sodium-ion batteries. *Carbon* **2019**, *146*, 78–87. [CrossRef]
26. Hwang, J.; Matsumoto, K.; Hagiwara, R. $Na_3V_2(PO_4)_3$ @Carbon Nanofibers: High Mass Loading Electrode Approaching Pra Sodium Secondary Batteries Utilizing Ionic Liquid Electrolytes. *ACS Appl. Energy Mater.* **2019**, *2*, 2818–2827. [CrossRef]
27. Tang, K.; Tian, H.; Zhang, Y.; Cai, Y.; Du, H.; Zhu, M.; Yao, X.; Su, Z. Multilevel carbon composite construction of NASICON- $NaVPO_4F$/C/CNT cathode material for enhanced-performance sodium-ion batteries. *J. Mater. Chem. C* **2025**, *13*, 6605– [CrossRef]
28. Wang, Q.; Peng, D.; Chen, Y.; Xia, X.; Liu, H.; He, Y.; Ma, Q. A facile surfactant-assisted self-assembly of $LiFePO_4$/grap composites with improved rate performance for lithium ion batteries. *J. Electroanal. Chem.* **2018**, *818*, 68–75. [CrossRef]
29. Samigullin, R.R.; Zakharkin, M.V.; Drozhzhin, O.A.; Antipov, E.V. Thermal Stability of NASICON-Type $Na_3V_2(PO_4)_3$ $Na_4VMn(PO_4)_3$ as Cathode Materials for Sodium-ion Batteries. *Energies* **2023**, *16*, 3051. [CrossRef]
30. Lim, S.Y.; Kim, H.; Shakoor, R.A.; Jung, Y.; Choi, J.W. Electrochemical and Thermal Properties of NASICON Struct $Na_3V_2(PO_4)_3$ as a Sodium Rechargeable Battery Cathode: A Combined Experimental and Theoretical Study. *J. Electrochem* **2012**, *159*, A1393–A1397. [CrossRef]

Koleva, V.; Tushev, T.; Harizanova, S.; Kukeva, R.; Shipochka, M.; Markov, P.; Stoyanova, R. Multi-electron redox reactions with iron and vanadium ions at a mixed phosphate–sulfate electrode during sodium intercalation. *Mater. Adv.* **2024**, *5*, 8599–8614. [CrossRef]

Essehli, R.; Alkhateeb, A.; Mahmoud, A.; Boschini, F.; Yahia, H.B.; Amin, R.; Belharouak, I. Optimization of the compositions of polyanionic sodium-ion battery cathode $NaFe_{2-x}V_x(PO_4)(SO_4)_2$. *J. Power Sources* **2020**, *469*, 228417–228426. [CrossRef]

Li, S.-F.; Hou, X.-K.; Gu, Z.-Y.; Meng, Y.-F.; Zhao, C.-D.; Zhang, H.-X.; Wu, X.-L. Sponge-like $NaFe_2PO_4(SO_4)_2$@rGO as a high-performance cathode material for sodium-ion batteries. *New J. Chem.* **2021**, *45*, 4854–4860. [CrossRef]

Pati, J.; Dhaka, R.S. Mixed polyanionic $NaFe_{1.6}V_{0.4}(PO_4)(SO_4)_2$@CNT cathode for sodium-ion batteries: Electrochemical diffusion kinetics and distribution of relaxation time analysis at different temperatures. *J. Power Sources* **2024**, *609*, 234646–234659. [CrossRef]

Zboril, R.; Mashlan, M.; Papaefthymiou, V.; Hadjipanayis, G. Thermal decomposition of $Fe_2(SO_4)_3$: Demonstration of Fe_2O_3 polymorphism. *J. Radioanal. Nucl. Chem.* **2003**, *255*, 413–417. [CrossRef]

Jiang, Y.; Yang, Z.; Li, W.; Zeng, L.; Pan, F.; Wang, M.; Wei, X.; Hu, G.; Gu, L.; Yu, Y. Nanoconfined carbon-coated $Na_3V_2(PO_4)_3$ particles in mesoporous carbon enabling ultralong cycle life for sodium-ion batteries. *Adv. Energy Mater.* **2015**, *5*, 1402104. [CrossRef]

Sun, F.; Gao, J.; Zhu, Y.; Chen, G.; Wu, S.; Qin, Y. Adsorption of SO_2 by typical carbonaceous material: A comparative study of carbon nanotubes and activated carbons. *Adsorption* **2013**, *19*, 959–966. [CrossRef]

Kundu, S.; Wang, Y.M.; Xia, W.; Muhler, M. Thermal stability and reducibility of oxygen-containing functional groups on multiwalled carbon nanotube surfaces: A quantitative highresolution XPS and TPD/TPR study. *J. Phys. Chem. C* **2008**, *112*, 16869–16878. [CrossRef]

Clark, D.T.; Dilks, A. ESCA applied to polymers. 23. RF glow discharge modification of polymers in pure oxygen and helium oxygen mixtures. *J. Polym. Sci. Polym. Chem.* **1979**, *17*, 957–976. [CrossRef]

Harizanova, S.; Tushev, T.; Koleva, V.; Stoyanova, R. Carbon-based composites with mixed phosphate-pyrophosphates with improved electrochemical performance at elevated temperature. *Materials* **2023**, *16*, 6546. [CrossRef] [PubMed]

Augustyn, V.; Simon, P.; Dunn, B. Pseudocapacitive oxide materials for high-rate electrochemical energy storage. *Energy Environ. Sci.* **2014**, *7*, 1597–1614. [CrossRef]

Veleva, S.; Marinova, D.; Harizanova, S.; Koleva, V.; Lefterova, E.; Shipochka, M.; Dimitrov, O.; Stoyanova, A.; Stoyanova, R. Mixing approaches in enhancing the capacitive performance of rGO-based hybrid electrodes. *Materials* **2025**, *18*, 2460. [CrossRef]

Affi, J.; Handayani, M.; Anggoro, M.A.; Esmawan, A.; Sabarman, H.; Satriawan, A.; Shalannanda, W.; Siburian, R.; Anshori, I. Electrochemical and capacitive behavior of reduced graphene oxide from green reduction of graphene oxide by urea for supercapacitor electrodes. *J. Mater. Sci. Mater. Electron.* **2023**, *34*, 1638. [CrossRef]

Kachmar, A.; Tobis, M.; Frąckowiak, E. Electrochemical capacitor based on reduced graphene oxide/NiS_2 composite. *ChemElectroChem* **2022**, *9*, e202200834. [CrossRef]

Schulz, N.; Hausbrand, R.; Wittich, C.; Dimesso, L.; Jaegermann, W. XPS-surface analysis of SEI layers on Li-ion cathodes: Part II. SEI-composition and formation inside composite electrodes. *J. Electrochem. Soc.* **2018**, *165*, A833–A846. [CrossRef]

Kalapsazova, M.; Kostov, K.; Zhecheva, E.; Stoyanova, R. Hybrid Li/Na ion batteries: Temperature-induced reactivity of three-layered oxide (P3 $Na_{2/3}Ni_{1/3}Mg_{1/6}Mn_{1/2}O_2$) toward lithium ionic liquid electrolytes. *Front. Chem.* **2020**, *8*, 600140–600151. [CrossRef]

Shipitsyn, V.; Antrasian, N.; Soni, V.; Mu, L.; Ma, L. Fundamentals and perspectives of electrolyte additives for non-aqueous Na-ion batteries. *Energy Mater.* **2023**, *3*, 300038–300075. [CrossRef]

Jin, Y.; Le, P.M.L.; Gao, P.; Xu, Y.; Xiao, B.; Engelhard, M.H.; Cao, X.; Vo, T.D.; Hu, J.; Zhong, L.; et al. Low-solvation electrolytes for high-voltage sodium-ion batteries. *Nat Energy* **2022**, *7*, 718–725. [CrossRef]

Kukeva, R.; Vassilev, G.; Kalapsazova, M.; Rasheev, H.; Tadjer, A.; Simova, S.; Stoyanova, R. Electrochemical oxidation and reduction of sodium electrolytes tracked by in-situ/ex-situ magnetic resonance spectroscopy and computational modelling. *Electrochim. Acta* **2025**, *526*, 146191. [CrossRef]

Kim, T.; Choi, W.; Shin, H.-C.; Choi, J.-Y.; Kim, J.M.; Park, M.S.; Yoon, W.S. Applications of voltammetry in lithium ion battery research. *J. Electrochem. Sci. Technol.* **2020**, *11*, 14–25. [CrossRef]

molecules

Article

Tailoring Electrocatalytic Properties of sp^2-Bonded Carbon Nanoforms Through Doping

Paweł Szroeder [1,*], Agnieszka Banaszak-Piechowska [1] and Ihor Sahalianov [2]

[1] Faculty of Physics, Kazimierz Wielki University, Powstańców Wielkopolskich 2, 85-090 Bydgoszcz, Poland; agnb@ukw.edu.pl

[2] Laboratory of Organic Electronics, Department of Science and Technology, Linköping University, SE-60174 Norrköping, Sweden; ihor.sahalianov@liu.se

* Correspondence: psz@ukw.edu.pl; Tel.: +48-668-847-464

Abstract: The symmetry of the valence and conduction bands in graphene and carbon nanotubes allows for easy modification of the electronic structure, which is correlated with their electrocatalytic activity. Modifying the electronic structure of the sp^2-bonded nanocarbons by substituting carbon atoms with electron donors/acceptors and through covalent functionalization can facilitate heterogeneous electron transfer (HET), which is beneficial for designing carbon-based, high-performance electrocatalysts. Based on the Gerischer–Marcus model, we discuss how we can match the density of π-electron states (DOS) of a nanocarbon electrode to the redox potential of redox species using electron and hole doping. Along with the results, this article provides guidance on how to match the properties of nanocarbons to specific electroactive analytes, oxygen reduction reaction (ORR), hydrogen evolution reaction (HER), and oxygen evolution reaction (OER).

Keywords: graphene; carbon nanotubes; density of π-electron states; heterogeneous electron transfer; intrinsic defects; donor doping; acceptor doping; functionalization

1. Introduction

An efficient electrocatalyst should be characterized by its ability to lower the energy barrier of desired electrochemical reactions and hinder undesired side reactions. Suitable electrode materials in electrocatalysis should also be low-cost and environmentally friendly. These expectations are met by carbon-based electrocatalysts, which exhibit high activity and strong structure operability [1–8]. Among carbon-based materials, sp^2-bonded nanocarbons with reduced dimensionality, such as two-dimensional (2D) graphene and one-dimensional (1D) carbon nanotubes (CNTs), have attracted much research attention due to their utility in electronic and structural engineering, which has opened up the possibility of tailoring their properties through lattice deformation, doping, and functionalization [9–14].

Low-dimensional sp^2-bonded carbons combine two features that are key to their use as electrocatalysts: (1) the arrangement of carbon atoms in the form of a two-dimensional, honeycomb-like, crystalline lattice and (2) the electronic properties determined by π-electrons. As for the first feature, carbon atoms in the hexagonal carbon lattice are bound together by strong σ bonds, which provide high chemical stability. In addition, the hexagonal carbon lattice, due to its flexibility, can form planar (graphene), spheroidal (fullerenes, carbon onions), cylindrical (CNTs), and conical (carbon nanocones) structures, whose common feature is a large specific surface area up to 2630 $m^2 \cdot g^{-1}$ [15].

As for the second feature, π-electrons in the hexagonal carbon lattice form a two-dimensional gas of massless Dirac fermions, which flow in the hexagonal plane with a

 https://doi.org/10.3390/molecules3006

Fermi velocity of 10^6 m s^{-1} [16]. As a consequence, charge carriers' mobility reaches a value of 2.5×10^5 cm^2 V^{-1} s^{-1}, which translates into relatively high in-plane electrical conductivity of 100 MS m^{-1} [17]. On the other hand, the in-plane π electrons in unperturbed sp^2-bonded structures show poor activity in electrocatalytic processes [18]. Therefore, realizing the full application potential of sp^2-bonded carbon nanoforms in electrocatalysis requires modification of their π-electronic structure, which could stimulate the π-electrons to contribute to the heterogeneous electron transfer (HET). The most obvious strategy is doping, which leads to changes in π-electron or hole concentration. Several approaches have been proposed to dope sp^2-bonded carbons, which include (i) doping using intrinsic defects [19], (ii) substitutional doping with heteroatoms [20,21], (iii) covalent functionalization [22], and (iv) adsorption-based (noncovalent) doping [23].

Intrinsic defects are the crystalline order perturbations without the presence of foreign atoms. Point defects (vacancies, sp^3 sites) and line defects (dislocations, edges) have an influence on the electronic structure as they lead to local rehybridization of σ and π bonds [24]. In particular, missing orbitals at the vacancy sites give rise to creating additional midgap bands, which contribute to the density of states (DOS) at the Fermi level [25]. A similar effect is produced by sp^3 defects, which create π-orbital vacancies [26]. The band structure of the pristine graphene with line defects also possesses flat bands, which lead to the high DOS near the Fermi level [27].

Topological defects, such as n-sided polygons ($n \neq 6$), which substitute hexagons in the lattice, induce not only a local curvature of a flat lattice but also produce a finite DOS at the Fermi level that increases with the curvature. In turn, heptagon–pentagon pairs and Stone–Walls defects give rise to the modulation of the DOS [28].

The commonly used substitutional dopants are neighbors of carbon atoms in the periodic table, boron and nitrogen, which produce acceptor-like and donor-like states, respectively [29]. Boron atoms attract delocalized π-electrons from carbon atoms that lead to the lowering of the Fermi level with respect to the Dirac point. The population of the antibonding π^* electron states increases, which translates directly into the concentration of holes (p doping) [30]. Nitrogen atoms provide electrons to the conduction band (n doping), which results in the shift of the Fermi level above the Dirac point [31,32].

The π-electronic band structure of graphene functionalized with -OH, -COOH, and -NH_2 functional groups depends on their configuration on different sites of the sp^2-bonded carbon lattice. In general, functional groups produce a flat band near the Dirac point and additionally cause a slight down-shift of the Fermi level with respect to the Dirac point (hole doping). However, in some specific configurations, functional groups can open a band gap [33]. Hole doping induced through functionalization is confirmed experimentally in graphene and nanotubes [34,35].

Another method for modifying the electron structure of sp^2-bonded nanocarbons is molecular-dipole-induced doping. Molecules adsorbed on the surface do not distort the carbon crystalline lattice, but they create high-density electron–hole puddles [36]. Noncovalent doping can be realized through adsorption of strong acceptor (p-doping) or donor (n-doping) molecules [23]. Adsorption of guest molecules on sp^2-bonded carbons is ubiquitous. On graphene exposed to ambient air, hydrocarbon contaminants adsorb very quickly, which significantly reduces its electrocatalytic activity [37,38]. It has also been shown that molecules of organic dispersants used for the preparation of nanocarbon electrodes adsorb permanently on their surface, giving the doping effect [39].

There are a number of publications in which Fermi level engineering in carbon materials is discussed in terms of electrocatalysis [40–45]. In light of all of the existing knowledge, it is an undeniable fact that disorder and doping are necessary to improve the kinetics of the HET on π-electronic carbon systems. In this paper, we present the results of calculating electrode reaction curves on graphene and carbon nanotubes in which the DOS is modified through disorder, heteroatom doping, and functionalization. As redox probes, we use hexacyanoferrate (II)/(III) and hexaammineruthenium (II)/(III), anionic and cationic redox pairs commonly used to assess the electrocatalytic activity of electrode materials. The sp^2-bonded carbon nanoforms are promising electrode materials in fuel cells and batteries [46], as well as in electrochemical water splitting [47]. Therefore, we also discuss the effect of heteroatom doping and functionalization on the performance of oxygen reduction reaction (ORR), hydrogen evolution reaction (HER), and oxygen evolution reaction (OER). In further parts of the paper, we confront the results obtained with the experimental data available in the literature. In this work, we emphasize Raman spectroscopy, which is an invaluable tool for determining both disorder [48–52] and doping [53–55] in carbons.

2. Results

2.1. Density of π-Electronic States of Electrodes and Alignment of the Redox States

DOS is a key parameter of any electrode material, as it determines the ability to transfer an electron of a certain energy between the electrode and the redox species. Figure 1 shows the DOS of pristine graphene and a metallic and semiconducting carbon nanotube calculated using tight-binding (TB) approximation. On the right side of Figure 1, we show the Dirac cones that occur in the band structure of π-electrons at the vertices of graphene's first Brillouin zone. A characteristic feature of the DOS of sp^2-bonded carbons with reduced dimensionality is the occurrence of van Hove singularities, i.e., non-smooth points (kinks and spikes), which originate from critical and saddle points in the electronic band structure.

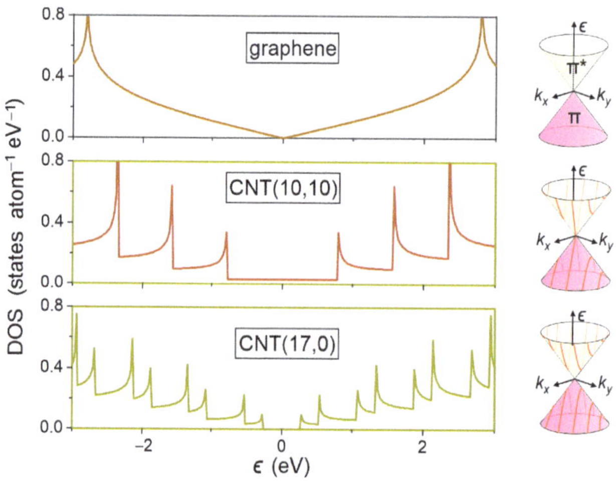

Figure 1. DOS of graphene and a metallic CNT(10,10) and semiconducting carbon nanotube CNT(17,0) calculated using TB approximation. On the right side, Dirac cones for bonding π- and antibonding π^*-electronic bands are shown. In CNTs, allowed states in the (k_x, k_y) space form series of dispersion curves marked as red lines. In metallic CNT, dispersion curves intersect the Dirac point. Zero in the energy scale refers to the Fermi level.

In 2D graphene, van Hove singularity is formed at $\epsilon = 0$, where the apices of Dirac cones formed by bonding π- and antibonding π^*-electronic states merge together. At this energy, the kink in the DOS is formed with a density of states equal to zero. The DOS increases smoothly as we move toward higher and lower energies up to a value of ± 2.7 eV, where another van Hove singularity in the form of spikes with a high DOS occurs. These sharp maxima correspond to states near the center of the graphene's first Brillouin zone, whose dispersion relation is flat. Due to the quantum confinement in quasi 1D carbon nanotubes, the allowed states in the (k_x, k_y) space form a series of one-dimensional sections with corresponding dispersion curves for π and π^* electrons, marked in Figure 1. Each curve has a saddle point, which corresponds to the van Hove singularity in the DOS. Thus, a consequence of the limited dimensionality of nanotubes is a larger number of van Hove singularities than in 2D graphene.

While the DOS of a semiconducting CNT(17,0) nanotube at the Fermi level is zero, the DOS of a metallic nanotube is 0.029 states atom^{-1} eV^{-1}. For comparison, gold has a DOS at the Fermi level of 0.28 states atom^{-1} eV^{-1}, and highly oriented pyrolytic graphite (HOPG) has been reported to have a minimum of 0.0022 states atom^{-1} eV^{-1} [56].

We used in the calculations negatively charged $Fe(CN)_6{}^{3-/4-}$ and positively charged $Ru(NH_3)_6{}^{3+/2+}$ redox couples, which are commonly used to evaluate the electrocatalytic activity of electrode material. The position of the redox potentials relative to the electron bands of the nanocarbon electrode is shown in Figure 2. The electrochemical scale is converted to the absolute energy scale according to Trassati's formula: ϵ [eV] $= -4.5\,\text{eV} - eE$ (vs. SHE) [57]. Unoccupied redox states of the electron acceptor (oxidized form, $W_{ox}(\epsilon)$, and filled states of the electron donor, $W_{red}(\epsilon)$) are separated by doubled reorganization energy, λ. In an anodic reaction, electron transfer occurs from the filled states of the reduced form of the redox couple to empty states of the electrode. In the cathodic reaction, the opposite reaction takes place.

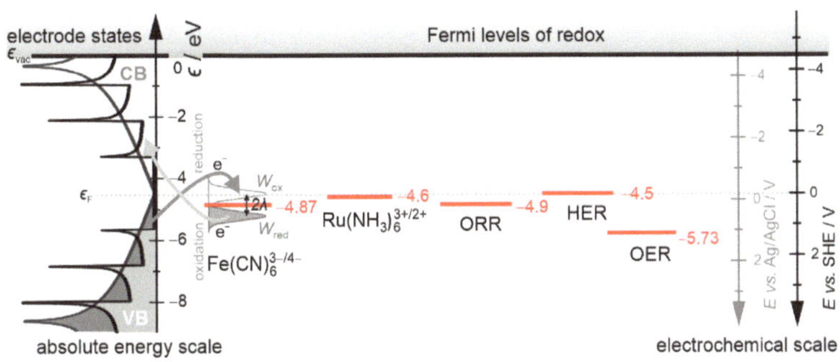

Figure 2. Alignment of the standard reduction potentials of the $Fe(CN)_6{}^{3-}$, $Ru(NH3)_6{}^{3+}$, ORR, HER, and OER with respect to the position of valence (VB) and conduction (CB) bands of graphene and carbon nanotubes. The $E°$ in V vs. SHE are converted to absolute energy scale. The $\epsilon_{F,redox}$ values are given in eV vs. vacuum.

Another calculation concerns the rate of ORR with a four-electron O_2 reduction pathway: $O_2 + 2H_2O + 4e^- \rightarrow 4OH^-$. The four-electron pathway ensures high current densities and high onset potentials, which are fundamentally important in air cathodes of fuel cells and metal–air batteries [46]. The standard reduction potential of this reaction is $E° = 0.402$ V vs. SHE [58]. After conversion to the energy scale, we obtain $\epsilon_{F,redox} = -4.902$ eV vs. vacuum. In turn, the efficiency of water electrolysis depends on the performance of HER, which occurs on the cathode, and OER occurring on the anode [59]. The cathodic HER pathway is given by $2H^+(aq) + 2e^- \rightarrow H_2(g)$ with $E° = 0.0$ V vs. SHE, while the anodic OER occurs according to the pathway $2H_2O\,(l) \rightarrow O_2(g) + 4H^+(aq) + 4e^-$ with $E° = 1.23$ V vs. SHE. The standard potentials correspond to $\epsilon_{F,redox} = -4.5$ eV and $\epsilon_{F,redox} = -5.73$ eV vs. vacuum, respectively. The positions of the ORR, HER, and OER standard potentials relative to the DOS of graphene and CNT are shown in Figure 2.

The Gerischer–Marcus theory states that electron transfer is not limited to states at the Fermi level but occurs between all states forming the quasi-valence and conduction band of the electrode. For this reason, the DOS of π-electrons plays a key role in redox reactions at sp^2-bonded nanocarbon electrodes.

2.2. HET Reaction Rate

2.2.1. Graphene

In Figure 3, we have juxtaposed the DOS of the modified graphene with simulations of the anodic and cathodic reaction rates calculated based on the Gerischer–Marcus model.

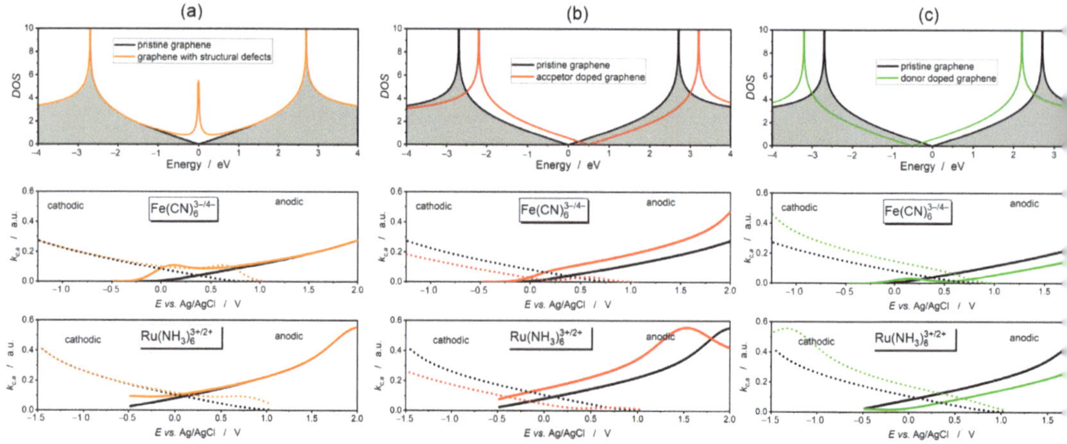

Figure 3. DOS of modified graphene with corresponding cathodic (dotted line) and anodic (continuous line) reaction rate curves for $Fe(CN)_6{}^{3-/4-}$ and $Ru(NH3)_6{}^{3+/2+}$ redox couples. (**a**) Graphene with structural defects. (**b**) Graphene doped with acceptor dopant; hole concentration of 1.7×10^{13} cm^{-2}. (**c**) Graphene doped with donor dopant; electron concentration of 1.7×10^{13} cm^{-2}. The gray color indicates the π-electron states of unmodified graphene. Zero in the energy scale refers to the Fermi level.

Intrinsic defects, such as graphene edges, vacancies, Stone–Walles defects, etc., induce additional localized states near the Dirac point, which contribute to the DOS of graphene (Figure 3a). The consequence is an increase in the anodic and cathodic reaction rates of the $Fe(CN)_6{}^{3-/4-}$ redox couple relative to unmodified graphene at electrode potentials ranging between −0.25 V and 1.0 V vs. Ag/AgCl. The cathodic reaction curve shows a local maximum at the potential of ~+0.5 V vs. Ag/AgCl. A similar maximum is seen in the cathodic curve at the potential of ~+0. 1 V vs. Ag/AgCl. In the case of the $Ru(NH3)_6{}^{3+/2+}$ redox couple, a similar effect occurs; however, no clear maxima are seen. This is a consequence of the higher reorganization energy of the $Ru(NH_3)_6{}^{3+/2+}$ redox couple.

Acceptor and donor doping of graphene results in an up-shift and a down-shift of electronic states, respectively. An upward shift of the electron bands on the energy scale caused by acceptor doping causes a downward shift of the anodic and cathodic reaction rate curves (on the electrochemical scale). Donor doping, on the other hand, leads to a shift of the bands toward lower energies. This is accompanied by an upward shift in the cathodic and anodic reaction curves.

2.2.2. Carbon Nanotubes

The substitutional boron and nitrogen doping of the nanotubes also significantly affects the shape of the reaction rate curves. As we show in Figure 4, the introduction of boron atoms (electron acceptor) results in the creation of new states within the quasi-band gap below the Fermi level and up-shift of the levels in the quasi-conduction band. At a boron concentration of 3 at%, the bands shift upward by about 1 eV. In contrast, replacing carbon atoms with nitrogen atoms (electron donors) leads to a downward shift of the bands.

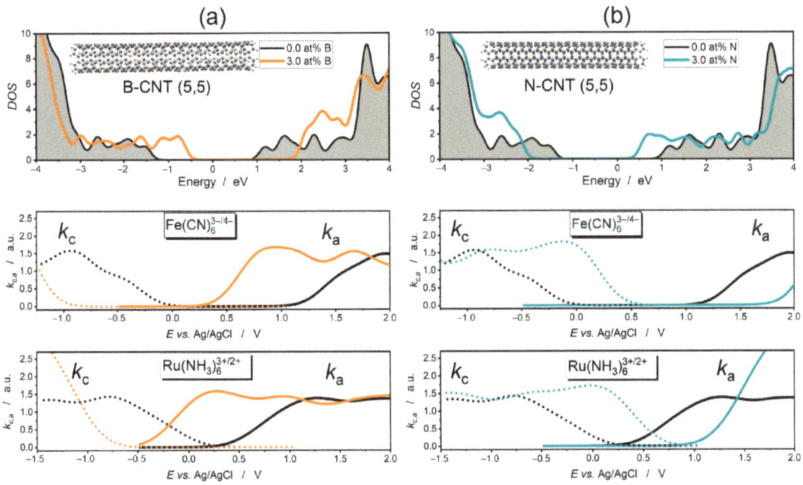

Figure 4. DOS of boron- and nitrogen-doped carbon nanotubes with corresponding cathodic (dotted line) and anodic (continuous line) reaction rate curves for Fe(CN)$_6{}^{3-/4-}$ and Ru(NH3)$_6{}^{3+/2+}$ redox couples. Atomic concentration of dopants—3 at%. (**a**) B-CNT(5,5). (**b**) N-CNT(5,5). The gray color indicates the π-electron states of unmodified CNT. Zero in the energy scale refers to the Fermi level.

As with graphene, the upward shift of the electron bands causes the anodic and cathodic reaction curves to shift significantly to lower potentials, while a downward shift of the electron bands leads to an upward shift of the electrode reaction curves.

Covalent functionalization with -OH and -COOH groups produces an effect similar to the substitution of carbon atoms by acceptor (boron) dopant atoms. An important difference, however, is that functionalization generates bands below the Fermi level, while levels in the quasi-conduction band shift up only slightly. The consequence is a decrease in the quasi-band gap. As shown in Figure 5, carboxyl groups contribute more to the generation of hole states below the Fermi level than hydroxyl groups. This is due to the fact that the -COOH group demonstrates a pronounced electron withdrawal capability compared to -OH due to the presence of two oxygens and stabilization of partial negative charge by carbon atom. The elevated propensity for carboxyl groups to suck electrons out of the sp^2 carbon lattice is also confirmed by first-principles calculations [60].

Curves of the cathodic reaction rates, k_c, which reflect the electron transfer from the filled conduction quasi-bands of the functionalized CNTs to empty states of the oxidized form of redox couples, are only slightly shifted toward lower potentials of the electrode. In the case of the carboxylated CNTs, the intensity of k_c, which is a measure of the cathodic current density during the electrochemical reaction, also decreases. On the other hand, the curves of the anodic reaction rate, k_a, shift to lower potentials, and their intensities increase significantly due to the abundance of unoccupied states above the top of the valence band produced through functionalization with -OH and -COOH groups. This effect is more spectacular in the case of carboxylated CNTs, in which many more new states are formed above the top of the valence band. Thus, functionalization with -OH and COOH groups transforms carbon nanotubes into electrodes that have a high capacity to oxidize electroactive species.

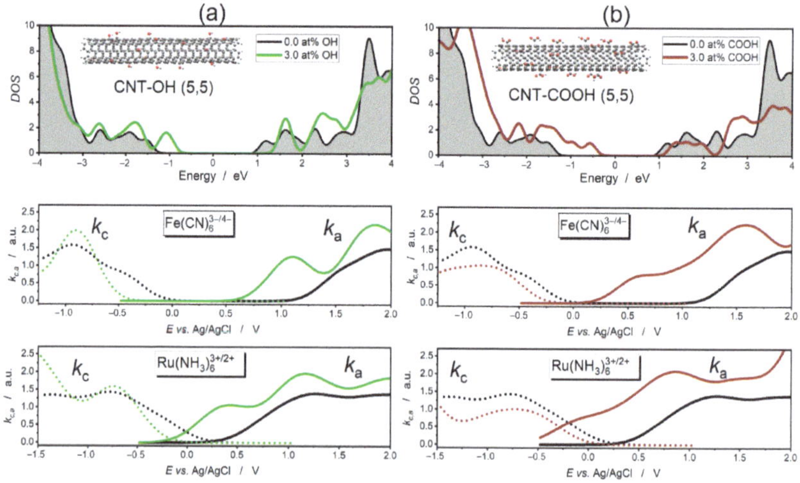

Figure 5. DOS of covalently functionalized carbon nanotubes with corresponding cathodic (dotted line) and anodic (continuous line) reaction rate curves for $Fe(CN)_6{}^{3-/4-}$ and $Ru(NH3)_6{}^{3+/2+}$ redox couples. Atomic concentration of -OH and -COOH groups—3 at%. (**a**) Hydroxylated CNT(5,5). (**b**) Carboxylated CNT(5,5). The gray color indicates the π-electron states of unmodified CNT. Zero in the energy scale refers to the Fermi level.

2.3. ORR, HER, and OER

The curves of cathodic ORR and HER on pristine, heteroatom-doped, and functionalized CNTs are compared in Figure 6a,b. The curves reflect the role of the conductivity quasi-band in the performance of the cathodic reaction. Acceptor doping induced through the substitution of carbon atoms with boron atoms shifts the conductivity quasi-band of the CNTs toward higher energies (Figure 4a), resulting in the absence of electron states that could participate in electron transfer from the electrode to O_2 (ORR) and H^+ (HER). Therefore, the onset of the cathodic reaction on the B-CNT electrode is shifted toward lower potentials by as much as 0.9 V compared to the pristine CNT electrode. The reverse is true for acceptor-doped nanotubes. By shifting the conduction quasi-band downward in N-CNTs (Figure 4b), more electron states from the conduction band can participate in electron transfer. Therefore, the onset of the cathodic reaction is shifted toward higher potentials of the N-CNT electrode by 0.5 V compared to pristine CNT.

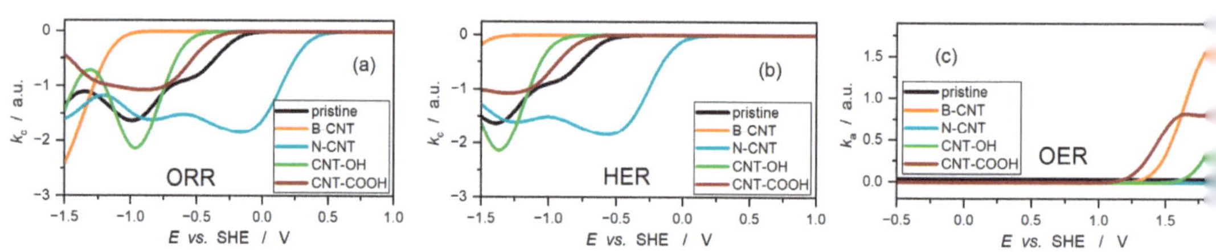

Figure 6. (**a**,**b**) are cathodic reaction rate curves, k_c, of the ORR and HER calculated for pristine, heteroatom-doped, and functionalized CNTs. (**c**) Anodic reaction rate curves, k_a, for OER for pristine and modified CNTs.

Functionalization with -OH and -COOH groups leads to only a slight shift of the conduction quasi-band edge toward higher energies (Figure 5a,b). The onset of the ORR and HER shifts toward lower potentials by only 0.2 V in CNT-COOH and by 0.4 V in CNT-OH relative to pristine CNT.

On the other hand, states from the quasi-valence band contribute to the performance of the anodic OER, which is shown in Figure 6c. Due to the high standard potential of OER (1.23 V vs. SHE), pristine CNTs show negligible OER catalytic activity. Even more overstimulating is the introduction of donor dopant, which has the effect of shifting the top of the valence band to lower energies (Figure 4b). For this reason, in the k_a curves of pristine CNTs and N-CNTs, the onset of the anodic reaction is not observed in the range of the calculated electrode potentials. In contrast, any modification of the electron structure leading to the formation of additional electron states in the valence quasi-band contributes to lowering the onset of the OER reaction. As we show in Figure 6c, substitution of carbon atoms by boron is very effective. Also, functionalization with -OH and -COOH groups contributes to lowering the OER onset potential, with the carboxylation of nanotubes yielding better results.

3. Discussion

There are a number of tools to track doping- and functionalization-induced changes in the structure of sp^2 carbons, such as X-ray fluorescence spectroscopy (XPS), Fourier transform infrared (FTIR) spectroscopy, or Raman spectroscopy. Based on the XPS results, both the presence of the binding states, which correspond to the specific functional groups, can be identified [35], and the heteroatom content can be assessed [20]. XPS is used for surface chemical characterization of materials, so analytical results are often affected by impurities in the form of hydrocarbons and oxides permanently adsorbed from the air. While FTIR is an invaluable technique in identifying chemical functional groups in bulk carbon nanoforms [39], it does not provide much information about the properties of sp^2-bonded aromatic carbon backbone. Raman spectroscopy is the most versatile spectroscopic technique when it comes to assessing the effects of heteroatom doping and functionalization on the electronic structure and, consequently, on the electrocatalytic properties of sp^2-bonded carbon nanoforms. From Raman spectra, information on the dimensionality of carbon nanoforms, the concentration of defects, and the concentration of holes or electrons can be extracted simultaneously. This is an essential advantage over XPS and FTIR, which, in practice, should be used as a complement to Raman spectroscopy.

In Figure 7, we show an example spectrum of partially reduced graphene oxide, where the main Raman features of the sp^2-bonded carbons are highlighted. The D band at ~1340 cm^{-1} is the first-order hexagon breathing mode, and it is inactive in a perfect lattice. The G band at ~1580 cm^{-1} comes from the in-plane C-C bond stretching vibrations of a hexagonal carbon lattice. The 2D band at ~2680 cm^{-1} is the second-order hexagon breathing mode [61,62].

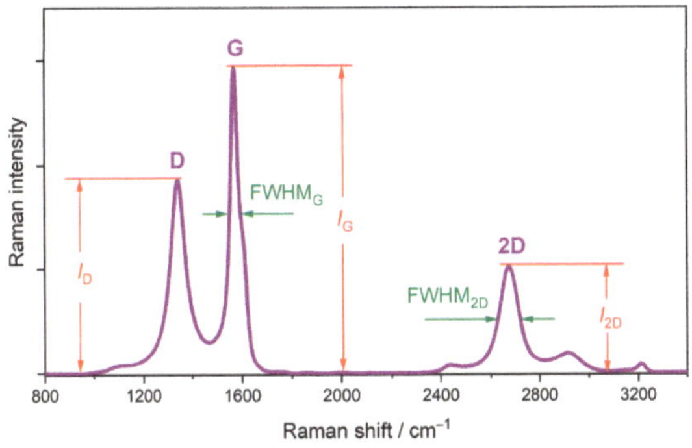

Figure 7. Raman spectrum of partially reduced graphene oxide, showing how to determine such parameters as I_D, I_G, I_{2D}, and FWHM.

The I_D/I_G ratio delivers information about the disorder. In the case of point defects, it is possible to estimate the defect density from I_D/I_G [63,64]. The presence of intrinsic defects also causes a decrease in 2D band intensity. This allows the crystalline quality to be assessed on the basis of the I_{2D}/I_G ratio. Additionally, doping results in increasing the full width at half maximum (FWHM) of all Raman features. Electron and hole doping shifts the position of the G band to higher frequencies [65]. The position of the 2D band allows us to distinguish between electron and hole doping. Upon increasing the charge transfer into the π^* band (donor doping), the G mode softens, and the position of the 2D band shifts two frequencies lower. An opposite effect produces the introduction of holes into the π band; the 2D mode stiffens, and the 2D band shifts to higher frequences [65–67].

The correlation between intrinsic defects and HET has been studied in graphene produced through laser beam irradiation of carbon precursors [68]. Graphene obtained through irradiation with a UV laser beam showed a lower I_{2D}/I_G ratio than graphene produced using an IR laser beam. The lower crystalline quality of the graphene translated into enhanced HET kinetics. For both $Fe(CN)_6^{3-/4-}$ and $Ru(NH_3)_6^{3+/2+}$ redox couples, the HET standard rate constant was about twice as high on the UV-produced graphene with more intrinsic defects. In turn, one paper [69] reported a strong correlation between the I_D/I_G ratio and the improved HET kinetics manifested by reduced redox peak separation of cyclic voltammograms of the $Fe(CN)_6^{3-/4-}$ redox couple, which was observed in reduced graphene oxides. A comprehensive study on the effect of point defects in graphene generated by Ar^+ irradiation on electrocatalytic activity is provided by Ref. [70]. In that study, FcMeOH was used as an electrochemical probe. The standard rate constant increased by more than one order of magnitude when the point defect density estimated from I_D/I_G increased from 4.2×10^{10} to 5.7×10^{12} cm^{-1}.

The influence of heteroatom doping on electrocatalytic activity was investigated in boron- and nitrogen-reduced graphene oxides through electrochemical impedance spectroscopy [71]. Impedance spectra of the $Fe(CN)_6^{3-/4-}$ redox couple show a significant decrease in electron transfer resistance for both N-doped and B-doped graphene, with a greater decrease occurring for acceptor doping. This behavior is explained in Figure 3. An upward shift of the electron bands (boron doping) is more conducive to electron transfer than a downward shift of the bands (nitrogen doping) because the Fermi level of the redox couple, $\epsilon_{F,redox}$, is below the Dirac point (the Fermi level of pure graphene, Figure 2). We might expect that for a redox couple with a higher $\epsilon_{F,redox}$, (lower standard potential), the donor dopant would be more effective.

A similar effect is produced by heteroatom doping of carbon nanotubes. On boron-doped CNTs, the HET kinetics of $Fe(CN)_6^{3-/4-}$ is slightly better than on nitrogen-doped CNTs and much better than on pristine CNTs [72]. In both of the case studies referred to, the carbon sp^2 network was heavily defected, so the subtle G-band shift phenomenon in the Raman spectra could not be observed.

A strong influence of covalent functionalization with -OH group on the HET kinetics has been reported in Ref. [35]. Hydroxylation of carbon nanotubes results in enhanced HET kinetics between the $Fe(CN)_6^{3-/4-}$ redox couple and the electrode, whereas for the $Ru(NH_3)_6^{3+/2+}$ redox couple the HET standard rate constant decreased radically. Raman spectra of pristine CNTs and OH-CNTs have confirmed the effect of donor doping, and a blue-shift of the G- and 2D-band has been observed.

In Ref. [73], it has been shown that strongly N-doped CNTs synthesized from ethylenediamine are an excellent catalyst of ORR, which manifests itself in a shift of the ORR onset potential toward higher potential and a significant reduction of the current density compared to pristine CNTs. Strong donor doping is confirmed in that work by the shift of the G-mode to lower frequencies. The enhanced HER performance of N-CNTs, relative to pristine CNTs, was in turn shown in Ref. [74], where donor doping is also confirmed by Raman data. On the other hand, enhanced catalytic activity of B-CNTs toward OER has been reported in [75], where a reduction in the OER onset potential and a marked increase in the anodic current density on the B-CNT electrode relative to pristine CNTs were observed. The positive effect of carboxylation of nanotubes on OER performance is reported in Ref. [76]. In that work, it was shown that carboxylation of the nanotubes caused a downward shift in the onset potential of the OER by 0.4 V.

4. Materials and Methods

The π-electron DOS of graphene was calculated using the nearest neighbor TB model [77,78], where zero energy is referenced as the Fermi level, ϵ_F. Calculation of graphene's DOS with structural defects was based on the assumption that disorder induces a resonant impurity band near the Fermi level and that it is characterized by the Thomas–Porter distribution [79]. In the calculation of graphene's DOS subjected to substitutional doping with acceptor or donor atoms on the DOS, we relied on the assumption that holes or electrons injected into the graphene lattice by acceptor or donor dopants cause a shift of the Dirac point by $\delta\epsilon_{DP} = \pm\sqrt{\pi}\hbar v_F\sqrt{n}$. Here, the plus sign refers to acceptor (hole) doping, while the minus sign refers to donor (electron) doping. The Fermi velocity $\hbar v_F = 6.73$ eV Å is an experimental parameter taken from the literature, and n is the carrier concentration (holes or electrons) [80].

The electronic structure of carbon nanotubes was studied through density functional theory (DFT). A pristine nanotube CNT(5) made of 420 carbon atoms was considered as an initial model. We used the ωB97XD [81] exchange-correlation functional and 6-31G(d) basis set, which has previously proven to be a reasonable combination for studying the electron structure of large organic systems [82–84]. Numerical simulations were carried out with the Gaussian16 computational package [85]. The nanotube's DOS was extracted from Gaussian output data by using GaussSum software (Version 3.0, https://gausssum.sourceforge.net/, accessed on 7 March 2025) [86].

The cathodic reaction rate, k_c, and the anodic reaction rate, k_a, on an electrode with a specific DOS were calculated as a function of electrode potential, E, using the Gerischer–Marcus integrals [87–89]:

$$k_c(E) \propto \int_{-\infty}^{\infty} f(\epsilon - eE)\ \mathrm{DOS}(\epsilon - eE)\ W_{ox}(\epsilon)\ d\epsilon, \tag{1a}$$

$$k_a(E) \propto \int_{-\infty}^{\infty} [1 - f(\epsilon - eE)]\ \mathrm{DOS}(\epsilon - eE)\ W_{red}(\epsilon)\ d\epsilon. \tag{1b}$$

Equation (1a) describes the HET of electrons from the occupied states expressed as the product of the Fermi–Dirac distribution, $f(\epsilon - eE)$, and the density of electronic states, $\mathrm{DOS}(\epsilon - eE)$, to the unoccupied states of the oxidized form of the redox couple, $W_{ox}(\epsilon)$. In the anodic reaction, electrons are transferred from filled states of the reduced form of the redox couple represented by $W_{red}(\epsilon)$ to empty states of the electrode given by the product of $[1 - f(\epsilon - eE)]$ and $\mathrm{DOS}(\epsilon - eE)$.

Distribution of the unoccupied redox states of the electron acceptor (the oxidized form of the redox couple) and the occupied states of the electron donor (the reduced form of the redox couple) in the electrolyte is given by Gaussian distributions:

$$W_{ox}(\epsilon) = \frac{1}{4\pi\lambda kT}\exp\left(-\frac{(\epsilon - \epsilon_{\mathrm{F,redox}} - \lambda)^2}{4kT\lambda}\right), \tag{2a}$$

$$W_{red}(\epsilon) = \frac{1}{4\pi\lambda kT}\exp\left(-\frac{(\epsilon - \epsilon_{F,redox} + \lambda)^2}{4kT\lambda}\right), \quad (2b)$$

where $\epsilon_{F,redox}$ is the Fermi energy of the redox species obtained through the conversion of its standard potential to an absolute energy scale and λ is the reorganization energy of the redox species. Standard potentials of $Fe(CN)_6^{3-/4-}$ and $Ru(NH_3)_6^{3+/2+}$ are +0.37 V and +0.1 V vs. SHE, respectively [90]. Their equivalent values on the energy scale are −4.87 eV and −4.6 eV, respectively. Reorganization energies, λ, of the $Fe(CN)_6^{3-/4-}$ and $Ru(NH_3)_6^{3+/2+}$ redox probes are 0.35 eV and 0.8 eV, respectively [91]. In calculating the $k_c(E)$ of ORR and HER, we assumed the values of standard reduction potentials of 0.402 V vs. SHE and 0.0 V vs. SHE [58], which on the energy scale correspond to values of −4.902 eV and −4.5 eV vs. vacuum, respectively. The $k_a(E)$ of OER was calculated assuming a standard reaction potential of 1.23 V ($\epsilon_{F,redox} = -5.73$ eV). In the calculation of ORR, HER, and OER rates, a value of $\lambda = 0.26$ eV was assumed [92]. HET through the nanocarbon electrode–electrolyte interface depends on alignment between the Fermi level of the nanocarbon electrode, ϵ_F, and the distribution of the electron states of redox species in the electrolyte [11,12,93,94]. In calculations, the ϵ_F in graphene and carbon nanotubes was set to −4.5 eV. It is an averaged value derived from the work function of graphene and various nanotubes [95].

5. Conclusions

Upon analyzing the cathodic and anodic reaction curves, we draw the following conclusions, which also provide guidance on how to match nanocarbon electrodes to specific electrochemical reactions:

1. If we want both anodic and cathodic reactions to occur at lower electrode potentials, we use acceptor-doped nanocarbons as the electrode material. This is particularly true for OER, which should start at the lowest possible anode potential;
2. If we want both anodic and cathodic reactions to occur at higher electrode potentials, we use donor doping. This guideline is particularly applicable to ORR and HER catalysis, where we are concerned that the onset of the ORR and HER reaction should occur at the highest possible cathode potential;
3. If we want the anodic reaction to primarily occur at lower electrode potentials, we use functionalized nanocarbons (functionalization with a -COOH group is more effective than -OH). This tip, in particular, is applicable to OER catalysis;
4. If we do not care about the selectivity of the reaction and focus only on the efficiency, we use electrodes with intrinsic defects (point defects, edges, etc.).

Indeed, the above rules apply to Fermi level engineering of sp^2-bonded nanocarbons, which is an effective tool for matching the electrode material to the redox potential of a specific electroactive medium. This translates into all possible electrochemical applications, including electrochemical sensors, fuel cells, metal–air batteries (ORR), and electrochemical water splitting (HER and OER). Referring to the selected case studies, we show that the doping effect can be easily monitored in nanocarbons using Raman spectroscopy.

Author Contributions: Conceptualization, P.S. and A.B.-P.; methodology, P.S., A.B.-P. and I.S.; software, P.S. and I.S.; validation, P.S.; formal analysis, P.S.; investigation, P.S., A.B.-P. and I.S.; data curation, P.S., A.B.-P. and I.S.; writing—original draft preparation, P.S.; writing—review and editing, P.S., A.B.-P. and I.S.; supervision, P.S.; project administration, P.S.; funding acquisition, P.S. All authors have read and agreed to the published version of the manuscript.

Funding: The research was funded by the Ministry of Science and Higher Education under the program 'Regional Initiative of Excellence' in years 2024–2027, Project No. RID/SP/0048/2024/01.

Institutional Review Board Statement: Not applicable.

Informed Consent Statement: Not applicable.

Data Availability Statement: The data presented in this study are available from the corresponding author upon reasonable request.

Acknowledgments: The authors would like to thank Przemysław Ziółkowski for providing the Raman data used in this study.

Conflicts of Interest: The authors declare no conflicts of interest.

Abbreviations

The following abbreviations are used in this manuscript:

HET	Heterogeneous electron transfer
DOS	Density of states
CNT	Carbon nanotube
SHE	Standard hydrogen electrode
ORR	Oxygen reduction reaction
HER	Hydrogen evolution reaction
OER	Oxygen evolution reaction

References

Walcarius, A. Electrocatalysis, sensors and biosensors in analytical chemistry based on ordered mesoporous and macroporous carbon-modified electrodes. *TrAC Trends Anal. Chem.* **2012**, *38*, 79–97. [CrossRef]

Paul, R.; Zhu, L.; Chen, H.; Qu, J.; Dai, L. Recent advances in carbon-based metal-free electrocatalysts. *Adv. Mater.* **2019**, *31*, 1806403. [CrossRef]

Kamedulski, P.; Lukaszewicz, J.P.; Witczak, L.; Szroeder, P.; Ziolkowski, P. The importance of structural factors for the electrochemical performance of graphene/carbon nanotube/melamine powders towards the catalytic activity of oxygen reduction reaction. *Materials* **2021**, *14*, 2448. [CrossRef]

Wang, J.; Kong, H.; Zhang, J.; Hao, Y.; Shao, Z.; Ciucci, F. Carbon-based electrocatalysts for sustainable energy applications. *Prog. Mater. Sci.* **2021**, *116*, 100717. [CrossRef]

Macchi, S.; Denmark, I.; Le, T.; Forson, M.; Bashiru, M.; Jalihal, A.; Siraj, N. Recent advancements in the synthesis and application of carbon-based catalysts in the ORR. *Electrochem* **2021**, *3*, 1–27. [CrossRef]

Das, S.; Ghosh, S.; Kuila, T.; Murmu, N.C.; Kundu, A. Biomass-Derived advanced carbon-based electrocatalysts for oxygen reduction reaction. *Biomass* **2022**, *2*, 155–177. [CrossRef]

Gwóźdź, M.; Brzęczek-Szafran, A. Carbon-based electrocatalyst design with phytic acid—A versatile biomass-derived modifier of functional materials. *Int. J. Mol. Sci.* **2022**, *23*, 11282. [CrossRef]

Yadav, S.; Ahmad, A.; Priyadarshini, M.; Dubey, B.K.; Ghangrekar, M.M. Transition towards renewable nano-carbon-based electrocatalysts in electrochemical and bio-electrochemical technologies. *FlatChem* **2024**, *44*, 100623. [CrossRef]

Radchenko, T.; Tatarenko, V.; Sagalianov, I.; Prylutskyy, Y.; Szroeder, P.; Biniak, S. On adatomic-configuration-mediated correlation between electrotransport and electrochemical properties of graphene. *Carbon* **2016**, *101*, 37–48. [CrossRef]

Sagalianov, I.Y.; Radchenko, T.M.; Prylutskyy, Y.I.; Tatarenko, V.A.; Szroeder, P. Mutual influence of uniaxial tensile strain and point defect pattern on electronic states in graphene. *Eur. Phys. J. B* **2017**, *90*, 112. [CrossRef]

Szroeder, P.; Sagalianov, I.Y.; Radchenko, T.M.; Tatarenko, V.A.; Prylutskyy, Y.I.; Strupiński, W. Effect of uniaxial stress on the electrochemical properties of graphene with point defects. *Appl. Surf. Sci.* **2018**, *442*, 185–188. [CrossRef]

Szroeder, P.; Sahalianov, I.; Radchenko, T.; Tatarenko, V.; Prylutskyy, Y. The strain-and impurity-dependent electron states and catalytic activity of graphene in a static magnetic field. *Opt. Mater.* **2019**, *96*, 109284. [CrossRef]

Talla, J.A.; Salman, S.A. Electronic structure tuning and band gap engineering of carbon nanotubes: Density functional theory. *Nanosci. Nanotechnol. Lett.* **2015**, *7*, 381–386. [CrossRef]

Qian, L.; Xie, Y.; Zhang, S.; Zhang, J. Band engineering of carbon nanotubes for device applications. *Matter* **2020**, *3*, 664–695. [CrossRef]

Sur, U.K. Graphene: A rising star on the horizon of materials science. *Int. J. Electrochem.* **2012**, *2012*, 237689. [CrossRef]

Novoselov, K.S.; Geim, A.K.; Morozov, S.V.; Jiang, D.; Katsnelson, M.I.; Grigorieva, I.V.; Dubonos, S.V.; Firsov, A.A. Two-dimensional gas of massless Dirac fermions in graphene. *Nature* **2005**, *438*, 197–200. [CrossRef]

17. Rizzi, L.; Zienert, A.; Schuster, J.; Köhne, M.; Schulz, S.E. Electrical conductivity modeling of graphene-based conductor mate *ACS Appl. Mater. Interfaces* **2018**, *10*, 43088–43094. [CrossRef]
18. Mazánek, V.; Luxa, J.; Matějková, S.; Kučera, J.; Sedmidubský, D.; Pumera, M.; Sofer, Z. Ultrapure graphene is a poor electro lyst: Definitive proof of the key role of metallic impurities in graphene-based electrocatalysis. *ACS Nano* **2019**, *13*, 1574– [CrossRef]
19. Zhang, C.; Chen, S. Defect-and dopant-controlled carbon nanotubes fabricated by self-assembly of graphene nanoribbons. *Res.* **2015**, *8*, 2988–2997. [CrossRef]
20. Liu, Y.; Shen, Y.; Sun, L.; Li, J.; Liu, C.; Ren, W.; Li, F.; Gao, L.; Chen, J.; Liu, F.; et al. Elemental superdoping of graphene carbon nanotubes. *Nat. Commun.* **2016**, *7*, 10921. [CrossRef]
21. Ha, S.; Choi, G.B.; Hong, S.; Kim, D.W.; Kim, Y.A. Substitutional boron doping of carbon materials. *Carbon Lett.* **2018**, *27*, 1–
22. Bekyarova, E.; Sarkar, S.; Wang, F.; Itkis, M.E.; Kalinina, I.; Tian, X.; Haddon, R.C. Effect of covalent chemistry on the electr structure and properties of carbon nanotubes and graphene. *Acc. Chem. Res.* **2013**, *46*, 65–76. [CrossRef] [PubMed]
23. Hu, T.; Gerber, I.C. Theoretical study of the interaction of electron donor and acceptor molecules with graphene. *J. Phys. Che* **2013**, *117*, 2411–2420. [CrossRef]
24. Banhart, F.; Kotakoski, J.; Krasheninnikov, A.V. Structural defects in graphene. *ACS Nano* **2011**, *5*, 26–41. [CrossRef]
25. Farjam, M. Visualizing the influence of point defects on the electronic band structure of graphene. *J. Phys. Condens. Matter* *26*, 155502. [CrossRef]
26. Wehling, T.; Katsnelson, M.; Lichtenstein, A. Impurities on graphene: Midgap states and migration barriers. *Phys. Rev. B-Con Matter Mater. Phys.* **2009**, *80*, 085428. [CrossRef]
27. Berger, D.; Ratsch, C. Line defects in graphene: How doping affects the electronic and mechanical properties. *Phys. Rev. B* *93*, 235441. [CrossRef]
28. Cortijo, A.; Vozmediano, M.A. Effects of topological defects and local curvature on the electronic properties of planar grap *Nucl. Phys. B* **2007**, *763*, 293–308. [CrossRef]
29. Fujimoto, Y. Formation, energetics, and electronic properties of graphene monolayer and bilayer doped with heteroatoms. *Condens. Matter Phys.* **2015**, *2015*, 571490. [CrossRef]
30. Liu, J.; Liang, T.; Tu, R.; Lai, W.; Liu, Y. Redistribution of π and σ electrons in boron-doped graphene from DFT investiga *Appl. Surf. Sci.* **2019**, *481*, 344–352. [CrossRef]
31. Tison, Y.; Lagoute, J.; Repain, V.; Chacon, C.; Girard, Y.; Rousset, S.; Joucken, F.; Sharma, D.; Henrard, L.; Amara, H.; Electronic interaction between nitrogen atoms in doped graphene. *ACS Nano* **2015**, *9*, 670–678. [CrossRef] [PubMed]
32. Joucken, F.; Tison, Y.; Le Fèvre, P.; Tejeda, A.; Taleb-Ibrahimi, A.; Conrad, E.; Repain, V.; Chacon, C.; Bellec, A.; Girard, Y.; Charge transfer and electronic doping in nitrogen-doped graphene. *Sci. Rep.* **2015**, *5*, 1–10. [CrossRef] [PubMed]
33. Tonel, M.; Lara, I.V.; Zanella, I.; Fagan, S.B. The influence of the concentration and adsorption sites of different chemical gro on graphene through first principles simulations. *Phys. Chem. Chem. Phys.* **2017**, *19*, 27374–27383. [CrossRef]
34. Singla, R.; Kottantharayil, A. Stable hydroxyl functionalization and p-type doping of graphene by a non-destructive p chemical method. *Carbon* **2019**, *152*, 267–273. [CrossRef]
35. Szroeder, P.; Ziółkowski, P.; Sahalianov, I.; Madajski, P.; Trzcinski, M. The Hydroxylated Carbon Nanotubes as the Hole Oxida System in Electrocatalysis. *Materials* **2024**, *17*, 3532. [CrossRef]
36. Sreeprasad, T.; Berry, V. How do the electrical properties of graphene change with its functionalization? *Small* **2013**, *9*, 341– [CrossRef]
37. Velický, M.; Bissett, M.A.; Toth, P.S.; Patten, H.V.; Worrall, S.D.; Rodgers, A.N.J.; Hill, E.W.; Kinloch, I.A.; Novoselov, Georgiou, T.; et al. Electron transfer kinetics on natural crystals of MoS_2 and graphite. *Phys. Chem. Chem. Phys.* **2015** 17844–17853. [CrossRef]
38. Li, Z.; Wang, Y.; Kozbial, A.; Shenoy, G.; Zhou, F.; McGinley, R.; Ireland, P.; Morganstein, B.; Kunkel, A.; Surwade, S.P.; et al. E of airborne contaminants on the wettability of supported graphene and graphite. *Nat. Mater.* **2013**, *12*, 925–931. [CrossRef]
39. Szroeder, P.; Ziółkowski, P.; Mosińska, L.; Trykowski, G. Boosting electrochemical performance of single-walled carbon nano three-dimensional architectures through appropriate selection of organic dispersant. *Diam. Relat. Mater.* **2024**, *148*, 11 [CrossRef]
40. Asefa, T.; Huang, X. Heteroatom-Doped carbon materials for electrocatalysis. *Chem. Eur. J.* **2017**, *23*, 10703–10713. [CrossRe
41. Hu, C.; Dai, L. Doping of carbon materials for metal-free electrocatalysis. *Adv. Mater.* **2019**, *31*, 1804672. [CrossRef]
42. Zhao, R.; Chen, Y.; Huang, S. Doping engineering on carbons as electrocatalysts for oxygen reduction reaction. *Fundam. Res.* *1*, 807–823. [CrossRef]
43. He, Q.; Qiao, S.; Zhou, Y.; Vajtai, R.; Li, D.; Ajayan, P.M.; Ci, L.; Song, L. Carbon Nanotubes-Based Electrocatalysts: Struct Regulation, Support Effect, and Synchrotron-Based Characterization. *Adv. Funct. Mater.* **2022**, *32*, 2106684. [CrossRef]
44. Chattopadhyay, J.; Pathak, T.S.; Pak, D. Heteroatom-doped metal-free carbon nanomaterials as potential electrocatalysts. *Mole* **2022**, *27*, 670. [CrossRef] [PubMed]

Zhou, Q.; Shi, M.; Wu, M.; Zhao, N.; Shi, P.; Zhu, Y.; Wang, A.; Ye, C.; Lin, C.-T.; Fu, L. Optimizing Graphene Dopants for Direct Electrocatalytic Quantification of Small Molecules and Ions. *Catalysts* **2023**, *14*, 8. [CrossRef]

Wang, D.-W.; Su, D. Heterogeneous nanocarbon materials for oxygen reduction reaction. *Energy Environ. Sci.* **2014**, *7*, 576–591. [CrossRef]

Raza, A.; Hassan, J.Z.; Qumar, U.; Zaheer, A.; Babar, Z.U.D.; Iannotti, V.; Cassinese, A. Strategies for robust electrocatalytic activity of 2D materials: ORR, OER, HER, and CO2RR. *Mater. Today Adv.* **2024**, *22*, 100488. [CrossRef]

Dillon, R.; Woollam, J.A.; Katkanant, V. Use of Raman scattering to investigate disorder and crystallite formation in as-deposited and annealed carbon films. *Phys. Rev. B* **1984**, *29*, 3482. [CrossRef]

Ferrari, A.C.; Robertson, J. Interpretation of Raman spectra of disordered and amorphous carbon. *Phys. Rev. B* **2000**, *61*, 14095. [CrossRef]

Pimenta, M.; Dresselhaus, G.; Dresselhaus, M.S.; Cançado, L.G.; Jorio, A.; Saito, R. Studying disorder in graphite-based systems by Raman spectroscopy. *Phys. Chem. Chem. Phys.* **2007**, *9*, 1276–1290. [CrossRef]

Dresselhaus, M.; Jorio, A.; Filho, A.G.S.; Saito, R. Defect characterization in graphene and carbon nanotubes using Raman spectroscopy. *Philos. Trans. R. Soc. A Math. Phys. Eng. Sci.* **2010**, *368*, 5355–5377. [CrossRef]

Rebelo, S.L.; Guedes, A.; Szefczyk, M.E.; Pereira, A.M.; Araújo, J.P.; Freire, C. Progress in the Raman spectra analysis of covalently functionalized multiwalled carbon nanotubes: Unraveling disorder in graphitic materials. *Phys. Chem. Chem. Phys.* **2016**, *18*, 12784–12796. [CrossRef] [PubMed]

Rao, A.M.; Eklund, P.C.; Bandow, S.; Thess, A.; Smalley, R.E. Evidence for charge transfer in doped carbon nanotube bundles from Raman scattering. *Nature* **1997**, *388*, 257–259. [CrossRef]

McGuire, K.; Gothard, N.; Gai, P.; Dresselhaus, M.; Sumanasekera, G.; Rao, A. Synthesis and Raman characterization of boron-doped single-walled carbon nanotubes. *Carbon* **2005**, *43*, 219–227. [CrossRef]

Suzuki, S.; Hibino, H. Characterization of doped single-wall carbon nanotubes by Raman spectroscopy. *Carbon* **2011**, *49*, 2264–2272. [CrossRef]

Brownson, D.A.; Kampouris, D.K.; Banks, C.E. Graphene electrochemistry: Fundamental concepts through to prominent applications. *Chem. Soc. Rev.* **2012**, *41*, 6944–6976. [CrossRef]

Trasatti, S. The absolute electrode potential: An explanatory note (Recommendations 1986). *Pure Appl. Chem.* **1986**, *58*, 955–966. [CrossRef]

Bard, A.J.; Parsons, R.; Jordan, J. *Standard Potentials in Aqueous Solution*; Routledge: Abingdon-on-Thames, UK, 2017.

Zubair, M.; Hassan, M.M.U.; Mehran, M.T.; Baig, M.M.; Hussain, S.; Shahzad, F. 2D MXenes and their heterostructures for HER, OER and overall water splitting: A review. *Int. J. Hydrogen Energy* **2022**, *47*, 2794–2818. [CrossRef]

Liang, S.; Gao, P.; Wang, A.; Zeng, C.; Bao, G.; Tian, D. Insights into the influence of functional groups on the properties of graphene from first-principles calculations. *J. Phys. Org. Chem.* **2022**, *35*, e4387. [CrossRef]

Ferrari, A.C. Raman spectroscopy of graphene and graphite: Disorder, electron–phonon coupling, doping and nonadiabatic effects. *Solid State Commun.* **2007**, *143*, 47–57. [CrossRef]

Ferrari, A.C.; Meyer, J.C.; Scardaci, V.; Casiraghi, C.; Lazzeri, M.; Mauri, F.; Piscanec, S.; Jiang, D.; Novoselov, K.S.; Roth, S.; et al. Raman spectrum of graphene and graphene layers. *Phys. Rev. Lett.* **2006**, *97*, 187401. [CrossRef] [PubMed]

Lucchese, M.M.; Stavale, F.; Ferreira, E.H.M.; Vilani, C.; Moutinho, M.V.O.; Capaz, R.B.; Achete, C.A.; Jorio, A. Quantifying ion-induced defects and Raman relaxation length in graphene. *Carbon* **2010**, *48*, 1592–1597. [CrossRef]

Cançado, L.G.; Jorio, A.; Ferreira, E.H.M.; Stavale, F.; Achete, C.A.; Capaz, R.B.; Moutinho, M.V.d.O.; Lombardo, A.; Kulmala, T.S.; Ferrari, A.C. Quantifying defects in graphene via Raman spectroscopy at different excitation energies. *Nano Lett.* **2011**, *11*, 3190–3196. [CrossRef]

Basko, D.; Piscanec, S.; Ferrari, A. Electron-electron interactions and doping dependence of the two-phonon Raman intensity in graphene. *Phys. Rev. B-Condens. Matter Mater. Phys.* **2009**, *80*, 165413. [CrossRef]

Das, A.; Pisana, S.; Chakraborty, B.; Piscanec, S.; Saha, S.K.; Waghmare, U.V.; Novoselov, K.S.; Krishnamurthy, H.R.; Geim, A.K.; Ferrari, A.C.; et al. Monitoring dopants by Raman scattering in an electrochemically top-gated graphene transistor. *Nat. Nanotechnol.* **2008**, *3*, 210–215. [CrossRef]

Ott, A.K.; Ferrari, A.C. Raman spectroscopy of graphene and related materials. In *Encyclopedia of Condensed Matter Physics*, 2nd ed.; Chakraborty, T., Ed.; Elsevier: Amsterdam, The Netherlands, 2024; pp. 233–247.

Santos, N.F.; Pereira, S.O.; Moreira, A.; Girão, A.V.; Carvalho, A.F.; Fernandes, A.J.S.; Costa, F.M. IR and UV Laser-Induced Graphene: Application as Dopamine Electrochemical Sensors. *Adv. Mater. Technol.* **2021**, *6*, 2100007. [CrossRef]

Chee, S.Y.; Poh, H.L.; Chua, C.K.; Šaněk, F.; Sofer, Z.; Pumera, M. Influence of parent graphite particle size on the electrochemistry of thermally reduced graphene oxide. *Phys. Chem. Chem. Phys.* **2012**, *14*, 12794–12799. [CrossRef]

Zhong, J.-H.; Zhang, J.; Jin, X.; Liu, J.-Y.; Li, Q.; Li, M.-H.; Cai, W.; Wu, D.-Y.; Zhan, D.; Ren, B. Quantitative correlation between defect density and heterogeneous electron transfer rate of single layer graphene. *J. Am. Chem. Soc.* **2014**, *136*, 16609–16617. [CrossRef]

71. Tian, H.; Sofer, Z.; Pumera, M.; Bonanni, A. Investigation on the ability of heteroatom-doped graphene for biorecogn *Nanoscale* **2017**, *9*, 3530–3536. [CrossRef]
72. Ye, J.; Lu, J.; Ma, T.; Wen, D. Untangling the effects of doping carbon with diverse heteroatoms on the bioelectrochemist glucose oxidase. *Anal. Chem.* **2023**, *95*, 7685–7692. [CrossRef]
73. Higgins, D.C.; Meza, D.; Chen, Z. Nitrogen-doped carbon nanotubes as platinum catalyst supports for oxygen reduction rea in proton exchange membrane fuel cells. *J. Phys. Chem. C* **2010**, *114*, 21982–21988. [CrossRef]
74. Davodi, F.; Tavakkoli, M.; Lahtinen, J.; Kallio, T. Straightforward synthesis of nitrogen-doped carbon nanotubes as highly a bifunctional electrocatalysts for full water splitting. *J. Catal.* **2017**, *353*, 19–27. [CrossRef]
75. Cheng, Y.; Tian, Y.; Fan, X.; Liu, J.; Yan, C. Boron doped multi-walled carbon nanotubes as catalysts for oxygen reduction rea and oxygen evolution reactionin in alkaline media. *Electrochim. Acta* **2014**, *143*, 291–296. [CrossRef]
76. Zhang, X.; Zhang, W.; Dai, J.; Sun, M.; Zhao, J.; Ji, L.; Chen, L.; Zeng, F.; Yang, F.; Huang, B.; et al. Carboxylated carbon nano with high electrocatalytic activity for oxygen evolution in acidic conditions. *InfoMat* **2022**, *4*, e12273. [CrossRef]
77. Hobson, J.; Nierenberg, W. The statistics of a two-dimensional, hexagonal net. *Phys. Rev.* **1953**, *89*, 662. [CrossRef]
78. Castro Neto, A.H.; Guinea, F.; Peres, N.M.; Novoselov, K.S.; Geim, A.K. The electronic properties of graphene. *Rev. Mod.* **2009**, *81*, 109–162. [CrossRef]
79. Huang, B.-L.; Chang, M.-C.; Mou, C.-Y. Density of states of graphene in the presence of strong point defects. *Phys. Rev. B Con Matter Mater. Phys.* **2010**, *82*, 155462. [CrossRef]
80. Gierz, I.; Riedl, C.; Starke, U.; Ast, C.R.; Kern, K. Atomic hole doping of graphene. *Nano Lett.* **2008**, *8*, 4603–4607. [CrossRef
81. Chai, J.-D.; Head-Gordon, M. Long-range corrected hybrid density functionals with damped atom–atom dispersion correc *Phys. Chem. Chem. Phys.* **2008**, *10*, 6615–6620. [CrossRef]
82. Sahalianov, I.; Singh, S.K.; Tybrandt, K.; Berggren, M.; Zozoulenko, I. The intrinsic volumetric capacitance of conducting poly Pseudo-capacitors or double-layer supercapacitors? *RSC Adv.* **2019**, *9*, 42498–42508. [CrossRef]
83. Sahalianov, I.; Hynynen, J.; Barlow, S.; Marder, S.R.; Müller, C.; Zozoulenko, I. UV-to-IR absorption of molecularly p-d polythiophenes with alkyl and oligoether side chains: Experiment and interpretation based on density functional theory. *J. Chem. B* **2020**, *124*, 11280–11293. [CrossRef]
84. Sahalianov, I.; Say, M.G.; Abdullaeva, O.S.; Ahmed, F.; Glowacki, E.; Engquist, I.; Berggren, M.; Zozoulenko, I. Volum double-layer charge storage in composites based on conducting polymer PEDOT and cellulose. *ACS Appl. Energy Mater.* **20** 8629–8640. [CrossRef]
85. Frisch, M.J.; Trucks, G.W.; Schlegel, H.B.; Scuseria, G.E.; Robb, M.A.; Cheeseman, J.R.; Scalmani, G.; Barone, V.; Petersson, Nakatsuji, H.; et al. *Gaussian 16 Rev. C.01*; Gaussian, Inc.: Wallingford, CT, USA, 2016.
86. O'boyle, N.M.; Tenderholt, A.L.; Langner, K.M. Cclib: A library for package-independent computational chemistry algorith *Comput. Chem.* **2008**, *29*, 839–845. [CrossRef]
87. Gerischer, H. Über den Ablauf von Redoxreaktionen an Metallen und an Halbleitern. *Z. Für Phys. Chem.* **1960**, *26*, 223 [CrossRef]
88. Heller, I.; Kong, J.; Heering, H.A.; Williams, K.A.; Lemay, S.G.; Dekker, C. Individual single-walled carbon nanotub nanoelectrodes for electrochemistry. *Nano Lett.* **2005**, *5*, 137–142. [CrossRef] [PubMed]
89. Szroeder, P. Electron transfer kinetics at single-walled carbon nanotube paper: The role of band structure. *Phys. E Low-Dir Syst. Nanostruct.* **2011**, *44*, 470–475. [CrossRef]
90. Bard, A.J.; Faulkner, L.R.; White, H.S. *Electrochemical Methods: Fundamentals and Applications*; John Wiley & Sons: Hoboker USA, 2022.
91. Zhang, G.; Cuharuc, A.S.; Güell, A.G.; Unwin, P.R. Electrochemistry at highly oriented pyrolytic graphite (HOPG): Lower for the kinetics of outer-sphere redox processes and general implications for electron transfer models. *Phys. Chem. Chem.* **2015**, *17*, 11827–11838. [CrossRef]
92. Sakaushi, K.; Kumeda, T.; Hammes-Schiffer, S.; Melander, M.M.; Sugino, O. Advances and challenges for experiment and th for multi-electron multi-proton transfer at electrified solid–liquid interfaces. *Phys. Chem. Chem. Phys.* **2020**, *22*, 19401–1 [CrossRef]
93. Gerischer, H. The impact of semiconductors on the concepts of electrochemistry. *Electrochim. Acta* **1990**, *35*, 1677–1699. [Cross
94. Szroeder, P.; Górska, A.; Tsierkezos, N.; Ritter, U.; Strupiński, W. The role of band structure in electron transfer kineti low-dimensional carbon. *Mater. Und Werkst.* **2013**, *44*, 226–230. [CrossRef]
95. Chakrapani, V.; Sumanasekera, G.U.; Abeyweera, B.; Sherehiy, A.; Angus, J.C. Electrochemically induced p-type conductivi carbon nanotubes. *ECS Solid State Lett.* **2013**, *2*, M57. [CrossRef]

molecules

MDPI

cle

gnin-Derived Activated Carbon as Electrode Material for igh-Performance Supercapacitor

nghao Pan [1,2], Yongfeng Ji [2], Suxia Ren [2], Tingzhou Lei [2,*] and Lili Dong [2,*]

1 School of Environmental Science and Engineering, Changzhou University, Changzhou 213164, China; s22030857045@smail.cczu.edu.cn
2 Institute of Urban & Rural Mining, Changzhou University, Changzhou 213164, China; 13222666378@163.com (Y.J.); rensuxia@cczu.edu.cn (S.R.)
* Correspondence: leitingzhou@163.com (T.L.); dongli2050@cczu.edu.cn (L.D.)

Abstract: Utilizing lignin-derived activated carbon in supercapacitors has emerged as a promising approach to alleviating environmental pollution and promoting the high-value utilization of byproducts in the papermaking industry. In this study, activated carbons (LACs) were prepared using a simple one-step KOH activation approach and by employing enzymatic hydrolysis lignin (EHL). The impact of the KOH activation parameters on the microstructure and capacitive performance of the LACs was investigated by varying the KOH/EHL ratio and activation temperature. The optimized sample $LAC_{800\text{-}4}$ showed an interconnected porous structure with a high surface area of 2285 m^2/g, abundant micropores, and a small number of mesopores, which makes it a suitable electrode material for supercapacitors. The sample $LAC_{800\text{-}4}$ demonstrated a high specific capacitance of 291.3 F/g in a three-electrode system. Under a symmetrical supercapacitor electrode system, the specific capacitance of the $LAC_{800\text{-}4}$ electrode reached 186.8 F/g at 0.5 A/g. After 10,000 cycles at 20 A/g, the capacitance retention rate remained at 96.1%. The symmetrical supercapacitor also demonstrated a superior energy density of 6.5 Wh/kg. This work provides valuable insights into the transformation of low-value natural biomass derivatives into environmentally friendly, high-performing supercapacitor electrode materials.

Keywords: lignin; activated carbon; KOH activation; supercapacitors

1. Introduction

In recent years, the use of carbon-based supercapacitors in energy storage and conversion devices has garnered significant attention due to their high energy density, rapid charge and discharge rates, and superior safety performance [1,2]. Based on the formation mechanism of double-layer capacitance, the carbon material electrode should contain abundant micropores to provide a high surface area, enough active sites for electron storage, and an appropriate amount of mesopores and macropores to supply fast transmission and diffusion channels for the electrolyte [3,4]. Therefore, preparing activated carbon materials that meet the above requirements has gained considerable attention.

In recent years, biomass has been recognized as an ideal precursor for activated carbon due to its renewability, chemical stability, low cost, and unique structure. Wu et al. [5] prepared honeycomb-like carbon foam from wheat flour via one-step carbonization. The unique structure provided the carbon foam with a significant capacitive performance, with a high specific capacitance of 473 F/g at 0.5 A/g and a high energy density of 29.3 63.5 Wh/kg. Chen et al. [6] used rice husk as a raw material to prepare hierarchical porous carbon. The contribution of different rice husk components, including cellulose, hemicellulose, lignin, and SiO_2, to the formation of pores was investigated. Lignin mainly contributes to the formation of mesopores and micropores, which is crucial in obtaining a high specific capacitance. The optimized sample showed a specific capacitance of 51.4 F/g at 0.5 A/g in the symmetric supercapacitor, while its capacitance retention was as high as 96.2 % at 20 A/g. Studies have shown that biomass-derived carbon materials could be effectively employed in supercapacitors. Among the numerous biomass precursors developed, lignin has attracted significant attention due to its high carbon content, numerous self-doping heteroatoms, rigid chemical structure, and large and cheap output.

Lignin is the most abundant aromatic polymer in the world. Approximately 70 million tons of lignin are produced from pulping byproducts annually, but almost 98% of that lignin is burned as low-quality fuel or discarded directly as waste [7,8]. The conversion of lignin into carbon-based electrode materials not only alleviates environmental pollution but also realizes the high-value utilization of lignin. In recent years, some methods for the preparation of lignin-based carbon materials for use in supercapacitors have been developed. Zhang et al. [9] prepared carbon materials from lignin with and without KOH. The carbon material (LC) obtained without KOH showed a large texture with a small surface area, while the hierarchical porous carbon (LHPC) prepared with KOH showed a high surface area of 907 m^2/g. Benefiting from the hierarchical porous structure, the LHPC sample presented a superior captive performance, with a specific capacitance of 165.0 F/g in 1M H_2SO_4 at 0.05 A/g. Shi et al. [10] fabricated porous carbon from enzymatically hydrolyzed lignin and urea through a KOH activation procedure. The influences of activation temperatures and elemental doping on the microstructure and capacitive performance of the synthesized ONS-HPC materials were investigated and the ONS-HPC-600 electrode material presented the best electrochemical performance. It can be concluded that the activation parameters strongly influence the microstructure and capacitive performance of carbon materials. However, systematic discussions regarding the influence of the activation parameters on the pore structure and electrochemical performance of lignin-derived activated carbon have rarely been reported.

In this study, activated carbons (LACs) were prepared from enzymatic hydrolysis lignin and KOH using a facile one-step method. The lignin-derived activated carbon was characterized to analyze the mechanism implicated in the formation of the porous structure of LACs. The intrinsic regular relationship between the structure and electrochemical performance of LACs was further discussed. This work will provide a valuable reference for the fabrication of lignin-derived carbon electrodes for supercapacitors.

2. Results and Discussion

Scanning electron microscopy (SEM) was used to characterize the morphology and microstructure of the as-prepared samples. As shown in Figure 1a, LAC_{800-2} exhibited a bulk structure, in which only the surface was etched. This may be attributed to the low quantity of the KOH added, which contributed to the low etching degree and relatively complete carbon skeleton. As the KOH/EHL mass ratio increased, the LAC_{800-3} and LAC_{800-4} formed a well-developed porous structure. Generally speaking, an appropriate porous structure should contain a large number of micropores to provide enough active sites for ion storage, a certain amount of mesopores to provide short and convenient ion transfer channels, and large pores to provide buffer layers for electrolyte ions [11]. Thus, LAC_{800-3} and LAC_{800-4}, which have interconnected porous structures, may exhibit a significant capacity for electrochemical energy storage due to the improved electron and ion transport efficiency. However, the carbon skeleton was over-etched due to the excessive use of KOH, as indicated in Figure 1d [12]. The activation temperature also has a strong impact on the porous structure of carbon materials. The $LAC_{600\text{-}4}$ sample showed few pores on its surface due to the low activation temperature used, namely 600 °C; this hindered the process of KOH activation. As the activation temperature increased to 700 and 800 °C, a porous structure was achieved due to the suitable etching of KOH (Figure 1c,f). However, using an overly high activation temperature of 900 °C led to the collapse of $LAC_{900\text{-}4}$'s structure (Figure 1g). The TEM image shows that $LAC_{800\text{-}4}$ had a rough surface with numerous pores. The HRTEM image indicated the typical porous amorphous structure of $LAC_{800\text{-}4}$ [13].

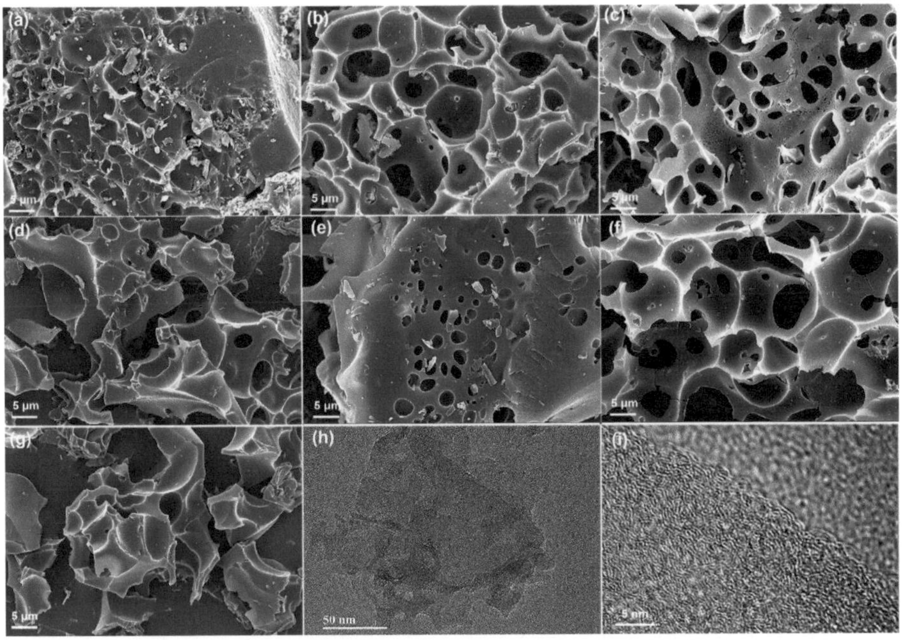

Figure 1. SEM images of (**a**) $LAC_{800\text{-}2}$, (**b**) $LAC_{800\text{-}3}$, (**c**) $LAC_{800\text{-}4}$, (**d**) $LAC_{800\text{-}5}$, (**e**) $LAC_{600\text{-}4}$, (**f**) $LAC_{700\text{-}4}$, and (**g**) $LAC_{900\text{-}4}$; (**h**) TEM and (**i**) HRTEM images of $LAC_{800\text{-}4}$.

To study the crystal structure of the samples, XRD and Raman tests were performed, as shown in Figure 2. The samples prepared with various KOH contents presented two typical diffraction peaks at 25 and 43 degrees, which were ascribed to the (002) and (100) crystal planes of carbon (Figure 2a) [14]. The broad and low peaks indicate the low crystallinity and typical amorphous carbon structure of the $LAC_{800\text{-}2}$, $LAC_{800\text{-}3}$, $LAC_{800\text{-}4}$, and $LAC_{800\text{-}5}$ samples [15]. As the KOH content increased, the (002) crystal plane showed a gradual reduction in diffraction peak intensity. This suggests that the structure of LAC_s was etched by KOH; this reduced its regularity, creating visible pores or faults and lowering the degree of graphitization [16]. The Raman spectra degree was further studied to investigate the graphitization of the $LAC_{800\text{-}2}$, $LAC_{800\text{-}3}$, $LAC_{800\text{-}4}$, and $LAC_{800\text{-}5}$ samples. According to Figure 2b, two peaks at 1330 and 1585 cm^{-1} appeared in all the Raman spectra, which were ascribed to the defect-induced D-band and ideal graphite-derived G band, respectively. The degree of irregular arrangement in the LACs was estimated using the I_D/I_G ratio, which was calculated via Gaussian function fitting. The I_D/I_G values of $LAC_{800\text{-}2}$, $LAC_{800\text{-}3}$, $LAC_{800\text{-}4}$, and $LAC_{800\text{-}5}$ were calculated to be 0.88, 0.90, 0.93, and 0.96, respectively. These results suggest that the degree of graphitization decreases as the amount of KOH activator increases. The XRD patterns of LAC_{600-4}, LAC_{700-4}, LAC_{800-4}, and LAC_{900-4} (Figure 2c) all showed relatively wide diffraction peaks, indicating that the samples possessed amorphous structures [17]. The Raman spectra further confirmed this finding. With the increase in the activation temperature, the ratio of I_D/I_G gradually increased, indicating that the degree of graphitization decreased. This is mainly because the graphitization of carbon materials occurs at temperatures above 1000 °C. When the activation temperature is 600–900 °C and the mass ratio of KOH to EHL is high, the activation of KOH plays a dominant role, leading to an increase in the degree of disorder in the samples [18].

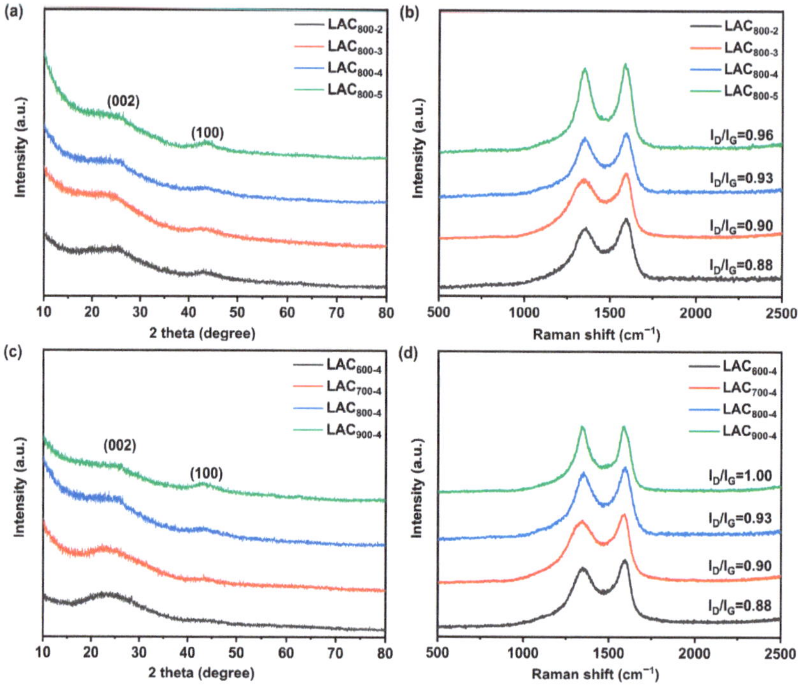

Figure 2. XRD patterns (**a**) and Raman spectra (**b**) of LAC_{800-2}, LAC_{800-3}, LAC_{800-4}, and LAC_{800-5}; XRD patterns (**c**) and Raman spectra (**d**) of LAC_{600-4}, LAC_{700-4}, LAC_{800-4}, and LAC_{900-4}.

XPS spectra were selected to further determine the chemical composition and valence states of LAC_{800-4}. As shown in Figure 3a, the survey XPS spectrum of LAC_{800-4} revealed the presence of C and O, which were the major elements. Four sub-peaks at 284.7 eV, 285.6 eV, 286.5 eV, and 288.9 eV were fitted in the C1s spectrum, and these were attributed to the sp^2 C=C, sp^3 C-C, C-O, and C=O bands, respectively [19]. In the high-resolution O 1s spectrum, there were three characteristic peaks at 531.4, 533.3, and 534.9 eV, which were attributed to C=O, C-O-C, and C=O-OH or water, respectively [20]. The presence of the C=O group promoted the wettability of the LAC_{800-4} surface, which could reduce the charge impedance of the supercapacitor. The oxygen-containing functional groups could also serve as the active sites to enhance the capacitive performance via a reversible redox reaction [21].

N_2 adsorption–desorption analysis was used to further investigate the microstructures of the LACs, as illustrated in Figure 4. All the LACs presented I-type isothermal curves and H4-type hysteresis rings. The isothermal curve of LAC_{800-2} exhibited a significant increase in $P/P_0 < 0.02$, indicating the existence of a large number of micropores [21]. In addition to the sharp rise in $P/P_0 < 0.01$, a rising stage in the range of $0.02 < P/P_0 < 0.4$ appeared for the LAC_{800-3}, LAC_{800-4}, and LAC_{800-5} samples. As the KOH/EHL mass ratios increased, the H4-type hysteresis ring became larger, indicating that the number of mesopores increased. The slight rise in the range of $0.9 < P/P_0 < 1.0$ reveals the presence of some macropores [22]. The N_2 adsorption–desorption isotherm of samples under different activation temperatures exhibited similar characteristics (Figure 4c). In particular, the LAC_{900-4} sample exhibited a larger hysteresis loop and wider rising stage, suggesting an increased percentage of mesopores. These findings show that the LACs had a hierarchical porous structure, and the pore size distribution curves further confirmed these results (Figure 4b,d). All the LACs exhibited pore size distributions in the range of 0.6–2 nm and 2–6 nm, implying the existence of both micropores and mesopores.

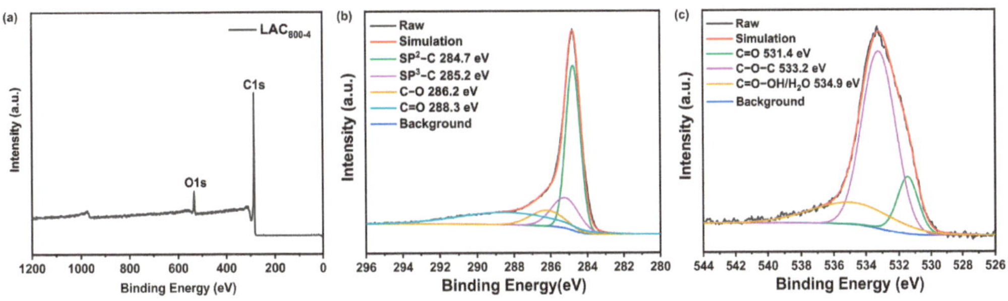

Figure 3. Full XPS spectrum of LAC_{800-4} (**a**); peak fitting spectra of C 1s (**b**) and O 1s (**c**) of LAC_{800-4}.

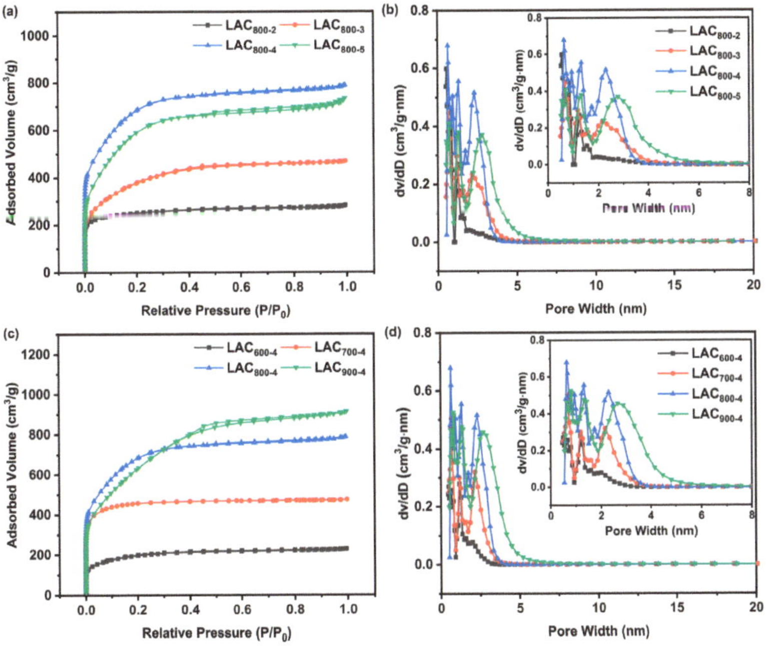

Figure 4. N_2 adsorption–desorption curves (**a**) and pore size distribution curves (**b**) of LAC_{800-2}, LAC_{800-3}, LAC_{800-4}, and LAC_{800-5}; N_2 adsorption–desorption curves (**c**) and pore size distribution curves (**d**) of LAC_{600-4}, LAC_{700-4}, LAC_{800-4}, and LAC_{900-4}.

It was found that the KOH/EHL mass ratio strongly influences the microstructures of LACs. A relatively low ratio of KOH/EHL (2:1) results in insufficient activation of the carbon skeleton, which only occurred on the surface of LAC_{800-2} (as presented in the SEM image). Along with the increase in the KOH/EHL mass ratio, the N_2 adsorption capacity of LAC_{800-3} and LAC_{800-4} gradually increased. The LAC_{800-4} sample exhibited the highest surface area of 2285 m^2/g. This is because the appropriate KOH activation led to a reasonable pore size distribution, with a large number of micropores and some mesopores [23]. However, the addition of excess KOH led to the collapse of the porous structure, thereby decreasing the surface area and widening the pore size distribution of LAC_{800-5}. As illustrated in Figure 4c,d, the activation temperature is another factor that significantly affects the activation procedure. KOH activation is difficult to achieve at a low activation temperature of 600 °C, resulting in the low surface area observed in LAC_{600-4}. The N_2 adsorption capacity gradually increased as the activation temperature increased to 800 °C, implying that there was an increase in the surface area and number of micropores. The pore size distribution range also extended due to the more intensive reaction between the carbon skeleton and KOH at higher activation temperatures. However, strong KOH etching occurred at a high activation temperature of 900 °C, leading to a decrease in the number of micropores and an increase in the number of mesopores. In summary, the LACs prepared from lignin possessed an interconnected porous structure with an abundance of micropores and an appropriate number of mesopores and macropores. The dominating micropores in the electrode material provide a high surface area and abundant active sites for electron storage, while the mesopores and macropores enable fast transmission and diffusion channels for electrolytes [24].

The electrochemical properties of the LAC_{800-2}, LAC_{800-3}, LAC_{800-4}, and LAC_{800-5} samples were tested in a three-electrode system, as displayed in Figure 5. Under a scan rate of 10 mV/s, all the CV curves of LAC_s showed a quasi-rectangular shape, indicating their double-layer characteristics and capacity for a rapid electrode response [25]. Compared with the other samples, LAC_{800-4} presented the largest integrated area, showing the best capacitive performance. The GCD curves of LAC_S at 1 A/g were further examined, as illustrated in Figure 5b. All the GCD curves presented an approximate isosceles triangular shape, revealing the typical electric double-layer capacitance and reversibility of the electrode materials. As the KOH/EHL ratio increased, the specific capacitance of the LAC_S increased first and subsequently declined when the KOH/EHL ratio surpassed 3:1. The specific capacitances of LAC_{800-2}, LAC_{800-3}, LAC_{800-4}, and LAC_{800-5} are 168.8, 207.5, 291.3, and 246.3 F/g, respectively. The resistance of the prepared porous carbon electrodes is characterized by the electrochemical impedance spectroscopy (EIS) measurement, as illustrated in Figure 5c. The Nyquist plot curves of the four samples showed a semicircle at high frequencies and a straight line at low frequencies, corresponding to the interfacial charge–transfer resistance (R_{ct}) and Warburg resistance ($Z_{w)}$, respectively. The intercept of the curve on the real axis in the high-frequency region corresponds to the equivalent series resistance (R_s) [26]. The LAC_{800-4} sample exhibited the smallest R_s and R_{ct} values of 0.42 and 0.53 Ω, demonstrating the superior charge transport properties. As discussed above, the LAC_{800-2} sample presented the smallest surface area, so it could not supply enough active sites for ion storage. As the KOH usage increased, a large number of micropores formed, which led to an increase in the surface area and specific capacitance. However, excessive KOH usage causes the collapse of the structure, which leads to a decreased specific capacitance. It can be concluded that the capacitive performance of lignin-based activated carbon can be adjusted by varying the KOH/EHL ratio.

Figure 6 shows the electrochemical performance of the LAC_{600-4}, LAC_{700-4}, LAC_{800-4}, and LAC_{900-4} samples under different activation temperatures (600–900 °C). It is known that a rectangular CV curve indicates the typical capacitive behavior of electrode materials. As shown in Figure 6a, the LAC_{800-3} sample presented a quasi-rectangular CV curve. However, the CV curves of other samples, especially LAC_{600-4}, deviate from the rectangle. In particular, the LAC_{800-4} electrode exhibits the highest current density and largest closed area, implying the best supercapacitance performance. The GCD curves showed the typical electrical double-layer capacitance according to the near-isosceles triangle shape (Figure 6b) [27]. At a current density of 1 A/g, the highest specific capacitance, namely 291.3 F/g, was calculated for LAC_{800-4}; meanwhile, LAC_{600-4}, LAC_{700-4}, and LAC_{900-4} exhibited specific capacitances of 92.5, 209.3, and 230.0 F/g, respectively. The activation temperature was also found to significantly influence the impedance (Figure 6c). The trends in R_s and R_{ct} were in the order of $LAC_{800-4} < LAC_{900-4} < LAC_{700-4} < LAC_{600-4}$. The LAC_{800-4} sample exhibited the lowest R_s and R_{ct} values and the largest slope in the low-frequency region. At a low activation temperature of 600 °C, KOH was unable to activate the sample effectively, which resulted in a small number of micropores and a low specific capacitance [28]. However, a high activation temperature of 900 °C led to the over-etching of the carbon skeleton, thereby resulting in a decrease in the specific surface area. In addition, the enlarged micropores led to a decrease in the specific capacitance. Therefore, a moderate activation temperature (T = 800 °C) needs to be used alongside the ideal microporous carbon material, thereby maximizing the performance of supercapacitors.

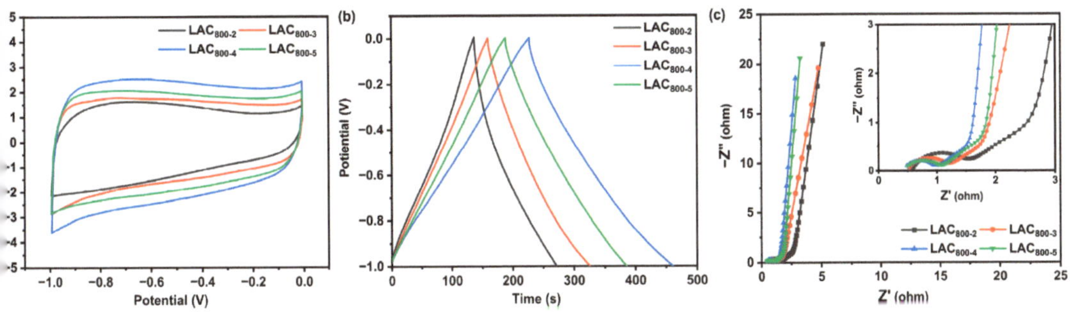

Figure 5. CV curves at 10 mV/s (**a**), GCD curves at 1 A/g (**b**), and Nyquist plots (**c**) of LAC_{800-2}, LAC_{800-3}, LAC_{800-4}, and LAC_{800-5}.

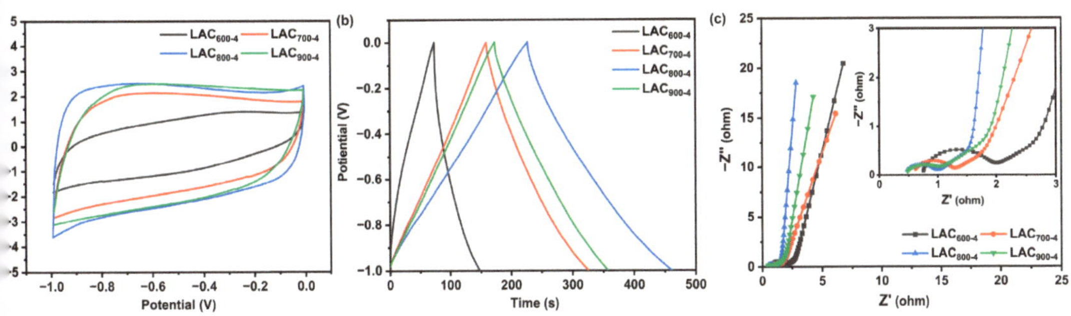

Figure 6. CV curves at 10 mV/s (**a**), GCD curves at 1 A/g (**b**), and Nyquist plots (**c**) of LAC_{600-4}, LAC_{700-4}, LAC_{800-4}, and LAC_{900-4}.

As the optimally prepared sample, the $LAC_{800\text{-}4}$-based symmetrical supercapacitor was tested in a two-electrode system to evaluate the capacitive performance practically. Figure 7a shows the CV curves of the symmetric supercapacitor at different scan rates, from 10 to 200 mV/s. All the CV curves demonstrated a near-ideal rectangular shape. The CV curves showed no discernible distortion as the scanning rate increased, indicating a superior double-layer capacitive performance [29]. Figure 7b shows the GCD curves of the symmetric supercapacitor at different current densities. All the GCD curves demonstrated a symmetrical triangle shape, implying that the capacitance of the symmetric supercapacitor was reversible. The $LAC_{800\text{-}4}$ electrode showed a specific capacitance of 186.8 F/g at 0.5 A/g and 140 F/g at 20 A/g, and the capacitance retention was calculated to be 74.9%. Furthermore, the durability of the symmetric supercapacitor was examined by performing 10,000 cycles at 20 A/g, with only a 3.9% reduction in the capacitance observed; this suggests that the cycling stability was excellent (Figure 7c). The Ragone plots (Figure 7d) showed that the corresponding energy density and power density of symmetric supercapacitors are 6.5 Wh/kg and 125 W/kg, values that are superior to previous reports (Table 1) [20,23,30–32].

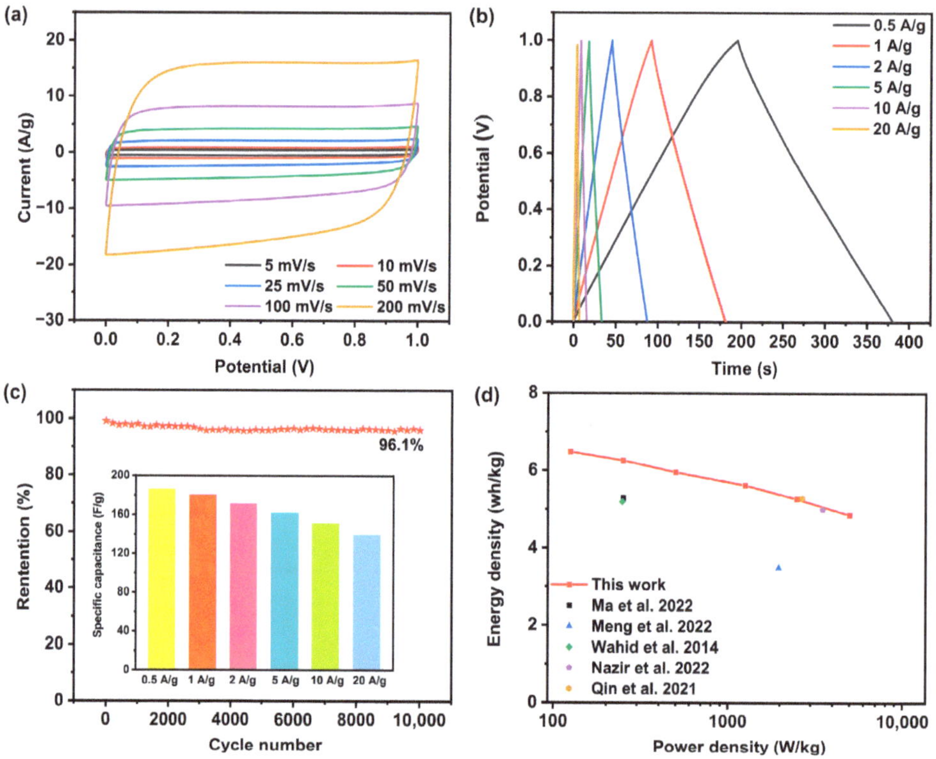

Figure 7. Electrochemical performance of the $LAC_{800\text{-}4}$ electrode in a two-electrode system in 6 M KOH: (**a**) CV curves at various scan rates. (**b**) GCD curves at various current densities. (**c**) Cycling stability at 20 A/g for 10,000 cycles. (**d**) Ragone plots and comparison with other reported results. Refs. [20,23,30–32].

Table 1. Comparison of the electrochemical performance of reported biomass-derived carbon materials.

aw Materials	Activator	S_{BET} (m^2/g)	Electrolyte	Capacitance (F/g)	Current Density (A/g)	Power Density (W/kg)	Energy Density (Wh/kg)	Ref.
Tea waste	KOH	1354	6 M KOH	98	1	5870	2.7	[20]
Chitosan	KOH	2787	6 M KOH	188	0.5	250.1	6.53	[23]
;arcane bagasse	KOH	1260	1 M H_2SO_4	225	1	3600	5	[30]
Banana peel	KOH	2452	1 M Na_2SO_4	138	0.5	2690	5.3	[31]
Gardenia	KOH	1637	6 M KOH	153	0.5	246.9	5.2	[32]
enzymatic drolysis lignin	KOH	2285	6 M KOH	186	0.5	125	6.5	This work

3. Experimental Section

3.1. Materials

Enzymatic hydrolysis lignin (EHL) was obtained from Shandong Longlive Biotechnology CO., Ltd., (Dezhou, China). The chemicals including KOH, HCl, and polytetrafluoroethylene were bought from Shanghai Macklin Biochemical Co., Ltd., (Shanghai, China). Acetylene black was purchased from Shenzhen Kejingzhida Technology Co., Ltd., (Shezhen, China). Deionized water was used throughout the experiments.

3.2. Preparation of LACs

First, 30 g enzymatic hydrolysis lignin (EHL) was fully mixed with a certain amount of KOH. The KOH/EHL mass ratios were R = 2, 3, 4, and 5. The resulting samples were loaded into nickel crucibles, and the nickel crucibles were coated with calcined coke to provide an oxygen-free atmosphere. These were first heated at 550 °C for 2 h and then raised to various activation temperatures T (T = 600, 700, 800, and 900 °C) for 3 h at a heating rate of 5 degrees Celsius per minute. The obtained carbon samples were purified, washed, and dried and named LAC_{800-2}, LAC_{800-3}, LAC_{800-4}, LAC_{800-5}, LAC_{600-4}, LAC_{700-4}, and LAC_{900-4}, respectively.

The material characterization and electrochemical measurements are described in detail in the Supporting Information.

4. Conclusions

In summary, EHL-derived activated carbons were prepared using a facile one-step KOH activation approach. The results indicated that the microstructure and capacitive performance of the as-prepared LACs were significantly affected by the KOH/EHL ratio and activation temperature. The optimized LAC_{800-4} sample presented an interconnected porous structure with a high surface area of 2285m^2/g and an appropriate pore size distribution. When assembled into a symmetrical supercapacitor, the LAC_{800-4} electrode exhibited a specific capacitance of 186.8 F/g at 0.5A/g. Furthermore, the LAC_{800-4} symmetric supercapacitor showed excellent cycling stability, with only a 3.9% reduction in capacitance observed when 10,000 cycles were performed at 20 A/g. The LAC_{800-4} symmetric supercapacitor displayed a high energy density of 6.5 Wh/kg in 6 M aqueous KOH when the power density was 125 W/kg.

Supplementary Materials: The following supporting information can be downloaded at https://www.mdpi.com/article/10.3390/molecules30010089/s1, Electrochemical measurements.

Author Contributions: Conceptualization, T.L.; Methodology, Y.J.; Data curation, C.P.; Writing—original draft, C.P.; Writing—review and editing, S.R. and L.D.; Funding acquisition, T.L. All authors have read and agreed to the published version of the manuscript.

Funding: We acknowledge the financial support from the National Key Research and Development Program of China (2022YFB4201904).

Institutional Review Board Statement: Not applicable.

Informed Consent Statement: Not applicable.

Data Availability Statement: Data are contained within the article or Supplementary Materials.

Conflicts of Interest: The authors declare no conflicts of interest.

References

1. Bhat, M.Y.; Hashmi, S.A.; Khan, M.; Choi, D.; Qurashi, A. Frontiers and recent developments on supercapacitor's mate design, and applications: Transport and power system applications. *J. Energy Storage* **2023**, *58*, 106104. [CrossRef]
2. Zhang, W.L.; Yin, J.; Wang, C.W.; Zhao, L.; Jian, W.B.; Lu, K.; Lin, H.B.; Qiu, X.Q.; Alshareef, H.N. Lignin derived porous car Synthesis methods and supercapacitor applications. *Small Methods* **2021**, *5*, 2100896. [CrossRef] [PubMed]
3. Wang, T.; Liu, Z.G.; Li, P.F.; Wei, H.Q.; Wei, K.X.; Chen, X.R. Lignin-derived carbon aerogels with high surface are supercapacitor applications. *Chem. Eng. J.* **2023**, *466*, 143118. [CrossRef]
4. Dong, L.; Pan, C.; Ji, Y.; Ren, S.; Lei, T. Corncob-derived activated carbon as electrode material for high-performance supercapa *Materials* **2024**, *17*, 4341. [CrossRef]
5. Wu, X.; Jiang, L.; Long, C.; Fan, Z. From flour to honeycomb-like carbon foam: Carbon makes room for high energy de supercapacitors. *Nano Energy* **2015**, *13*, 527–536. [CrossRef]
6. Chen, Z.; Wang, X.; Xue, B.; Li, W.; Ding, Z.; Yang, X.; Qiu, J.; Wang, Z. Rice husk-based hierarchical porous carbon for performance supercapacitors: The structure-performance relationship. *Carbon* **2020**, *161*, 432–444. [CrossRef]
7. Jiang, M.; Xu, W.; Du, X.; Yang, X.; Wang, F.; Zhou, Y.; Pan, Y.; Lu, Y. An N, P, O-doped porous carbon electrode material der from a lignin-modified chitosan xerogel for a supercapacitor. *Mater. Today Sustain.* **2023**, *22*, 100372. [CrossRef]
8. Zhang, W.; Guo, T.; Liu, Y.; Zhang, X.; Zou, B.; Zhao, C.; Suo, H.; Wang, H.; Zhao, X. Electrocatalytic performance of ca layer and spherical carbon/carbon cloth composites towards hydrogen evolution from the direct electrolysis of bunsen rea product. *Chem. Res. Chin. Univ.* **2024**, *40*, 109–118. [CrossRef]
9. Zhang, W.; Lin, H.; Lin, Z.; Yin, J.; Lu, H.; Liu, D.; Zhao, M. 3D hierarchical porous carbon for supercapacitors prepared lignin through a facile template-free method. *Chemsuschem* **2015**, *8*, 2114–2122. [CrossRef]
10. Shi, F.Y.; Tong, Y.; Li, H.S.; Li, J.J.; Cong, Z.Y.; Zhai, S.R.; An, Q.D.; Wang, K. Synthesis of oxygen/nitrogen/sulfur cod hierarchical porous carbon from enzymatically hydrolyzed lignin for high-performance supercapacitors. *J. Energy Storage* *52*, 104992. [CrossRef]
11. Tian, J.; Liu, C.; Lin, C.; Ma, M. Constructed nitrogen and sulfur codoped multilevel porous carbon from lignin for performance supercapacitors. *J. Alloys Compd.* **2019**, *789*, 435–442. [CrossRef]
12. Chinnadurai, D.; Kim, H.-J.; Karupannan, S.; Prabakar, K. Multiscale honeycomb-structured activated carbon obtained nitrogen-containing mandarin peel: High-performance supercapacitors with significant cycling stability. *New J. Chem.* **201** 3486–3492. [CrossRef]
13. Zhang, W.; Zhao, M.; Liu, R.; Wang, X.; Lin, H. Hierarchical porous carbon derived from lignin for high performance superca tor. *Colloids Surf. A* **2015**, *484*, 518–527. [CrossRef]
14. Wan, X.; Shen, F.; Hu, J.; Huang, M.; Zhao, L.; Zeng, Y.; Tian, D.; Yang, G.; Zhang, Y. 3-D hierarchical porous carbon from oxid lignin by one-step activation for high-performance supercapacitor. *Int. J. Biol. Macromol.* **2021**, *180*, 51–60. [CrossRef] [Pub
15. Feng, P.; Wang, H.; Huang, P.; Zhong, L.; Gan, S.; Wang, W.; Niu, L. Nitrogen-doped lignin-derived porous carbon supercapacitors: Effect of nanoporous structure. *Chem. Eng. J.* **2023**, *471*, 144817. [CrossRef]
16. Klose, M.; Reinhold, R.; Logsch, F.; Wolke, F.; Linnemann, J.; Stoeck, U.; Oswald, S.; Uhlemann, M.; Balach, J.; Markowski, J.; Softwood Lignin as a Sustainable Feedstock for Porous Carbons as Active Material for Supercapacitors Using an Ionic Li Electrolyte. *ACS Sustain. Chem. Eng.* **2017**, *5*, 4094–4102. [CrossRef]
17. Chang, Z.Z.; Yu, B.J.; Wang, C.Y. Lignin-derived hierarchical porous carbon for high-performance supercapacitors. *J. Solid Electrochem.* **2016**, *20*, 1405–1412. [CrossRef]
18. Zhu, Q.; Zhao, D.; Cheng, M.; Zhou, J.; Owusu, K.A.; Mai, L.; Yu, Y. A new view of supercapacitors: Integrated supercapac *Adv. Energy Mater.* **2019**, *9*, 1901081. [CrossRef]
19. Guo, N.; Li, M.; Sun, X.; Wang, F.; Yang, R. Enzymatic hydrolysis lignin derived hierarchical porous carbon for supercapacito ionic liquids with high power and energy densities. *Green Chem.* **2017**, *19*, 2595–2602. [CrossRef]

Ma, Q.H.; Xi, H.T.; Cui, F.; Zhang, J.J.; Chen, P.; Cui, T.Y. Self-templating synthesis of hierarchical porous carbon with multi-heteroatom co-doping from tea waste for high-performance supercapacitor. *J. Energy Storage* **2022**, *45*, 103509. [CrossRef]
Zhai, S.; Li, K.; Li, C.; Zhai, C.; Han, Q.; Zhang, Z.; Fu, Y.; Li, X.; Jin, K.; Cai, Z.; et al. Lignin-derived N, S-codoped hierarchical porous carbons with high mesoporous rate for sustainable supercapacitive energy storage. *J. Storage* **2024**, *85*, 111036. [CrossRef]
Fu, F.; Yang, D.; Zhao, B.; Fan, Y.; Liu, W.; Lou, H.; Qiu, X. Boosting capacitive performance of N, S co-doped hierarchical porous lignin-derived carbon via self-assembly assisted template-coupled activation. *J. Colloid Interface Sci.* **2023**, *640*, 698–709. [CrossRef]
Meng, D.; Hu, Y.; Jing, Y.; Zhang, X.; Mahmud, S.; Su, S.; Zhu, J. One-step carbonization strategy of freeze-dried chitosan to prepare Nitrogen-Oxygen co-doped porous carbon supercapacitors with ultra-large specific surface area. *Fuel* **2022**, *320*, 124002. [CrossRef]
Park, S.; Song, J.; Lee, W.C.; Jang, S.; Lee, J.; Kim, J.; Kim, H.-K.; Min, K. Advances in biomass-derived electrode materials for energy storage and circular carbon economy. *Chem. Eng. J.* **2023**, *470*, 144234. [CrossRef]
Li, G.; Mao, K.; Liu, M.; Yan, M.; Zhao, J.; Zeng, Y.; Yang, L.; Wu, Q.; Wang, X.; Hu, Z. Achieving Ultrahigh Volumetric Energy Storage by Compressing Nitrogen and Sulfur Dual-Doped Carbon Nanocages via Capillarity. *Adv. Mater.* **2020**, *32*, 2004632. [CrossRef]
Zhang, D.Y.; He, Q.X.; Wang, D.M.; Zhao, W.; Wang, Z.W.; Zhou, Z.Y.; Zuo, L.G.; Li, L.Z.; Wei, W.Q. Preparation of nitrogen-doped lignin-based porous carbon materials and their application in a supercapacitor. *J. Mater. Sci.* **2024**, *59*, 19161–19174. [CrossRef]
Bai, Y.; Wang, Q.; Wang, J.; Zhang, S.; Wei, C.; Cao, L.; Zhang, S. In situ, nitrogen-doped porous carbon derived from mixed biomass as ultra-high-performance supercapacitor. *Nanomaterials* **2024**, *14*, 1368. [CrossRef] [PubMed]
Pang, J.; Zhang, W.; Zhang, H.; Zhang, J.; Zhang, H.; Cao, G.; Han, M.; Yang, Y. Sustainable nitrogen-containing hierarchical porous carbon spheres derived from sodium lignosulfonate for high-performance supercapacitors. *Carbon* **2018**, *132*, 280–293. [CrossRef]
Zhou, Q.Q.; Chen, Q.; Xu, W.J.; Wang, F.; Du, X.Q.; Zhou, Y.F.; Zhan, Y.L.; Jiang, M. Nitrogen and sulfur co-doped carbonized lignin nanotubes for supercapacitor applications. *Chem. Eng. J.* **2024**, *496*, 154126. [CrossRef]
Wahid, M.; Puthusseri, D.; Phase, D.; Ogale, S. Enhanced capacitance retention in a supercapacitor made of carbon from sugarcane bagasse by hydrothermal pretreatment. *Energy Fuels* **2014**, *28*, 4233–4240. [CrossRef]
Nazir, G.; Rehman, A.; Hussain, S.; Ikram, M.; Park, S.J. Supercapacitor performance based on nitrogen and sulfur co-doped hierarchically porous carbons: Superior rate capability and cycle stability. *Int. J. Energy Res.* **2022**, *46*, 15602–15616. [CrossRef]
Qin, C.; Wang, S.; Wang, Z.; Ji, K.; Wang, S.; Zeng, X.; Jiang, X.; Liu, G. Hierarchical porous carbon derived from Gardenia jasminoides Ellis flowers for high performance supercapacitor. *J. Energy Storage* **2021**, *33*, 102061. [CrossRef]

molecules

Article

Roughing Nitrogen-Doped Carbon Nanosheets for Loading of Monatomic Fe and Electroreduction of CO_2 to CO

Yuxuan Liu [1,2], Yufan Tan [2], Keyi Zhang [2], Tianqi Guo [3], Yao Zhu [1], Ting Cao [1], Haiyang Lv [1], Junpeng Zhu [2], Ze Gao [2], Su Zhang [4], Zheng Liu [1,*] and Juzhe Liu [2,*]

1 State Key Laboratory of Environmental Criteria and Risk Assessment, Chinese Research Academy of Environmental Sciences, Beijing 100012, China; liuyuxuan@ncepu.edu.cn (Y.L.); zhuyao@craes.org.cn (Y.Z.); caoting@craes.org.cn (T.C.); lvhaiyang@craes.org.cn (H.L.)
2 The Key Laboratory of Resources and Environmental System Optimization, Ministry of Education, College of Environmental Science and Engineering, North China Electric Power University, Beijing 102206, China; 18954531716@163.com (Y.T.); zhangkeyi@ncepu.edu.cn (K.Z.); zhujunpeng@ncepu.edu.cn (J.Z.); gaoze@ncepu.edu.cn (Z.G.)
3 International Institute for Interdisciplinary and Frontiers, Beihang University, Beijing 100191, China; guotianqi@buaa.edu.cn
4 School of Material Science and Engineering, China University of Petroleum (East China), Qingdao 266580, China; suzhangs@163.com
* Correspondence: liuzheng@craes.org.cn (Z.L.); liujuzhe@ncepu.edu.cn (J.L.)

Abstract: The catalyst is the pivotal component in CO_2 electroreduction systems for converting CO_2 into valuable products. Carbon-based single-atom materials (CSAMs) have emerged as promising catalyst candidates due to their low cost and high atomic utilization efficiency. The rational design of the morphology and microstructure of such materials is desirable but poses a challenge. Here, we employed different $Mg(OH)_2$ templates to guide the fabrication of two kinds of amorphous nitrogen-doped carbon nanosheet-supported Fe single atoms (FeSNC) with rough and flat surface structures. In comparison to flat FeSNC with saturated FeN_4 sites, the rough FeSNC (R-FeSNC) exhibited unsaturated FeN_{4-x} sites and contracted Fe-N bond length. The featured structure endowed R-FeSNC with superior capacity of catalyzing CO_2 reduction reaction, achieving an exceptional CO selectivity with Faradaic efficiency of 93% at a potential of −0.66 V vs. RHE. This study offers valuable insights into the design of CSAMs and provides a perspective for gaining a deeper understanding of their activity origins.

Keywords: single atom; CO_2 electroreduction; unsaturated coordination; amorphous; rough surface

1. Introduction

The electrocatalytic CO_2 reduction reaction (CRR) presents a promising approach to addressing global energy and environmental challenges by converting CO_2 to high-valuable chemicals [1–3]. Among various products, carbon monoxide (CO) stands out as a pivotal feedstock for the synthesis of multi-carbon chemicals [4–6]. Recently, substantial research efforts have been dedicated to developing cost-effective and high-selectivity catalysts for CO_2-to-CO conversion [7–9]. Carbon-based single-atom materials (CSAMs) with distinct structure models have drawn extensive attention for impressive performance and low costs, showing advantages over other common catalysts, e.g., noble metals or molecular materials [10–12]. The construction and optimization of such catalysts through morphology control and microstructure modulation represent a promising yet challenging area of exploration.

In terms of morphology control, typical strategies involve the construction of three-dimensional (3D) or two-dimensional (2D) structures, the introduction of pores on the surface, and the augmentation of surface rugosity and curvature [13–18]. For instance, the construction of a three-dimensional hierarchical Fe-N-C structure can effectively enhance mass transfer, facilitating the conversion of CO_2 molecules at single-atom sites [19]. Moreover, the curved carbon surface can provide abundant anchoring sites for single atoms and simultaneously tune their electronic structures, thereby optimizing electrocatalysis reaction energetics [20].

Apart from morphology control, the regulation of microstructures is also of great importance, primarily referring to the regulation on coordination environments of single-atom sites, which encompasses heteroatom doping and defect engineering, etc. [21–25]. The doping of nonmetallic elements (such as O, N, S and P) and the creation of defect sites can effectively modulate the coordination environment and electronic states of metal single atoms, which are closely correlated with their catalytic performance. For example, Zhang et al. designed an asymmetric atomic interface of CuN_3O/C which could reduce the Gibbs free energy of the CO* desorption step [26]. Zhao et al. created N-defects around Ni sites which can induce a stable pyridinic N dominant Ni-N_2 unsaturated coordination structure with enhanced kinetics toward the CO_2-to-CO conversion [27]. In addition, achieving favorable lattice strain through precise control of atomic coordination is also an effective strategy to enhance the electrocatalytic process [28]. For example, the strain regulation of atomically dispersed NiN_4 active sites in helical carbon could facilitate the generation of *COOH and suppress hydrogen evolution reaction (HER) [29]. On this basis, it is meaningful to develop effective synthetic strategies for modulating the structures of CSAMs and unravel their structure–activity relationships.

Herein, we fabricated flocculent and lamellar $Mg(OH)_2$ templates to direct the synthesis of FeSNC catalysts with rough and flat structures, respectively. Rough FeSNC (R-FeSNC) is featured with unsaturated FeN_{4-x} sites and contracted Fe-N bond lengths, which exhibited a considerable CO Faradaic efficiency (FE_{CO}) of 93% at −0.66 V vs. RHE and was superior to flat FeSNC (F-FeSNC). Our work provides an efficient strategy to tune the structure and catalytic property of CSAMs.

2. Results and Discussion

2.1. Material Characterization

The synthetic scheme is illustrated in Figure 1A. Initially, flocculent and lamellar $Mg(OH)_2$ templates are prepared in ethanol and aqueous solutions, respectively (Figures S1 and S2), which are subsequently blended with carboxymethyl chitosan, metal salts (ferrous acetate) and N-source (urea). After freeze-drying and grinding, the mixture underwent calcination. Subsequently, hydrochloric acid etching is employed to remove the templates, resulting in the final products. R-FeSNC and F-FeSNC can be obtained in synthesis procedures involving flocculent and lamellar $Mg(OH)_2$ templates, respectively. The X-Ray Diffraction (XRD) patterns in Figure S3 demonstrate the amorphous state of R-FeSNC and F-FeSNC since no sharp Bragg reflection peak is observed. The transmission electron microscopy (TEM) and scanning electron microscope (SEM) images of R-FeSNC show its two-dimensional structure with a rough surface (Figures 1B and S4). Atomically resolved high-angle annular dark-field scanning transmission electron microscopy (HAADF-STEM) imaging reveals the presence of isolated bright dots dispersed on the rough surface of the carbon sheet without metal clusters or particles, which correspond to monatomic Fe sites (Figure 1C). The corresponding energy-dispersive spectroscopy (EDS) elemental mapping images (Figure 1D) demonstrated the uniform distribution of C, N, and Cu on a R-FeSNC nanosheet. In comparison, F-FeSNC exhibits a tawo-dimensional structure with a relatively flat surface (Figures 1E and S5). The relevant HAADF-STEM image also verifies its single-atom distribution state (Figure 1F). The EDS mappings also show homogenous distribution of C, N, and Fe atoms for F-FeSNC in Figure 1G. According to the inductively coupled plasma optical emission spectrometry, the Fe contents of R-FeSNC and F-FeSNC were determined to be 1.3 and 1.0 wt%, respectively. The primary reason for the formation of different morphologies of R-FeSNC and F-FeSNC is attributed to the variation in the morphology of $Mg(OH)_2$ templates. Flocculent $Mg(OH)_2$, characterized by abundant edge sites and irregular surface undulations, significantly influences the carbon structure formation during the crosslinking and carbonization process when used as a hard template [30–32]. This resulted in the final carbon material inheriting some structural characteristics of the template. Consequently, after the template is removed, the R-FeSNC exhibits roughened features. In contrast, the surface of lamellar $Mg(OH)_2$ is relatively smooth and even, and the F-FeSNC obtained using it as a template correspondingly exhibits a flat surface.

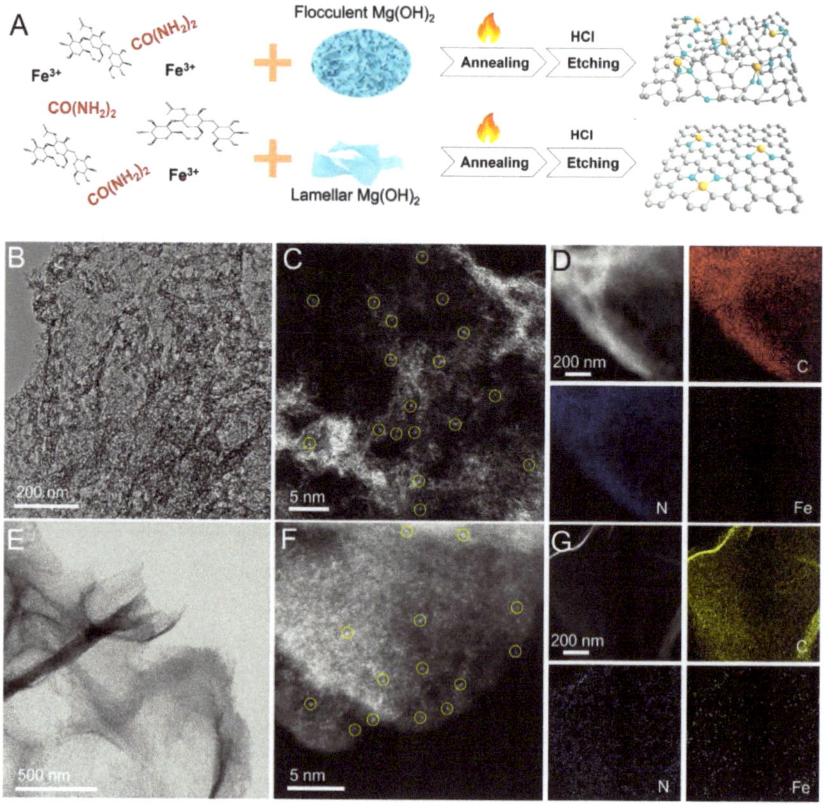

Figure 1. Synthesis and morphology characterization of R-FeSNC and F-FeSNC. (**A**) Schematic illustration of the synthetic route for R-FeSNC and F-FeSNC. (**B**) TEM and (**C**) HAADF-STEM images of R-FeSNC; the yellow circles mark single metal atoms. (**D**) Relevant EDS mapping images. (**E**) TEM and (**F**) HAADF-STEM images of F-FeSNC; the yellow circles mark single metal atoms. (**G**) Corresponding EDS mapping images.

The chemical state and coordination of R-FeSNC and F-FeSNC were investigated by X-ray absorption spectroscopy (XAS). The Fe K-edge XANES spectra of R-FeSNC and F-FeSNC exhibit nearly identical absorption edge positions, suggesting that the oxidation states of Fe in both catalysts are approximate. R-FeSNC shows a more dominated pre-peak at 7114 eV than that of F-FeSNC, indicative of its worse coordination symmetry. The Fe K-edge Fourier-transformed extended X-ray absorption fine structure (FT-EXAFS) spectra of R-FeSNC and F-FeSNC exhibit the characteristic peaks of the Fe-N bond at about 1.4 Å, and no Fe-Fe bond (2.1 Å) can be found [33,34]. This result further corroborates the single-atom dispersion of Fe, which is consistent with the HAADF-STEM results. However, the Fe-N peak of R-FeSNC is lower and left-shifted in comparison with F-FeSNC, suggesting their differential coordination environment. This fact can also be proven by EXAFS fitting and wavelet-transform contour plots (Figures 2C–F and S6 and Table S1). The Fe-N coordination numbers for R-FeSNC and F-FeSNC were determined to be approximately 3.5 and 4, respectively, while relevant mean bond lengths were estimated to be 1.99 Å and 2.01 Å, respectively. Apparently, R-FeSNC exhibits an unsaturated coordination configuration of FeN_{4-x} and shorter Fe-N bond length.

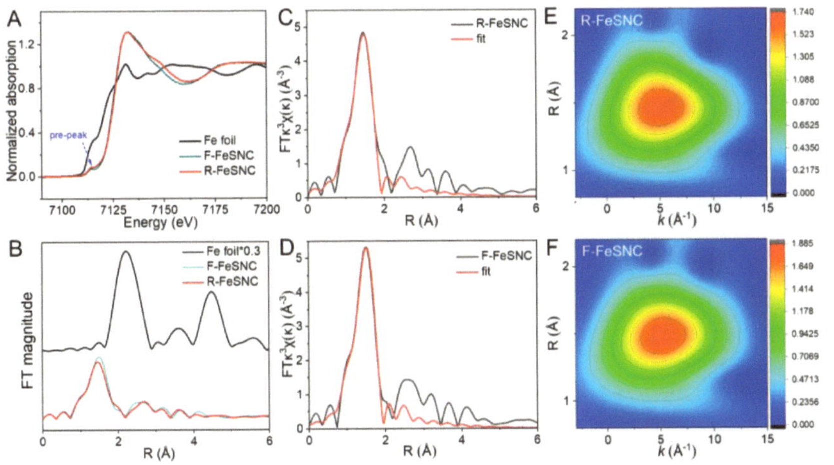

Figure 2. XAS characterization of R-FeSNC and F-FeSNC. (**A**) The normalized Fe K-edge XANES and (**B**) FT-EXAFS spectra of R-FeSNC and F-FeSNC, and the notation "*0.3" appended to "Fe foil" denotes that the peak intensity of the Fe K-edge EXAFS spectrum for the Fe foil has been multiplied by a factor of 0.3. The fitting results of EXAFS spectra for (**C**) R-FeSNC and (**D**) F-FeSNC. The wavelet transforms of (**E**) R-FeSNC and (**F**) F-FeSNC.

The electron and coordination state of R-FeSNC was further studied by X-ray photoelectron spectroscopy (XPS). The peaks at 712.7 eV, 726.1 eV and 719.2 eV are attributed to Fe^{3+}, suggesting that the supported Fe single atoms are trivalent [35]. To identify the type of nitrogen, peak decomposition analysis was conducted on N1s (Figure 3B), revealing peaks at 398.8 eV, 399.6 eV, 400.6 eV, and 401.2 eV corresponding to pyridinic nitrogen, Fe-N bonds, pyrrolic nitrogen, and graphitic nitrogen, respectively [36].

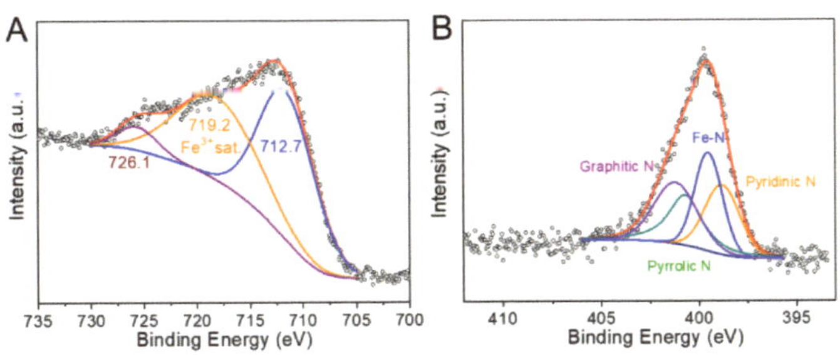

Figure 3. XPS studies of R-FeSNC. (**A**) Fe 2p and (**B**) N1s XPS spectra of R-FeSNC.

2.2. Electrocatalytic Performance

The CRR performance tests of R-FeSNC and F-FeSNC were carried out in a flow cell. The CRR electrocatalytic activity was confirmed by performing linear sweep voltammetry (LSV), feeding CO_2 or Ar. The larger current density of R-FeSNC with CO_2 feeding suggests its intrinsic catalytic capacity. The linear sweep voltammetry (LSV) result in Figure 4A demonstrates that the R-FeSNC electrode exhibited a higher total current density in the potential range from −0.46 V to −1.06 V, vs. RHE compared with F-FeSNC, suggesting its superior electrocatalytic activity.

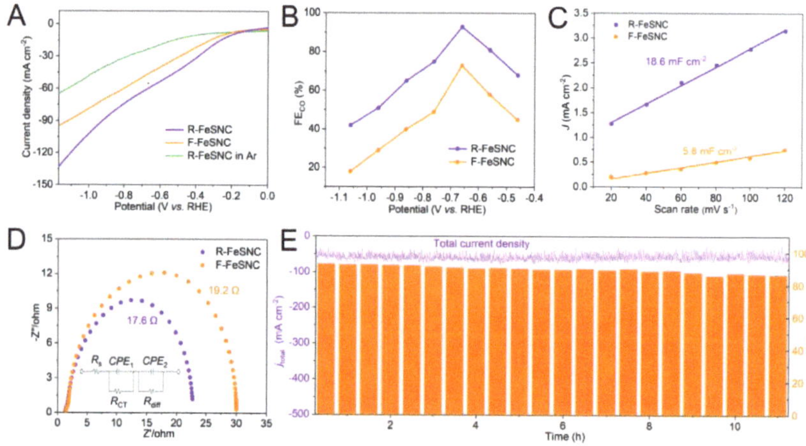

Figure 4. Electrocatalytic performance. (**A**) LSV curves of R-FeSNC and F-FeSNC with CO_2 and Ar feeding. (**B**) The CO FEs of R-FeSNC and F-FeSNC at different potentials. (**C**) C_{dl} fitting curves. (**D**) EIS data and equivalent circuit diagram fitting. (**E**) Stability test of R-FeSNC at a potential of −0.66 V vs. RHE.

The selectivity of two electrocatalysts was evaluated using FE analysis. It is found that the FE_{CO} of R-FeSNC and F-FeSNC is potential-dependent and exhibited a volcano trend (Figure 4B). R-FeSNC achieved an optimal FE_{CO} of 93% at −0.66 V vs. RHE with a total current density of 60 mA cm^{-2}, which is apparently advantageous over F-FeSNC with FE_{CO} of 66% and current density of 45 mA cm^{-2} at same potential. Accordingly, the partial CO current density of R-FeSNC is calculated to be 56 mA cm^{-2}, which is nearly double that of F-FeSNC~29 mA cm^{-2}. This result demonstrated the superior selectivity and activity of R-FeSNC for CRR. The CO_2RR performance of R-FeSNC is comparable to state-of-the-art electrocatalysts (Table S2).

The electrocatalytic active surface areas (ECSAs) of two catalysts were evaluated by assessing their double-layer capacitances, which were represented by the linear slopes obtained from plotting the differences in charging current densities against the scan rates (Figures 4C and S7). R-FeSNC demonstrates a linear slope value of 18.6 mF cm^{-2} 3-times higher than that of F-FeSNC (5.8 mF cm^{-2}). These strongly suggest that the rough surface endows R-FeSNC with a larger ECSA and more accessible active sites than F-FeSNC. Brunauere–Emmette–Teller (BET) adsorption and desorption curves and specific surface areas of R-FeSNC and F-FeSNC are provided in Figure S8. Accordingly, the specific surface area of R-FeSNC is 1120.846 m^2/g, which is larger than that of F-FeSNC (873.867 m^2/g). Electrochemical impedance spectroscopy (EIS) and relevant equivalent circuit diagram fittings were executed to investigate the charge transfer kinetics of two catalysts for CRR. As shown in Figure 4D, R-FeSNC exhibits a smaller circle radius and lower charge transfer resistance (R_{ct}) of 17.6 Ω than F-FeSNC (19.2 Ω) [37,38]. The electrocatalytic stability of R-FeSNC was determined by chronoamperometric measurements under a constant potential of −0.66 V vs. RHE as displayed in Figure 4E. R-FeSNC demonstrated good stability over 11 h electrolysis with almost unchanged total current density and slightly decreased FE_{CO}, which remained above 85%.

In order to explore the mechanism, the reaction intermediates generated over the two catalysts during CRR were monitored by in situ electrochemical attenuated total reflectance Fourier-transform infrared (ATR-FTIR) spectroscopy (Figure 5A,B). The peaks at ~2040 and ~1370 cm^{-1} can be assigned to linear-adsorbed *CO and *COOH species, respectively, which are key intermediates for electrochemical CO_2-to-CO transformation [39,40]. As applied potential increased, a more intense *COOH peak was observed for R-FeSNC than F-FeSNC, suggesting its facilitation of the process of CO_2 (g) + H^+ + e^- → *COOH. After that,

*COOH would cause protons and electrons to form *CO (*COOH + H^+ + e^- $\rightarrow$ *CO + H_2O) followed by *CO desorption (*CO $\rightarrow$ CO (g)). In addition, R-FeSNC exhibited less distinct *CO peaks than F-FeSNC, demonstrating that CO was easily desorbed from its surface. Meanwhile, R-FeSNC showed more obvious and differentiated peaks at 1660 and 1635 cm^{-1} which originate from the stretching of absorbed H_2O and HCO_3^- molecules [41]. The existence of H_2O and HCO_3^- molecules could contribute to the formation of H^+ via the dissociation processes ($HCO_3^- \rightarrow H^+ + CO_3^{2-}$ and $H_2O \rightarrow H^+ + OH^-$), thereby facilitating the formation of *COOH and *CO. Thus, the CO_2-to-CO electroreduction process should be more favorable on R-FeSNC than F-FeSNC.

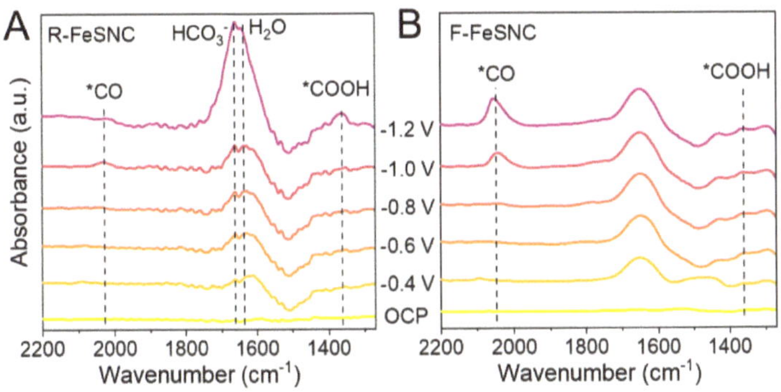

Figure 5. In situ ATR-FTIR spectra of (**A**) R-FeSNC and (**B**) F-FeSNC recorded at different applied potentials for CRR in a CO_2-saturated 0.5 M $KHCO_3$ solution. *CO and *COOH are adsorbed intermediates in the carbon dioxide reduction process.

As aforementioned, R-FeSNC exhibits higher activity and selectivity for the electrocatalytic reduction of CO_2 to CO, which is presumably closely related to its unique structural characteristics. Firstly, its rough surface can prevent the stacking of carbon nanosheets, thereby enhancing the solid–liquid contact area and consequently increasing the ECSA [42,43]. Secondly, the pits or channels on the rough surface can create specific localized environments, which may provide favorable conditions for the adsorption of H_2O and HCO_3^-, as evidenced by in situ ATR-FTIR. Thirdly, the amorphous rough surface, featuring abundant edges and dramatic structural fluctuations, can affect the coordination number and form of anchored Fe single atoms and N, leading to the formation of unsaturated FeN_{4-x} sites and shorter Fe-N bonds [44]. The unsaturated atomic coordination configuration and shortened bond distance can induce polarized charge distributions and an up-shifted total density of states towards the Fermi level which are conductive to stabilizing the key intermediate, i.e., *COOH, and accelerate charge transfers [45–47]. The characteristic microstructures should be conducive to optimizing the electronic and spatial structure of the active sites, thereby facilitating the kinetics of multi-step reactions.

3. Materials and Methods

3.1. Materials

All chemicals and reagents were analytical reagents (ARs). Ethanol (EtOH, C_2H_5OH) was obtained from Tianjin Okaibo Chemical Co., Ltd. (Tianjin, China). Potassium hydroxide (KOH) was purchased from Shanghai Macklin Biochemical Technology Co., Ltd. (Shanghai, China). Magnesium nitrate tetrahydrate ($\geq$99.98% metals basis), carboxymethyl chitosan, and iron nitrate 9-hydrate ($\geq$99.9% metals basis) were obtained from Shanghai Aladdin Biochemical Technology Co., Ltd. (Shanghai, China).

3.2. Catalyst Preparation

As is typical, 2.2 g of magnesium nitrate tetrahydrate was dissolved into 100 mL of ethanol and stirred for 0.5 h. After that, 1 g of NaOH was dissolved into 10 mL of water and then injected into the above solution immediately. The precipitate was centrifuged, washed, and dried overnight. Finally, flocculent $Mg(OH)_2$ was obtained. The lamellar $Mg(OH)_2$ was fabricated just by changing ethanol for dissolving magnesium nitrate tetrahydrate into water.

The rough R-FeSNC and F-FeSNC were fabricated for structure and performance comparison. For synthesis of R-FeSNC, 0.5 g of flocculent $Mg(OH)_2$ and 6.2 mg of $FeCl_3$ were dispersed into 30 mL of water by sonication for 0.5 h. After that, 0.5 g of carboxymethyl chitosan was dissolved into above suspension, which was then freeze-dried. R-FeSNC was synthesized by annealing the above-obtained powder at 900 °C for 1 h in an Ar atmosphere and acid pickling to remove templates. F-FeSNC can be synthesized following the above steps, except for changing flocculent $Mg(OH)_2$ to lamellar $Mg(OH)_2$.

3.3. *Catalyst Characterization*

X-ray diffraction (XRD, Rigaku SmartLab SE, Rigaku, Tokyo, Japan) was used to confirm crystal properties. X-ray photoelectron spectroscopy (XPS) was tested on the thermal ESCALAB 250xi spectrometer (Thermo Fisher, Waltham, MA, USA) and the peak binding energy of C1s 284.8 eV was used for calibration of all peaks. The microstructure of the catalyst was characterized by transmission electron microscopy (TEM, TALOS F200S, Thermo Fisher, Waltham, MA, USA) with an energy dispersive spectrometer (EDS). XAS spectra were obtained from the Beijing Synchrotron Radiation Facility at 1W1B (Beijing Electron–Positron Collider II, China). Fe foil was measured for energy calibration. BET surface area was measured with Quantachrome, Autosorb-iQ. The in situ electrochemical attenuated total reflectance Fourier transform infrared (ATR-FTIR, Shimadzu IRTracer-100, Kyoto, Japan) was carried out to obtain the chemical bonds or functional information of different materials.

3.4. *Preparation of Gas Diffusion Electrodes (GDEs)*

Carbon paper (Sinero YLS-30T) with a microporous layer was purchased from the Sinero Store (Suzhou Sinero Technology, Suzhou, China). 20 mg of catalysts (R-FeSNC and F-FeSNC) were dispersed in 4 mL ethanol with addition of 80 uL of 5 wt% Nafion solution. The ink was placed into the ultrasonic machine for one hour to ensure uniform dispersion. After that, the prepared ink was evenly sprayed on a GDE (3 cm $\times$ 8 cm) through the spray gun, and the weight of the sprayed catalyst was determined by the weight difference in GDE before and after airbrushing. The loading of catalyst was controlled at 0.3 mg cm^{-2}. The GDEs were cut to 1 cm $\times$ 3 cm and dried in a vacuum overnight at 60°C before using.

4. Conclusions

In summary, we manipulated the morphology of amorphous N-doped carbon-supported, isolated Fe atoms using diverse $Mg(OH)_2$ templates. Flocculent and lamellar $Mg(OH)_2$ templates lead to R-FeSNC and F-FeSNC, respectively. The R-FeSNC features a rough surface and unsaturated FeN_{4-x} sites. In CRR electrocatalysis, R-FeSNC demonstrated satisfactory selectivity and activity with an FE_{CO} of 93% at a current density of 60 mA cm^{-2}, superior to F-FeSNC with an FE_{CO} of 73% and a current density of 45 mA cm^{-2}. The enhanced performance of R-FeSNC can be attributed to well-designed morphology and microstructure that can increase ECSA, create specific localized environments and optimize active sites. This work paves the way to building highly efficient and selective CO_2RR catalysts by modulating the morphology and microstructure of CSAMs.

Supplementary Materials: The following supporting information can be downloaded at: https://www.mdpi.com/article/10.3390/molecules29235561/s1. Figure S1: The SEM images of flocculent $Mg(OH)_2$. Figure S2: The SEM images of lamellar $Mg(OH)_2$. Figure S3: The XRD patterns of R-FeSNC and F-FeSNC. Figure S4: The SEM image of R-FeSNC. Figure S5: The SEM image of F-FeSNC. Figure S6: EXAFS curves-fitting results of Fe K-edge of Fe foil. Figure S7: CV curves of (A) R-FeSNC and (B) F-FeSNC at varied scan rates (20, 40, 60, 80, 100 and 120 mV s-1). Figure S8: BET adsorption and desorption curves and surface area of (A) R-FeSNC and (B) F-FeSNC. Table S1: EXAFS fitting parameters extracted from the Fe K-edge. Table S2: Performance comparison of R-FeSNC with some advanced catalysts. References [7–9,48–54] are cited in the supplementary materials.

Author Contributions: Y.L., Y.T., K.Z., T.G., Y.Z., T.C., H.L., J.Z. and J.L., synthesis, characterization and writing; Y.T., Z.G. and Z.G., characterization and analysis; S.Z., J.L. and Z.L., conceptualization, writing and editing. All authors have read and agreed to the published version of the manuscript.

Funding: This work is supported by National Natural Science Foundation of China (No. 52302092), Fundamental Research Funds for the Central Universities (No. 2023MS057) and the Taishan Scholar Project of Shandong Province (tsqn202306131). The authors thank the staff at 1W1B of BSRF.

Institutional Review Board Statement: Not applicable.

Informed Consent Statement: Not applicable.

Data Availability Statement: Data are contained within the article and Supplementary Materials.

Conflicts of Interest: The authors declare no conflicts of interest.

erences

Shi, L.; Li, M.; You, B.; Liao, R. Theoretical Study on the Electro-Reduction of Carbon Dioxide to Methanol Catalyzed by Cobalt Phthalocyanine. *Inorg. Chem.* **2022**, *61*, 16549–16564. [CrossRef] [PubMed]

Kumar, A.; Prajapati, P.K.; Aathira, M.S.; Bansiwal, A.; Boukherroub, R.; Jain, S.L. Highly Improved Photoreduction of Carbon Dioxide to Methanol using Cobalt Phthalocyanine Grafted to graphitic carbon nitride as photocatalyst under visible light irradiation. *J. Colloid Interface Sci.* **2019**, *543*, 201–213. [CrossRef] [PubMed]

Park, J.; Jeong, C.; Na, M.; Oh, Y.; Lee, K.-S.; Yang, Y.; Byon, H.R. Subnanometer Cu Clusters on Porous Ag Enhancing Ethanol Production in Electrochemical CO_2 Reduction. *ACS Catal.* **2024**, *14*, 3198–3207. [CrossRef]

Tan, D.; Wulan, B.; Cao, X.; Zhang, J. Strong Interactions of Metal-Support for Efficient Reduction of Carbon Dioxide into Ethylene. *Nano Energy* **2021**, *89*, 106460. [CrossRef]

Tan, D.; Zhang, J.; Yao, L.; Tan, X.; Cheng, X.; Wan, Q.; Han, B.; Zheng, L.; Zhang, J. Multi-shelled CuO Microboxes for Carbon Dioxide Reduction to Ethylene. *Nano Res.* **2020**, *13*, 768–774. [CrossRef]

Mistry, H.; Varela, A.S.; Bonifacio, C.S.; Zegkinoglou, I.; Sinev, I.; Choi, Y.W.; Kisslinger, K.; Stach, E.A.; Yang, J.C.; Strasser, P.; et al. Highly Selective Plasma-Activated Copper Catalysts for Carbon Dioxide Reduction to Ethylene. *Nat. Commun.* **2016**, *7*, 12123. [CrossRef]

Al-Attas, T.A.; Marei, N.N.; Yong, X.; Yasri, N.G.; Thangadurai, V.; Shimizu, G.; Siahrostami, S.; Kibria, M.G. Ligand-Engineered Metal-Organic Frameworks for Electrochemical Reduction of Carbon Dioxide to Carbon Monoxide. *ACS Catal.* **2021**, *11*, 7350–7357. [CrossRef]

Dong, H.; Zhang, L.; Yang, P.; Chang, X.; Zhu, W.; Ren, X.; Zhao, Z.; Gong, J. Facet Design Promotes Electroreduction of Carbon Dioxide to Carbon Monoxide on Palladium Nanocrystals. *Chem. Eng. Sci.* **2019**, *194*, 29–35. [CrossRef]

Ma, M.; Trzesniewski, B.J.; Xie, J.; Smith, W.A. Selective and Efficient Reduction of Carbon Dioxide to Carbon Monoxide on Oxide-Derived Nanostructured Silver Electrocatalysts. *Angew. Chem.-Int. Ed.* **2016**, *55*, 9748–9752. [CrossRef]

Yang, H.; Lin, Q.; Zhang, C.; Yu, X.; Cheng, Z.; Li, G.; Hu, Q.; Ren, X.; Zhang, Q.; Liu, J.; et al. Carbon Dioxide Electroreduction on Single-Atom Nickel Decorated Carbon Membranes with Industry Compatible Current Densities. *Nat. Commun.* **2020**, *11*, 593. [CrossRef]

Lv, C.; Huang, K.; Fan, Y.; Xu, J.; Lian, C.; Jiang, H.; Zhang, Y.; Ma, C.; Qiao, W.; Wang, J.; et al. Electrocatalytic Reduction of Carbon Dioxide in Confined Microspace Utilizing Single Nickel Atom Decorated Nitrogen-Doped Carbon Nanospheres. *Nano Energy* **2023**, *111*, 108384. [CrossRef]

Hu, C.; Hong, X.; Liu, M.; Shen, K.; Chen, L.; Li, Y. Hierarchically Ordered Pore Engineering of Carbon Supports with High-Density Edge-Type Single-Atom Sites to Boost Electrochemical CO_2 Reduction. *Adv. Mater.* **2024**, 2409531. [CrossRef]

Karaiskakis, A.N.; Biddinger, E.J. Evaluation of the Impact of Surface Reconstruction on Rough Electrodeposited Copper-Based Catalysts for Carbon Dioxide Electroreduction. *Energy Technol.* **2017**, *5*, 901–910. [CrossRef]

Wang, C.; Chen, Y.; Su, D.; Man, W.; Lau, K.; Han, L.; Zhao, L.; Zhan, D.; Zhu, X. In situ Electropolymerized 3D Microporous Cobalt-Porphyrin Nanofilm for Highly Effective Molecular Electrocatalytic Reduction of Carbon Dioxide. *Adv. Mater.* **2023**, *35*, 2303179. [CrossRef] [PubMed]

He, Y.; Rao, H.; Song, K.; Li, J.; Yu, Y.; Lou, Y.; Li, C.G.; Han, Y.; Shi, Z.; Feng, S. 3D Hierarchical $ZnIn_2S_4$ Nanosheets with Rich Zn Vacancies Boosting Photocatalytic CO_2 Reduction. *Adv. Funct. Mater.* **2019**, *29*, 1905153. [CrossRef]

Gao, Y.; Li, S.; Gong, L.; Li, J.; Qi, D.; Liu, N.; Bian, Y.; Jiang, J. Unprecedented POSS-Linked 3D Covalent Organic Frameworks with 2-Fold Interpenetrated scu or sqc Topology Regulated by Porphyrin Center for Photocatalytic CO_2 Reduction. *Angew. Chem.-Int. Ed.* **2024**, *63*, e202404156. [CrossRef]

Yu, Z.; Li, Y.; Torres-Pinto, A.; LaGrow, A.P.; Diaconescu, V.M.; Simonelli, L.; Sampaio, M.J.; Bondarchuk, O.; Amorim, I.; Araujo, A.; et al. Single-atom Ir and Ru Anchored on Graphitic Carbon Nitride for Efficient and Stable Electrocatalytic/Photocatalytic Hydrogen Evolution. *Appl. Catal. B-Environ.* **2022**, *310*, 121318. [CrossRef]

18. Shang, H.; Wang, T.; Pei, J.; Jiang, Z.; Zhou, D.; Wang, Y.; Li, H.; Dong, J.; Zhuang, Z.; Chen, W.; et al. Design of a Single-A Indiumδ+-N_4 Interface for Efficient Electroreduction of CO_2 to Formate. *Angew. Chem.-Int. Ed.* **2020**, *59*, 22465–22469. [Cros
19. Jia, C.; Zhao, Y.; Song, S.; Sun, Q.; Meyer, Q.; Liu, S.; Shen, Y.; Zhao, C. Highly Ordered Hierarchical Porous Single-Ato Catalyst with Promoted Mass Transfer for Efficient Electroreduction of CO_2. *Adv. Energy Mater.* **2023**, *13*, 2302007. [CrossR
20. Wang, T.; Wang, J.; Lu, C.; Jiang, K.; Yang, S.; Ren, Z.; Zhang, J.; Liu, X.; Chen, L.; Zhuang, X.; et al. Single-Atom Anchored Cu Carbon Surface for Efficient CO_2 Electro-Reduction with Nearly 100% CO Selectivity and Industrially-Relevant Current De *Adv. Mater.* **2023**, *35*, 2205553. [CrossRef]
21. Huang, S.; Fang, Z.; Lu, C.; Zhang, J.; Sun, J.; Ji, H.; Zhu, J.; Zhuang, X. Well-defined Asymmetric Nitrogen/Carbon-Coordir Single Metal Sites for Carbon Dioxide Conversion. *J. Colloid Interface Sci.* **2024**, *675*, 683–688. [CrossRef] [PubMed]
22. Song, J.; Lei, X.; Mu, J.; Li, J.; Song, X.; Yan, L.; Ding, Y. Chlorine-Coordinated Unsaturated Ni-N_2 Sites for Efficient Electroche Carbon Dioxide Reduction. *Small* **2023**, *19*, 2304423. [CrossRef] [PubMed]
23. Liu, Y.; Deng, C.; Liu, F.; Dai, X.; Yang, X.; Chen, Y.; Wu, Z.; Guo, S.; Asefa, T.; Liu, Y. Coupling Electron Donors with Pr Repulsion via Pt-N_3-S Sites to Boost CO_2 Reduction in CO_2/H_2 Fuel Cell. *Nano Energy* **2024**, *126*, 109667. [CrossRef]
24. Gao, S.; Sun, Z.; Liu, W.; Jiao, X.; Zu, X.; Hu, Q.; Sun, Y.; Yao, T.; Zhang, W.; Wei, S.; et al. Atomic Layer Confined Vacancie Atomic-level Insights into Carbon Dioxide Electroreduction. *Nat. Commun.* **2017**, *8*, 14503. [CrossRef]
25. Zhang, E.; Wang, T.; Yu, K.; Liu, J.; Chen, W.; Li, A.; Rong, H.; Lin, R.; Ji, S.; Zheng, X.; et al. Bismuth Single Atoms Resulting Transformation of Metal-Organic Frameworks and their Use as Electrocatalysts for CO_2 Reduction. *J. Am. Chem. Soc.* **2019** 16569–16573. [CrossRef]
26. Song, P.; Hu, B.; Zhao, D.; Fu, J.; Su, X.; Feng, W.; Yu, K.; Liu, S.; Zhang, J.; Chen, C. Modulating the Asymmetric Atomic Inte of Copper Single Atoms for Efficient CO_2 Electroreduction. *ACS Nano* **2023**, *17*, 4619–4628. [CrossRef] [PubMed]
27. Jia, C.; Li, S.; Zhao, Y.; Hocking, R.K.; Ren, W.; Chen, X.; Su, Z.; Yang, W.; Wang, Y.; Zheng, S.; et al. Nitrogen Vacancy Ind Coordinative Reconstruction of Single-Atom Ni Catalyst for Efficient Electrochemical CO_2 Reduction. *Adv. Funct. Mater.* **202** 2107072. [CrossRef]
28. Yin, H.; Deng, Y.; He, Z.; Xu, W.; Hou, Z.; He, B.; Caha, I.; Cunha, J.; Karimi, M.; Yu, Z. Strain Engineering of Co_{SA}-N-C Cat Toward Enhancing the Oxygen Reduction Reaction Activity. *J. Colloid Interface Sci.* **2025**, *678*, 447–457. [CrossRef]
29. Li, M.; Zhang, D.; Wu, K.; Liu, Y.; Wang, P.; Cao, Y.; Yang, J. Local Compressive Strain Regulation of Atomically Dispersed N Sites for Enhancing CO_2 Electroreduction to CO. *Nanoscale* **2023**, *15*, 15700–15707. [CrossRef]
30. Tong, Y.; Wu, Y.; Liu, Z.; Yin, Y.; Sun, Y.; Li, H. Fabricating Multi-Porous Carbon Anode with Remarkable Initial Coulo Efficiency and Enhanced Rate Capability for Sodium-ion Batteries. *Chin. Chem. Lett.* **2023**, *34*, 107443. [CrossRef]
31. Li, M.; Li, G.; Ye, Q.; Deng, Y.; Chi, B.; Hua, Y.; Tian, X.; Rao, P. Fluorine and Sulfur Atoms Induced N-doped 3D Porous Ca Catalyzing Oxygen Reduction in the Zinc-air Battery. *Int. J. Hydrogen Energy* **2024**, *83*, 1107–1112. [CrossRef]
32. Zhu, Y.; Miyake, K.; Shu, Y.; Moroto, K.; Hirota, Y.; Uchida, Y.; Tanaka, S.; Zheng, T.; Katayama, M.; Inada, Y.; et al. Single At Co Coordinated with N in Microporous Carbon for Oxygen Reduction Reaction Obtained from Co/2-methylimidazole Anch to Y Zeolite as a Template. *Mater. Today Chem.* **2021**, *20*, 100410. [CrossRef]
33. Peng, X.; Wu, J.; Zhao, Z.; Wang, X.; Dai, H.; Xu, L.; Xu, G.; Jian, Y.; Hu, F. Activation of Peroxymonosulfate by Single-A Fe-g-C_3N_4 Catalysts for High Efficiency Degradation of Tetracycline via Nonradical Pathways: Role of High-Valent Iron Species and Fe-N_x Sites. *Chem. Eng. J.* **2022**, *427*, 130803. [CrossRef]
34. Gao, Y.; Duan, X.; Li, B.; Jia, Q.; Li, Y.; Fan, X.; Zhang, F.; Zhang, G.; Wang, S.; Peng, W. Fe Containing Template Derived At Fe-N-C to Boost Fenton-like Reaction and Charge Migration Analysis on Highly Active Fe-N_4 Sites. *J. Mater. Chem. A* **20** 14793–14805. [CrossRef]
35. Li, C.; Yang, W.; Guo, S.; Yang, Z.X.; Fida, H.; Liang, H. Efficient Degradation of Refractory Contaminants with Silver Ferri Persulfate Activation. *Funct. Mater. Lett.* **2019**, *12*, 1950083. [CrossRef]
36. Hu, K.; Xiao, Z.; Cheng, Y.; Yan, D.; Chen, R.; Huo, J.; Wang, S. Iron Phosphide/N, P-doped Carbon Nanosheets as Highly Effi Electrocatalysts for Oxygen Reduction Reaction over the whole pH Range. *Electrochim. Acta* **2017**, *254*, 280–286. [CrossRef]
37. Yu, K.; Sun, K.; Cheong, W.; Tan, X.; He, C.; Zhang, J.; Li, J.; Chen, C. Oxalate-Assisted Synthesis of Hollow Carbon Nano with Fe Single Atoms for Electrochemical CO_2 Reduction. *Small* **2023**, *19*, 2302611. [CrossRef]
38. Li, Z.; Wu, R.; Xiao, S.; Yang, Y.; Lai, L.; Chen, J.; Chen, Y. Axial Chlorine Coordinated Iron-Nitrogen-Carbon Single-A Catalysts for Efficient Electrochemical CO_2 reduction. *Chem. Eng. J.* **2022**, *430*, 132882. [CrossRef]
39. Wei, Z.; Ding, J.; Duan, X.; Chen, G.; Wu, F.; Zhang, L.; Yang, X.; Zhang, Q.; He, Q.; Chen, Z.; et al. Enhancing Sele Electrochemical CO_2 Reduction by In Situ Constructing Tensile-Strained Cu Catalysts. *ACS Catal.* **2023**, *13*, 4711–4718. [Cros
40. Zhu, S.; Jiang, B.; Cai, W.; Shao, M. Direct Observation on Reaction Intermediates and the Role of Bicarbonate Anions in Electrochemical Reduction Reaction on Cu Surfaces. *J. Am. Chem. Soc.* **2017**, *139*, 15664–15667. [CrossRef]
41. Wang, J.; Huang, Y.; Wang, Y.; Deng, H.; Shi, Y.; Wei, D.; Li, M.; Dong, C.; Jin, H.; Mao, S.; et al. Atomically Dispe Metal-Nitrogen-Carbon Catalysts with d-Orbital Electronic Configuration-Dependent Selectivity for Electrochemical CO_2-tc Reduction. *ACS Catal.* **2023**, *13*, 2374–2385. [CrossRef]
42. Yu, F.; Leung, P.; Xu, Q.; Mavrikis, S.; Nazarovs, P.; Shah, A.K.; Wang, L.; de León, C.P. Facile Spray-Printing of Hydrophobic Porous Gas Diffusion Electrodes Enabling Prolonged Electrochemical CO_2 Reduction to Ethylene. *J. Power Sources* **2023**, 233201. [CrossRef]

Kwon, T.; Prabhakaran, S.; Kim, D.; Kim, M.H.; Lee, Y. Engineering Hydrophobicity and High-index Planes of Gold Nanostructures for Highly Selective Electrochemical CO_2 Reduction to CO and Efficient CO_2 Capture. *Chem. Eng. J.* **2024**, *485*, 150045. [CrossRef]

Liu, Z.; Liu, Y.; Zhang, J.; Cao, T.; Sun, Z.; Liu, J.; Shang, H. Asymmetrically Coordinated Single Atom Cu Catalyst with Unsaturated C-Cu-N Structure for CO_2 Reduction to CO. *Nano Res.* **2024**, *17*, 3911–3918. [CrossRef]

Zhao, X.; Fang, R.; Wang, F.; Shen, Z.; Yang, X.; Kong, X.; Li, Y. Coordinatively Unsaturated Single-atom Cu Catalysts with Enhanced Catalytic Activity for Furfural Upgrading. *Chem. Eng. Sci.* **2023**, *280*, 119104. [CrossRef]

Xu, W.; Sun, Y.; Zhou, J.; Cao, M.; Luo, J.; Mao, H.; Hu, P.; Gu, H.; Zhai, H.; Shang, H.; et al. Coordinatively Unsaturated Single Co Atoms Immobilized on C_2N for Efficient Oxygen Reduction Reaction. *Nano Res.* **2023**, *16*, 2294–2301. [CrossRef]

Zhang, W.; Liu, D.; Liu, T.; Ding, C.; Chen, T.; Li, Y.; Liu, X.; Wang, L.; Li, C.; He, J.; et al. Coordinately Unsaturated Nickel Single Atom Electrocatalyst for Efficient CO_2 Conversion. *Nano Res.* **2023**, *16*, 10873–10880. [CrossRef]

Zhong, H.X.; Ghorbani-Asl, M.; Ly, K.H.; Zhang, J.C.; Ge, J.; Wang, M.C.; Liao, Z.Q.; Makarov, D.; Zschech, E.; Brunner, E.; et al. Synergistic Electroreduction of Carbon Dioxide to Carbon Monoxide on Bimetallic Layered Conjugated Metal-organic Frameworks. *Nat. Commun.* **2020**, *11*, 10.

Su, P.; Iwase, K.; Nakanishi, S.; Hashimoto, K.; Kamiya, K. Nickel-Nitrogen-Modified Graphene: An Efficient Electrocatalyst for the Reduction of Carbon Dioxide to Carbon Monoxide. *Small* **2016**, *12*, 6083–6089. [CrossRef]

Huang, P.; Chen, J.X.; Deng, P.L.; Yang, F.; Pan, J.; Qi, K.; Liu, H.F.; Xia, B.Y. Grain Refinement of Self-Supported Copper Electrode by Multiple-Redox Treatment for Enhanced Carbon Dioxide Electroreduction towards Carbon Monoxide Generation. *J. Catal.* **2020**, *381*, 608–614. [CrossRef]

Kim, J.; Kim, H.; Han, G.H.; Ahn, S.H. Solution-phase-reconstructed Zn-based Nanowire Electrocatalysts for Electrochemical Reduction of Carbon Dioxide to Carbon Monoxide. *Int. J. Energy Res.* **2021**, *45*, 7987–7997. [CrossRef]

Wijaya, D.T.; Haryanto, A.; Lim, H.; Jin, K.; Lee, C. Sub-2 nm Mixed Metal Oxide for Electrochemical Reduction of Carbon Dioxide to Carbon Monoxide. *J. Energy Chem.* **2023**, *84*, 303–310. [CrossRef]

Ham, Y.S.; Choe, S.; Kim, M.J.; Lim, T.; Kim, S.K.; Kim, J.J. Electrodeposited Ag Catalysts for the Electrochemical Reduction of CO_2 to CO. *Appl. Catal. B-Environ.* **2017**, *208*, 35–43. [CrossRef]

Gao, F.Y.; Hu, S.J.; Zhang, X.L.; Zheng, Y.R.; Wang, H.J.; Niu, Z.Z.; Yang, P.P.; Bao, R.C.; Ma, T.; Dang, Z.; et al. High-Curvature Transition-Metal Chalcogenide Nanostructures with a Pronounced Proximity Effect Enable Fast and Selective CO_2 Electroreduction. *Angew. Chem. Int. Edit.* **2020**, *59*, 8706–8712. [CrossRef]

molecules

MDPI

Article

Fabrication of Pb-Containing PtAu Nanoflowers via Galvanic Replacement Method for Electrocatalytical Oxidation of Methanol

Zhao Huang [1,2], Zhirou Zhang [1], Long Chao [1,*] and Xueen Jia [1,3,*]

1 Hunan Key Laboratory of Biomedical Nanomaterials and Devices, College of Life Sciences and Chemistry, Hunan University of Technology, Zhuzhou 412007, China; huang1020@hut.edu.cn (Z.H.); dorazhangzr@163.com (Z.Z.)
2 Key Laboratory of Chemical Biology and Traditional Chinese Medicine Research (Ministry of Education of China), College of Chemistry and Chemical Engineering, Hunan Normal University, Changsha 410081, China
3 Department of Physics, Umeå University, SE-901 87 Umeå, Sweden
* Correspondence: chaolong4617@163.com (L.C.); xueen.jia@umu.se (X.J.); Tel.: +86-731-2218-3913 (L.C. & X.J.)

Abstract: A Pb-containing PtAu nanoflower electrocatalyst was deposited on the cathode via galvanic replacement reaction in a double-cabin galvanic cell (DCGC) with a Cu plate as the anode, a multi-walled carbon nanotube (MWCNT) modified glassy carbon electrode (GCE) as the cathode, 0.1 M $HClO_4$ aqueous solution as the anolyte, and Pb^{2+}-containing Pt^{4+} salt and Au^{3+} salt mixed aqueous solution as the catholyte, respectively, and the electrocatalytic performance of the modified electrode toward methanol oxidation in the alkaline medium was investigated. Electrochemical studies reveal that the stripping of bulk Cu can induce underpotential deposition (UPD) of Pb on Pt during the galvanic replacement reaction, which affects the morphology and composition of Pb-containing PtAu nanoparticles. Under the optimal experimental conditions, a Pb-Pt_3Au_1/MWCNTs/GCE shows the highest activity and the best stability toward electrocatalytic oxidation of methanol in the alkaline medium, and the Pt active area-normalized specific electrocatalytic activity of Pb-Pt_3Au_1/MWCNTs/GCE is as high as 59.8 mA cm_{Pt}^{-2}. We believe that the method presented here of depositing highly active noble metal nanostructures by galvanic replacement reaction in a DCGC device is expected to be widely applied in the preparation of nanomaterials for their study in fuel cells and electrocatalysis.

Keywords: double-cabin galvanic cell; underpotential deposition; Pb-containing PtAu nanoflower; electrocatalytical oxidation of methanol

1. Introduction

Direct methanol fuel cells (DMFCs) have attracted great interest as a renewable energy supply due to their advantages of low operating temperature, easy operation, and high energy density [1]. Currently, Pt, as a commonly used methanol electrocatalyst, is widely used in DMFC research, but Pt is expensive, scarce, and easily deactivated by CO_{ads} intermediates generated by methanol electrooxidation, which limits its commercial application in DMFCs [2]. Therefore, the key to the development of DMFCs is to develop efficient, low-cost, and anti-poisoning methanol electrocatalysts. Bi (or tri)-component Pt-based electrocatalysts formed by Pt with other transition metals or noble metals can effectively reduce the Pt content and improve the activity and anti-poisoning ability of Pt. So far, various Pt-based multi-component electrocatalysts have been studied, such as Pt-Mn [3], Pt-Fe [4], Pt-Co [5], Pt-Ru [6], Pt-Ni [7], Pt-Cu [8], Pt-Pd [9], Pt-Ag [10], Pt-Mo [11], and PtRuFe [12]. These multi-component Pt-based electrocatalysts can significantly improve the electrocatalytic activity and stability of Pt through electronic effects and bifunctional effects between components.

As a common noble metal, Au exhibits good chemical stability and weak catalytic activity in its bulk state, but nanosized Au particles show excellent electrocatalytic performance for small molecules [13,14]. The bi-component PtAu electrocatalyst can effectively enhance the anti-poisoning capability and catalytic activity of Pt due to the synergistic effect between Pt and Au components [15]. Currently, people use various methods to prepare PtAu electrocatalysts with different morphologies and compositions. For example, Zhang et al. prepared nanoporous PtAu electrocatalysts via dealloying and found that they could efficiently electrocatalyze the oxidation of methanol under acidic conditions [16]. Luo et al. synthesized PtAu nanoparticles supported on carbon via colloidal synthesis and observed that the bifunctional effect of Au could enhance the electrocatalytic activity of Pt toward methanol under alkaline conditions [17]. In addition, doping the PtAu electrocatalyst with other base metals (such as Cu and Pb) can further enhance the catalytic performance of the PtAu-based tri-component electrocatalyst. For instance, Wang et al. synthesized Au-Cu-X (X = Pt, Pd, Ag) ternary metal nanorods using Au-Cu nanorods as a galvanic cell displacement template, and the synthesized nanorods were found to be highly effective in the reduction of 4-nitrophenol [18]. To date, extensive studies have been conducted on the preparation of PtAu electrocatalysts; previous synthesis methods also have disadvantages such as high cost, complex operations, and environmental pollution. The galvanic cell method, due to its low energy consumption, simplicity, and low cost, is a typical green preparation technology. For example, Yoshii et al. synthesized Pd-Co bimetallic nanocatalysts supported on C using Co nanoparticles supported on C as a displacement template and found that they had high catalytic activity and selectivity for the hydrogenation of styrene [19]. In the previous galvanic cell method for the synthesis of Pt-based electrocatalysts, the anode and cathode reactions of the galvanic cell occurred in the same solution. Both reactions at the anode and cathode of the galvanic cell may affect the composition of the deposited Pt-based electrocatalyst. In some cases, further purification of the product is inevitably required, which adds complexity to the process and increases production costs, both of which are undesirable. The use of a double-cabin galvanic cell (DCGC) with separate anode and cathode compartments allows for the one-step deposition of clean, multi-component Pt-based electrocatalysts at the cathode, simplifying the operational steps. To our knowledge, there are rarely reports on the direct preparation of high-activity PtAu-based tri-component nanoelectrocatalyst-modified electrodes at the cathode using the DCGC setup.

In this work, using a high-purity Cu plate as the anode and a glassy carbon electrode (GCE) modified with multiwalled carbon nanotubes (MWCNTs) as the cathode in the DCGC device, a Pb-containing PtAu nanoflower electrocatalyst was deposited at the cathode through galvanic replacement, and the electrocatalytic performance of the resulting modified electrode toward methanol electrooxidation in an alkaline environment was investigated. Electrochemical studies show that the dissolution of bulk Cu during the galvanic reaction can induce underpotential deposition (UPD) of Pb on the Pt surface, leading to the deposition of Pb-containing PtAu nanoflower electrocatalysts on MWCNTs. Pb-containing PtAu nanoflower electrocatalysts exhibit excellent activity toward methanol electrocatalytic oxidation and have a potential application value in DMFCs.

2. Results and Discussion

2.1. Fabrication and Characterization of Pb-Containing PtAu Nanoparticles

Figure 1 shows the discharging current (i_{cell}) and the potential of a MWCNT/GCE cathode ($E_{cathode}$) during the depositing of Pb-Pt_3Au_1 and Pt_3Au_1 on its surface. When the anode and cathode are not connected, the electrochemical potential (ECP) of the Cu plate is approximately 0.008 V in 0.1 M $HClO_4$ solution (anode part), while the ECP of the MWCNT/GCE surface in the mixed solution of noble metal salts is about 0.78 V (cathode part). Therefore, after connecting the anode and cathode of the DCGC device with a Cu wire, the ECP difference between the anode and cathode drives the galvanic replacement reaction, leading to the dissolution of bulk Cu (Cu_{bulk}) at the anode and the deposition of noble metal nanoparticles on the surface of the MWCNTs/GCE at the cathode. During the deposition of Pt_3Au_1 from 3.0 mM H_2PtCl_6 + 1.0 mM $HAuCl_4$ + 0.1 M $HClO_4$ catholyte (without the presence of $Pb(ClO_4)_2$; Figure 1 b), the steady-state potential and i_{cell} of MWCNTs/GCE are 0.097 V and 52 μA, respectively. However, during the deposition of Pb-Pt_3Au_1 from 3.0 mM H_2PtCl_6 + 1.0 mM $HAuCl_4$ + 5.0 mM $Pb(ClO_4)_2$ + 0.1 M $HClO_4$ (with the presence of $Pb(ClO_4)_2$; Figure 1 a), the steady-state potential and i_{cell} of MWCNTs/GCE increase to 0.121 V and 62 μA, respectively. The response for the deposition of Pb-Pt_3Au_1 is larger than that for the deposition of Pt_3Au_1, which perhaps suggests that the addition of $Pb(ClO_4)_2$ to the catholyte facilitates the deposition of Pt and Au.

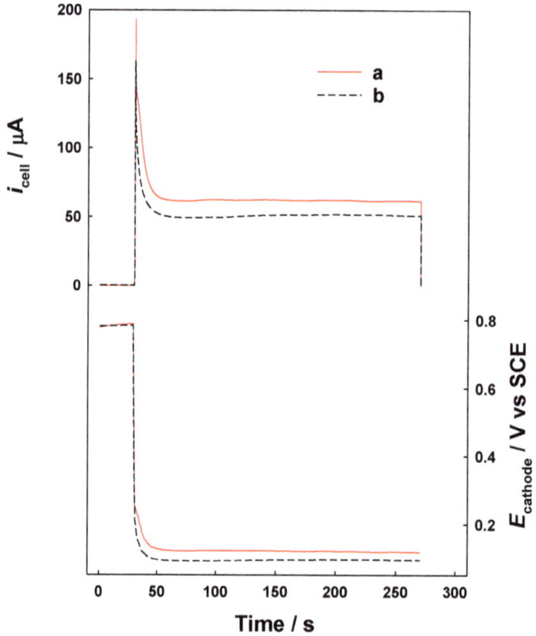

Figure 1. Simultaneous responses of i_{cell} and $E_{cathode}$ in the DCGC with a Cu plate as the anode, 0.1 M $HClO_4$ as the anolyte (stationary), and the MWCNTs/GCE as the cathode, while the catholyte (stationary) was 3.0 mM H_2PtCl_6 + 1.0 mM $HAuCl_4$ + 5.0 mM $Pb(ClO_4)_2$ + 0.1 M $HClO_4$ aqueous solution (a), for the deposition of Pb-Pt_3Au_1) and 3.0 mM H_2PtCl_6 + 1.0 mM $HAuCl_4$ + 0.1 M $HClO_4$ aqueous solution (b), for deposition of Pt_3Au_1), respectively. Short-circuit discharging of the DCGC (connecting the anode and cathode by a conducting Cu wire) was from 30 to 270 s.

Figure 2 shows the SEM images of the obtained Pt_3Au_1/MWCNTs/GCE and Pb-Pt_3Au_1/MWCNTs/GCE. For Pt_3Au_1/MWCNTs/GCE (Figure 2A,B), a large number of spherical nanoparticles with diameters of 40–240 nm are distributed on the surface of MWCNTs. However, in Pb-Pt_3Au_1/MWCNTs/GCE (Figure 2C,D), flower-like nanoparticles grow on the surface of MWCNTs, with diameters ranging from 110 to 450 nm, and the density of nanoparticles on the surface of Pb-Pt_3Au_1/MWCNTs/GCE is also lower than that of Pt_3Au_1/MWCNTs/GCE. EDS characterization (Figure 3) also proved that Pt and Au have been deposited on the surface of MWCNTs/GCE, and the atomic percentages of Pt and Au for Pt_3Au_1 nanoparticles were determined to be 53.9% and 46.1% (Figure 3A), respectively, which was unproportional to that of the catholyte solution (mole concentration ratio for $c_{H2PtCl6}$:c_{HAuCl4} = 3:1). Due to the higher ECP of Au versus that of Pt, the reaction kinetics for reducing $AuCl_4^-$ is anticipated to be significantly faster than that for reducing $PtCl_6^{2-}$, thus leading to the deposition of more Au in Pt_3Au_1 nanoparticles even at a high Pt/Au mole ratio in the catholyte. For Pb-Pt_3Au_1/MWCNTs/GCE, there are a small number of Pb atoms present in the obtained PtAu nanoparticles (Figure 3B), and the atomic percentages of Pt, Au, and Pb were determined to be 50.0%, 42.9%, and 7.1%, respectively, which indicated the occurrence of cathodic deposition of Pb from the $Pb(ClO_4)_2$-containing catholyte during the GRR deposition of $Pb\text{-}Pt_3Au_1$ nanoparticles, despite the higher activity of bulk lead (Pb_{bulk}) compared to Cu_{bulk} (anode here). The above results indicate that when $Pb(ClO_4)_2$ is added to the catholyte, the co-deposition of Pb on the cathode can influence the morphology and composition of PtAu nanoparticles, leading to the formation of Pb-containing PtAu nanoflower particles. Indeed, previous studies have also reported that the addition of transition metal ions (such as Cu^{2+}) can control the crystal structure and morphology for synthesizing noble metal nanostructures [20].

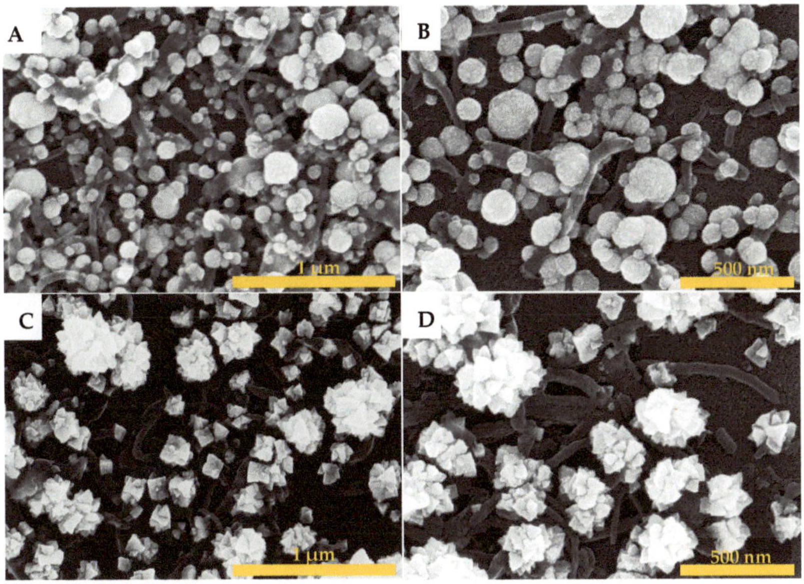

Figure 2. SEM of Pt_3Au_1/MWCNTs/GCE (**A**,**B**) and $Pb\text{-}Pt_3Au_1$/MWCNTs/GCE (**C**,**D**).

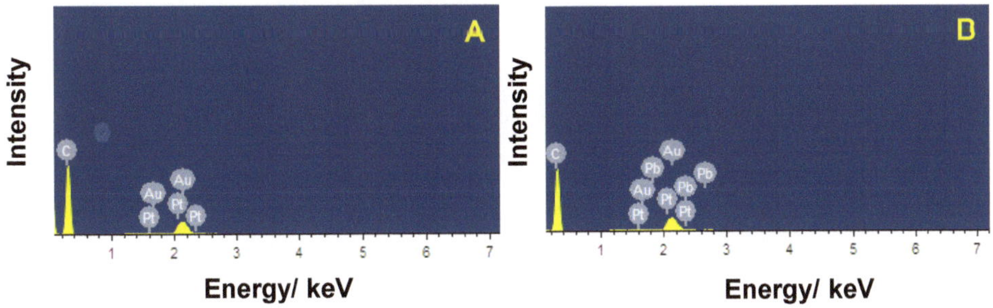

Figure 3. EDS spectra of Pt_3Au_1/MWCNTs/GCE (**A**) and $Pb\text{-}Pt_3Au_1$/MWCNTs/GCE (**B**), respectively.

Moreover, Figure S1 shows the X-ray diffraction (XRD) patterns of the as-synthesized Pt_3Au_1 and $Pb\text{-}Pt_3Au_1$ thin film sample on MWCNTs. The peaks at 37.1° and 42.4° are assigned to Au(111) and Pt(111), respectively, according to JCPDS 04-0784 of pure Pt and JCPDS 04-0802 of pure Au, and the peaks at 73.3° and 76.4° may correspond to (311) and (222) planes. In our case, no XRD peaks of Pb are observed on $Pb\text{-}Pt_3Au_1$ that may be the Pb of UPD is too small to produce an X-ray diffraction signal. X-ray photoelectron spectroscopy (XPS) was applied to characterize the Pt_3Au_1 and Pb-Pt_3Au_1 thin film sample also (as shown in Figure S2). The two 4f binding states of zero valence, at 84.4 ($4f_{5/2}$) and 87.5 eV ($4f_{7/2}$) for Au^0 and at 71.2 ($4f_{5/2}$) and 74.5 eV ($4f_{7/2}$) for Pt^0, were observed on both Pt_3Au_1 and Pb-Pt_3Au_1, respectively. Moreover, Pb may exist as Pb^{2+} in Pb-Pt_3Au_1, with the peak of Pb^{2+} $4f_{7/2}$ and $4f_{5/2}$ appearing at 139.6 eV and 144.6 eV, respectively, which is a large shift compared to pure Pb^0, indicating the electronic interaction between PtAu and Pb.

2.2. Electrochemical Study of the Mechanism of Pb Co-Deposition at the Cathode During the GRR Process in the DCGC Device

It is significant and intriguing to study the mechanism of Pb co-deposition at the cathode when depositing Pt and Au in the DCGC device with a Cu plate as the anode. Given that the ECP of Pb_{bulk} is lower than that of Cu_{bulk}, the dissolution of Cu_{bulk}, inducing the deposition of Pb_{bulk} during the GRR for depositing Pb-Pt_3Au_1 nanoparticles at the cathode, is thermodynamically impossible. Here, we studied the electrochemical deposition behavior of Pb on Pt or Au surfaces in cyclic voltammetry (CV) experiments. Figure 4 shows the CV curves for electroplated Pt-modified Au (Pt_{pla}/Au; as shown in Figure S3) and bare Au electrodes in a solution containing 3.0 mM $Pb(ClO_4)_2$ + 0.1 M $HClO_4$. For Pt_{pla}/Au, during the negative-going scan, the UPD of Pb on the Pt surface starts at 0.55 V [21]. The UPD of Pb on the Pt surface ($Pb^0{}_{UPD}$) significantly impedes the adsorption/desorption behavior of H on the Pt surface between 0 and −0.28 V. The UPD of Pb terminates at −0.45 V, where the bulk deposition of Pb ($Pb^0{}_{bulk}$) begins. For the bare Au electrode, during the negative-going scan, the UPD of Pb on Au starts at 0.06 V, and two distinct pairs of oxidation-reduction peaks appear at 0.01/−0.01 V and −0.23/−0.25 V, corresponding to the UPD behavior of Pb on the Au surface [22]. The UPD of Pb terminates at −0.45 V, where the bulk deposition of Pb occurs, and the dissolution peak of bulk Pb appears at −0.35 V. The CV experiments indicate that the deposition/dissolution potentials of $Pb^0{}_{UPD}$ are higher than those of $Pb^0{}_{bulk}$, i.e., the ECP of $Pb^{2+}/Pb^0{}_{UPD}$ is higher than that of $Pb^{2+}/Pb^0{}_{bulk}$, suggesting the thermodynamic possibility of the bulk dissolution of an active metal inducing the UPD of its ions (or other ions) on a relatively less active metal surface (such as Pt or Au).

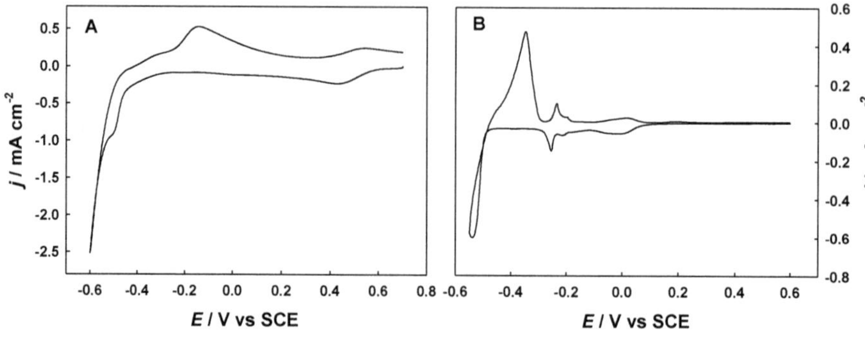

Figure 4. Cyclic voltammograms of Pt_{pla}/Au (**A**) and bare Au (**B**) in 3.0 mM $Pb(ClO_4)_2$ + 0.1 M $HClO_4$ aqueous solution, respectively. Scan rate: 20 mV s^{-1}.

In the DCGC device with a Cu plate as the anode, during the GRR for depositing Pb-Pt_3Au_1, the steady-state potential of the MWCNTs/GCE is 0.121 V, which falls within the potential range for the UPD of Pb on the Pt surface (0.55~−0.45 V) but is higher than the potential range for the UPD of Pb on the Au surface (0.06~−0.45 V). This indicates that during the deposition of Pb-Pt_3Au_1, Pb can undergo UPD on the Pt surface but not on the Au surface. Hence, the bulk dissolution of Cu at the anode can induce the UPD of Pb on Pt at the cathode in the DCGC device. The UPD effect of Pb on the Pt surface influences the growth process of PtAu nanoparticles, leading to the deposition of nanoflower structures on the MWCNT surface. Simultaneously, a small number of $Pb^0{}_{UPD}$ atoms are embedded within the PtAu nanoparticles, forming a Pb-containing PtAu nanoflower structure.

2.3. Electrocatalytical Oxidation of Methanol on Pb-Containing PtAu Nanoparticles in Alkaline Solution

Figure 5 shows the cyclic voltammograms (CVs) of Pb-Pt_3Au_1/MWCNTs/GCE, Pt_3Au_1/MWCNTs/GCE, and Pb-Pt/MWCNTs/GCE in a 1.0 M KOH (Figure 5A) and 1.0 M CH_3OH + 1.0 M KOH solution (Figure 5B,C). For Pb-Pt_3Au_1/MWCNTs/GCE (Figure 5A), the adsorption/desorption peaks of hydrogen on the Pt surface appear between −0.65 V and −1.0 V. The redox peaks at −0.13 V and −0.38 V correspond to the formation and reduction of Pt oxide [23]. Additionally, the redox peaks at 0.3 V and 0.03 V are attributed to the formation and reduction of Au oxide, which is consistent with the redox peaks of Au on the surface of Pb-Au/MWCNTs/GCE (inset of Figure 5A). The electrochemical characteristic peaks on the surfaces of Pt_3Au_1/MWCNTs/GCE and Pb-Pt/MWCNTs/GCE are similar to those of Pb-Pt_3Au_1/MWCNTs/GCE.

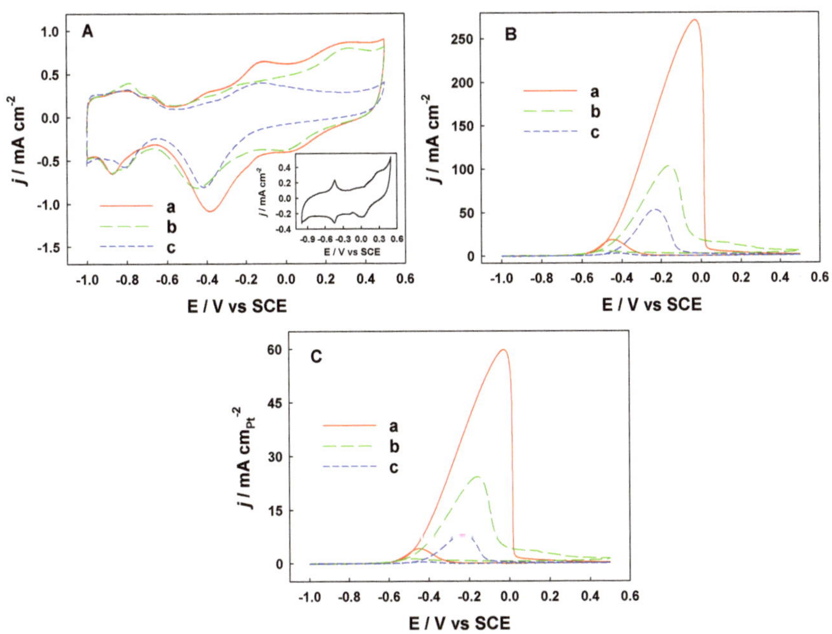

Figure 5. Cyclic voltammograms of Pb-Pt_3Au_1/MWCNTs/GCE (a), Pt_3Au_1/MWCNTs/GCE (b), and Pb-Pt/MWCNTs/GCE (c) in 1.0 M KOH aqueous solution (**A**) and 1.0 M CH_3OH + 1.0 M KOH aqueous solution (**B**,**C**), respectively. Currents in plans (**A**–**C**) are normalized to geometric area (**A**,**B**) and Pt active area (**C**), respectively. The inset in plan (**A**) is the CV of Pb-Au/MWCNTs/GCE in 1.0 M KOH aqueous solution. Scan rate: 50 mV s^{-1}.

For the electrocatalytic oxidation of methanol (Figure 5B), during the forward scan, the initial oxidation potential of methanol on the surface of Pb-Pt_3Au_1/MWCNTs/GCE is −0.64 V, which is shifted negatively by 30 mV and 130 mV compared to Pt_3Au_1/MWCNTs/GCE (−0.61 V) and Pb-Pt/MWCNTs/GCE (−0.51 V), respectively. The peak current for the oxidation of methanol on the surface of Pb-Pt_3Au_1/MWCNTs/GCE is 270 mA cm^{-2}, which is 2.62 times and 5.0 times higher than that of Pt_3Au_1/MWCNTs/GCE (103 mA cm^{-2}) and Pb-Pt/MWCNTs/GCE (54 mA cm^{-2}), respectively. Furthermore, we compared the electrochemical active area specific activity of Pt (mA $cm_{Pt}{}^{-2}$), as shown in Figure 5C. The peak current for the oxidation of methanol on the surface of Pb-Pt_3Au_1/MWCNTs/GCE is 59.8 mA cm_{Pt}^{-2}, which is 2.48 times and 7.29 times higher than that of Pt_3Au_1/MWCNTs/GCE (24.1 mA cm_{Pt}^{-2}) and Pb-Pt/MWCNTs/GCE (8.2 mA cm_{Pt}^{-2}), respectively.

The above results indicate that Pb-Pt_3Au_1/MWCNTs/GCE exhibits excellent electrocatalytic performance toward methanol, primarily due to the following reasons: (1) compared to the spherical structure of Pt_3Au_1 nanoparticles, the Pb-Pt_3Au_1 nanoflower particles have more surface defect sites and a higher number of exposed unsaturated Pt atoms, which enhance their catalytic activity [24]; (2) both the bifunctional effect between Pt and Au atoms [17] and the electronic effect between Pt and Pb atoms [25] contribute to enhancing the catalytic performance of the Pb-Pt_3Au_1 ternary metal catalyst.

Since the catalytic performance of noble metal electrocatalysts is related to their morphology, particle size, and composition [26], the modified electrodes with different Pt and Au contents were prepared by changing the concentration ratio of H_2PtCl_6 and $HAuCl_4$ in the catholyte, and the electrocatalytic performance of the obtained modified electrodes for methanol was studied, as shown in Figure 6. Increasing the concentration of $HAuCl_4$ in the catholyte results in an increase in the Au content on the electrode surface, accompanied by a decrease in Pt content, as evidenced by the increased oxidation-reduction peak of Au and the decreased oxidation-reduction peak of Pt (Figure 6A). The peak current density of methanol electrocatalytic oxidation also increases with the increased concentration of $HAuCl_4$, reaching a maximum value of 270 mA cm^{-2} at $c_{H2PtCl6}$: c_{HAuCl4} = 3:1 (Pb-Pt_3Au_1/MWCNTs/GCE; Figure 6B), and then decreases with the increased $HAuCl_4$ concentration. The electrochemical active area specific activity of Pt (Figure 6C) also reaches a maximum value of 59.8 mA cm_{Pt}^{-2} on Pb-Pt_3Au_1/MWCNTs/GCE. As the Au content on the electrode surface increases, the synergistic effect between Pt and Au atoms gradually intensifies. The activity of Pt atoms reaches its maximum at a ratio of $c_{H2PtCl6}$: c_{HAuCl4} = 3:1. As the Au content on the electrode surface continues to increase, the Pt content gradually decreases, as evidenced by the reduced oxidation-reduction peak of Pt observed in the above electrochemical characterization. The probability of Pt atoms being surrounded by Au atoms increases, and some Pt atoms may be separated by Au atoms, reducing the number of adjacent Pt atoms, which may hinder the adsorption and dissociation of methanol on the Pt surface and lead to a decrease in the activity of Pt atoms [27]. It has also been reported that methanol can only undergo electrooxidation on a surface with at least three adjacent Pt atoms [28].

The electrochemical stability of electrocatalysts is an important indicator for practical applications. We also used a potentiostatic method to evaluate the stability of different modified electrodes in catalyzing the oxidation of methanol in 1.0 M CH_3OH + 1.0 M KOH solution at −0.25 V, as shown in Figure 7. The Pb-Pt_3Au_1/MWCNTs/GCE exhibited the slowest rate of catalytic current decay, with a catalytic current of 8.19 mA cm_{Pt}^{-2} at 1800 s, which is significantly higher than that of Pb-Pt/MWCNTs/GCE (0.10 mA cm_{Pt}^{-2}), Pb-Pt_3Au_3/MWCNTs/GCE (0.83 mA cm_{Pt}^{-2}), Pb-Pt_3Au_5/MWCNTs/GCE (0.23 mA cm_{Pt}^{-2}), and Pb-Pt_3Au_7/MWCNTs/GCE (0.11 mA cm_{Pt}^{-2}). This indicates that the Pb-Pt_3Au_1 nanoelectrocatalyst can effectively remove toxic products such as CO_{ads} that are generated during the electrooxidation of methanol, maintaining the catalytic activity and electrochemical stability.

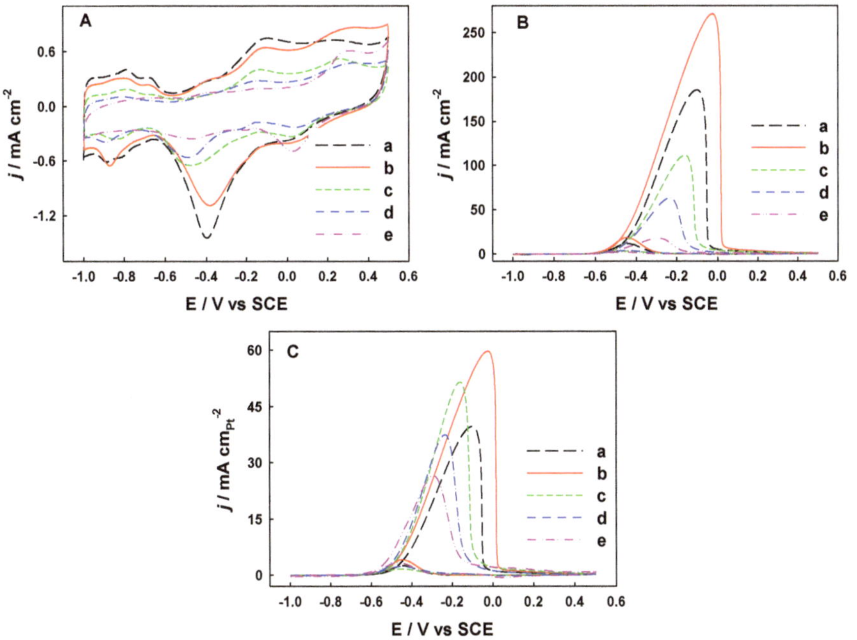

Figure 6. Cyclic voltammograms of Pb-$Pt_3Au_{0.5}$/MWCNTs/GCE (a), Pb-Pt_3Au_1/MWCNTs/GCE (b), Pb-Pt_3Au_3/MWCNTs/GCE (c), Pb-Pt_3Au_5/MWCNTs/GCE (d), and Pb-Pt_3Au_7/MWCNTs/GCE (e) in 1.0 M KOH aqueous solution (**A**) and 1.0 M CH_3OH + 1.0 M KOH aqueous solution (**B**,**C**), respectively. Currents in plans (**A**–**C**) are normalized to geometric area (**A**,**B**) and Pt active area (**C**), respectively. Scan rate: 50 mV s^{-1}.

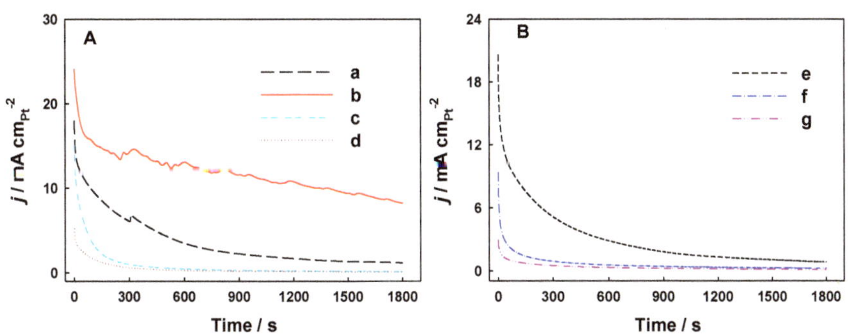

Figure 7. The *i*-t curves at −0.25 V of Pb-$Pt_3Au_{0.5}$/MWCNTs/GCE (a), Pb-Pt_3Au_1/MWCNTs/GCE (b), Pt_3Au_1/MWCNTs/GCE (c), Pb-Pt/MWCNTs/GCE (d) (**A**) and Pb-Pt_3Au_3/MWCNTs/GCE (e), Pb-Pt_3Au_5/MWCNTs/GCE (f), and Pb-Pt_3Au_7/MWCNTs/GCE (g) (**B**) in 1.0 M CH_3OH + 1.0 M KOH aqueous solution, respectively. Currents are normalized to Pt active area.

3. Materials and Methods

3.1. Instrumentation and Reagents

All electrochemical experiments were conducted on an Autolab PGSTAT 30 electrochemical workstation (Eco Chemie BV, Utrecht, The Netherlands) using a conventional three-electrode system. The working electrodes were either a glassy carbon electrode (GCE) with a diameter of 3 mm or a gold electrode (0.29 cm^2). The reference electrode was a saturated calomel electrode (SCE) equipped with a salt bridge, and the counter electrode was a carbon rod. All potentials were referenced to the SCE. The JEOL JSM-1230 high-resolution field-emission scanning electron microscope (SEM, Akishima, Japan) and the accompanying energy-dispersive X-ray spectroscopy (EDS) were used for morphological characterization and elemental composition analysis, respectively. X-ray diffraction (XRD) pattern was collected on a D8 Discover X-ray diffractometer (Bruker Co., Billerica, MA, USA). X-ray photoelectron spectroscopy (XPS) spectra were taken using an ESCA Escalab 220i XL (Thermo Fisher Scientific, Waltham, MA, USA) with a monochromated Al Kα X-ray source (1486.6 eV).

For the experiments conducted in the DCGC, when no external load was applied, the discharge current (i_{cell}) of the galvanic cell was monitored using the ECN module of the Autolab PGSTAT 30 electrochemical workstation [29] (Metrohm, Herisau, Switzerland). The accuracy of this setup was verified using a calibrated 1.5 V dry cell (error within 1 mV) under a 100 kΩ load; the dry cell voltage measured by the ECN setup was 1.501 V, with an i_{cell} of 15.01 μA, strictly adhering to Ohm's Law. The potentials of the anode and cathode in the galvanic cell setup were dynamically monitored using a high-impedance ($R \geq 10^{12}$ Ω) voltmeter, which was modified from a pH meter (Leici, Shanghai, China). All experiments were conducted at room temperature (26 $\pm$ 2 °C).

N,*N*-Dimethylformamide (DMF), $HAuCl_4$, H_2PtCl_6, and methanol were purchased from Tianjin Chemical Reagent Station (Tianjin, China). $Pb(ClO_4)_2$ and high-purity Cu plate (99.999%) were purchased from Alfa Aesar company (Tianjin, China). Multiwalled carbon nanotubes (MWCNTs; 60 nm outer diameter and 40 nm inner diameter on average) were purchased from Chengdu Organic Chemicals Co. Ltd. (Chengdu, China) and purified in concentrated acids before use. Other reagents were of analytical grade. All solutions were prepared using Milli-Q ultrapure water (Millipore, Burlington, MA, USA, >18 M cm).

3.2. Deposition of Pb-Containing PtAu Nanoparticles on MWCNTs/GCE in the DCGC Device

The GCE was sequentially polished with 1.0 μm and 0.05 μm alumina slurry on a chamois leather until a mirror-like finish was achieved. After each polishing step, the surface contaminants were washed off, and the electrode was cleaned in an ultrasonic water bath for 3 min each time; this process was repeated three times. Finally, the electrode was ultrasonically cleaned sequentially with 1:1 ethanol, 1:1 HNO_3, and distilled water. It was then subjected to cyclic voltammetry (CV) scans in 0.2 M $HClO_4$ solution (0 to 1.5 V, 30 mV s^{-1}) until reproducible voltammetric responses were observed. The resulting GCE was further subjected to CV scans in 2.0 mM $K_4Fe(CN)_6$ + 0.1 M Na_2SO_4 solution (−0.1 to 0.5 V, 50 mV s^{-1}). A peak-to-peak potential width of ca. 70 mV indicated that the electrode surface was clean. A 1 mg mL^{-1} dispersion of MWCNTs in DMF was prepared and sonicated for 15 min. Then, 5 μL of the MWCNT dispersion was cast onto the GCE and air-dried to form the MWCNTs/GCE.

For the deposition of Pb-containing PtAu nanoparticles on MWCNTs/GCE in a DCGC device based on the GRR principle, a Cu plate served as the anode, the anolyte was 0.1 M $HClO_4$ aqueous solution, MWCNTs/GCE served as the cathode, and the catholyte was 3.0 mM H_2PtCl_6 + 1.0 mM $HAuCl_4$ + 5.0 mM $Pb(ClO_4)_2$ + 0.1 M $HClO_4$ aqueous solution. The catholyte and anolyte were connected using a homemade U-shaped glass tube salt bridge filled with saturated KCl solution. Scheme 1 shows the schematic diagram of the total experimental setup, similar to that of a previous report [30]. After switching on the DCGC, the ECN and voltmeter were used to monitor the i_{cell} and the potential of MWCNTs/GCE (vs SCE), respectively. The total discharging time in the DCGC was 240 s. The resulting modified electrode was denoted as Pb-Pt_3Au_1/MWCNTs/GCE. Varying the concentration ratio of H_2PtCl_6 and $HAuCl_4$ in the catholyte can obtain modified electrodes with different Pt and Au content.

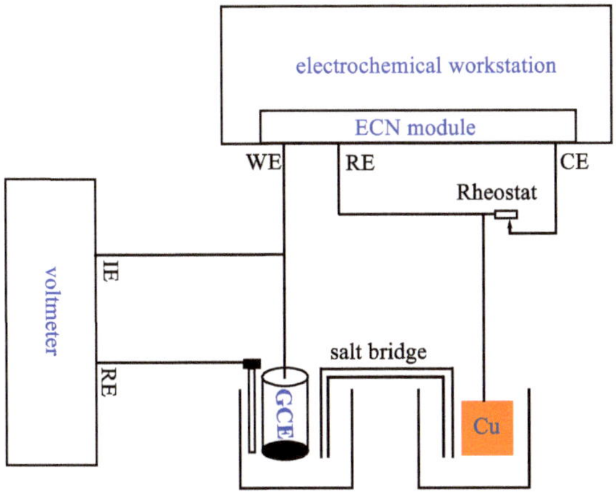

Scheme 1. Schematic diagram of the total experimental setup. GCE: glassy carbon electrode. IE: indicator electrode.

For comparative studies, we also prepared other modified electrodes in a similar procedure, and the abbreviation of the modified electrodes and the corresponding composition of catholytes are listed in Table 1.

Table 1. Abbreviation and the corresponding metal salt composition of catholyte for preparing the modified electrodes.

Abbreviation of Modified Electrodes	The Metal Salt Composition of Catholyte
Pb-Pt/MWCNTs/GCE	3.0 mM H_2PtCl_6 + 5.0 mM $Pb(ClO_4)_2$ + 0.1 M $HClO_4$
Pb-$Pt_3Au_{0.5}$/MWCNTs/GCE	5.0 mM $Pb(ClO_4)_2$ + 3.0 mM H_2PtCl_6 + 0.5 mM $HAuCl_4$ + 0.1 M $HClO_4$
Pb-Pt_3Au_1/MWCNTs/GCE	5.0 mM $Pb(ClO_4)_2$ + 3.0 mM H_2PtCl_6 + 1.0 mM $HAuCl_4$ + 0.1 M $HClO_4$
Pb-Pt_3Au_3/MWCNTs/GCE	5.0 mM $Pb(ClO_4)_2$ + 3.0 mM H_2PtCl_6 + 3.0 mM $HAuCl_4$ + 0.1 M $HClO_4$
Pb-Pt_3Au_5/MWCNTs/GCE	5.0 mM $Pb(ClO_4)_2$ + 3.0 mM H_2PtCl_6 + 5.0 mM $HAuCl_4$ + 0.1 M $HClO_4$
Pt_3Au_1/MWCNTs/GCE	3.0 mM H_2PtCl_6 + 1.0 mM $HAuCl_4$ + 0.1 M $HClO_4$
Pb-Au/MWCNTs/GCE	3.0 mM $HAuCl_4$ + 5.0 mM $Pb(ClO_4)_2$ + 0.1 M $HClO_4$

3.3. CV Experiment to Study the Electrochemical Deposition Behavior of Pb on Pt and Au Surfaces

The pretreatment process of the gold electrode was as follows. To remove contaminants from the electrode surface, the gold electrode was first cleaned with H_2SO_4 + H_2O_2 (v/v, 3:1) solution for 3 min and then subjected to CV scanning in 0.2 M $HClO_4$ solution (0~1.5 V, 30 mV s^{-1}) until a reproducible voltammetric response was achieved. The resulting gold electrode was then subjected to CV scans in 2.0 mM $K_4Fe(CN)_6$ + 0.1 M Na_2SO_4 solution (−0.1 to 0.5 V, 50 mV s^{-1}). A peak-to-peak potential width of ca. 70 mV indicated that the electrode had been properly cleaned.

The gold electrode was placed in a 3.0 mM H_2PtCl_6 + 0.1 M $HClO_4$ solution and electroplated at a constant potential of 0 V for 300 s to obtain a Pt-modified Au electrode (Pt_{pla}/Au). The obtained Pt_{pla}/Au electrode was placed in 0.1 M H_2SO_4 for CV characterization (−0.3–1.5 V, 50 mV s^{-1}). After Pt plating, a reduction peak of Pt oxide appeared at 0.35 V, and the adsorption/desorption peak of H on the Pt surface appeared at 0–0.28 V. The reduction peak of Au oxide at 0.85 V nearly disappeared, indicating that the Au electrode surface was fully covered with Pt (Figure S3). Then, using Pt_{pla}/Au as the working electrode, the electrochemical deposition behavior of Pb on the Pt_{pla}/Au surface was investigated by the CV method in 3.0 mM $Pb(ClO_4)_2$ + 0.1 M $HClO_4$ solution. Similarly, a bare gold electrode was used as the working electrode to study the electrochemical deposition behavior of Pb on the gold electrode surface.

3.4. Electrochemical Characterization of the Modified Electrodes and Electrocatalytic Oxidation of Methanol

The modified electrode was subjected to CV characterization in 1.0 M KOH (−1.0 to 0.5 V, 50 mV s^{-1}). The electrocatalytic oxidation of methanol was conducted in 1.0 M CH_3OH + 1.0 M KOH solution (−1.0 to 0.5 V, 50 mV s^{-1}). The stability of the modified electrode toward methanol electrocatalytic oxidation was evaluated using a potentiostatic method in 1.0 M CH_3OH + 1.0 M KOH solution at a potential of −0.25 V for 1800 s. The electrochemical active area of Pt was calculated from the hydrogen adsorption peak in 1.0 M KOH solution, using a conversion factor of 210 μC cm^{-2} [21]. All solutions were deoxygenated with high-purity nitrogen for at least 10 min.

4. Conclusions

In summary, a Pb-containing PtAu nanoflower electrocatalyst was deposited on the MWCNT/GCE cathode in a DCGC device. The electrochemical experimental results demonstrate that the bulk dissolution of Cu at the anode can induce the UPD of Pb on Pt at the cathode during the GRR process in the DCGC device. The UPD effect of Pb influenced the growth process of PtAu nanoparticles, leading to the formation of Pb-containing PtAu nanoflower structures. Under the optimal experimental conditions, the prepared Pb-Pt3Au1/MWCNTs/GCE showed the highest activity and stability for the electrocatalytic oxidation of methanol in an alkaline environment. The method of depositing high-activity electrocatalysts via GRR in the DCGC device is expected to have broad application value in electrocatalysis and fuel cell research.

Supplementary Materials: The following supporting information can be downloaded at: https://www.mdpi.com/article/10.3390/molecules29235492/s1. Figure S1. XRD patterns of the as-synthesized Pt_3Au_1 and Pb-Pt_3Au_1 sample. Figure S2. XPS spectra of the Au (4f), Pt 4f (a) and Pb 4f (b) regions for the as-synthesized Pt_3Au_1 and Pb-Pt_3Au_1 sample. Figure S3. CV curves at Pt_{pla}/Au and bare Au electrodes in 0.1 M H_2SO_4 aqueous solution. Scan rate: 50 mV s^{-1}.

Author Contributions: Conceptualization and methodology, Z.H., X.J. and L.C.; formal analysis, investigation, resources, and data curation, Z.H., Z.Z. and X.J.; writing—original draft preparation, Z.H.; writing, review, and editing, X.J. and L.C.; visualization, supervision, and project administration, X.J and L.C; funding acquisition, Z.H. and L.C. All authors have read and agreed to the published version of the manuscript.

Funding: This research was funded by a scientific research project of the Hunan Provincial Department of Education (23C0201), the Excellent Youth Program of Hunan Provincial Department of Education(23B0561), the Natural Science Foundation of Hunan Province (2022JJ50099, 2024JJ7137), and the Key Research and Development Program of Hunan Province (2022SK2009).

Institutional Review Board Statement: Not applicable.

Informed Consent Statement: Not applicable.

Data Availability Statement: Data are contained within the article and Supplementary Materials.

Conflicts of Interest: The authors declare no conflict of interest.

References

1. Berghian-Grosan, C.; Radu, T.; Biris, A.R.; Dan, M.; Voica, C.; Watanabe, F.; Biris, A.S.; Vulcu, A. Platinum nanoparticles cc by graphene layers: A low-metal loading catalyst for methanol oxidation in alkaline media. *J. Energy Chem.* **2020**, *40*, 8 [CrossRef]
2. Cao, L.; Scheiba, F.; Roth, C.; Schweiger, F.; Cremers, C.; Stimming, U.; Fuess, H.; Chen, L.; Zhu, W.; Qiu, X. Novel nanocomp $Pt/RuO_2{\cdot}xH_2O$/Carbon nanotube catalysts for direct methanol fuel cells. *Angew. Chem. Int. Ed.* **2006**, *45*, 5315–5319. [Cros [PubMed]

Lee, E.; Kim, S.; Jang, J.; Park, H.U.; Matin, M.A.; Kim, Y.T.; Kwon, Y.U. Effects of particle proximity and composition of Pt-M (M = Mn, Fe, Co) nanoparticles on electrocatalysis in methanol oxidation reaction. *J. Power Sources* **2015**, *294*, 75–81. [CrossRef]

Yang, Q.; Zhang, S.; Wu, F.; Zhu, L.; Li, G.; Chen, M.; Pei, A.; Feng, Y. Efficient and stable PtFe alloy catalyst for electrocatalytic methanol oxidation with high resistance to CO. *J. Energy Chem.* **2024**, *90*, 327–336. [CrossRef]

Zhang, L.; Zhang, X.; Chen, X.; Wang, A.; Han, D.; Wang, Z.; Feng, J. Facile solvothermal synthesis of $Pt_{71}Co_{29}$ lamellar nanoflowers as an efficient catalyst for oxygen reduction and methanol oxidation reactions. *J. Colloid. Interf. Sci.* **2019**, *536*, 556–562. [CrossRef]

Li, Z.; Ke, S.; Zheng, X.; Huang, Y.; Fu, W.; Wang, Y.; Nie, Y. Modulating d-orbital electronic configuration of PtRu via charge donation from Co-enriched core boosts methanol electrooxidation. *Chem. Eng. J.* **2024**, *493*, 152544–152552. [CrossRef]

Li, X.; Lei, H.; Yang, C.; Zhang, Q. Electrochemical fabrication of ultra-low loading Pt decorated porous nickel frameworks as efficient catalysts for methanol electrooxidation in alkaline medium. *J. Power Sources* **2018**, *396*, 64–72. [CrossRef]

Chen, R.; Gao, J.; Yang, J.; Zhang, F.; Wang, Q. Enhanced methanol electrooxidation catalysis via dual modulation of PtCu alloy and oxygen vacancies. *Fuel* **2024**, *371*, 131994–132002. [CrossRef]

Hanifah, M.F.R.; Jaafar, J.; Othman, M.; Ismail, A.F.; Rahman, M.A.; Yusof, N.; Aziz, F. One-pot synthesis of efficient reduced graphene oxide supported binary Pt-Pd alloy nanoparticles as superior electro-catalyst and its electro-catalytic performance toward methanol electro-oxidation reaction in direct methanol fuel cell. *J. Alloys Compd.* **2019**, *793*, 232–246. [CrossRef]

Eid, K. Rapid one-step aqueous synthesis of porous PtAg wavy nanochains for methanol electrooxidation with a high CO-tolerance. *J. Electroanal. Chem.* **2024**, *961*, 118207–118216. [CrossRef]

Morante-Catacora, T.Y.; Ishikawa, Y.; Cabrera, C.R. Sequential electrodeposition of Mo at Pt and PtRu methanol oxidation catalyst particles on HOPG surfaces. *J. Electroanal. Chem.* **2008**, *621*, 103–112. [CrossRef]

Ren, Y.; Askarov, S.; Zhang, Y.; Shi, D.; Wu, Q.; Chen, K.; Li, H. Nanoarchitectonics for modulation on the electronic structure of ultrafine PtRuFe nanowires as robust methanol electrooxidation catalysts. *J. Alloys Compd.* **2024**, *978*, 173442–173450. [CrossRef]

Baruch-Soto, M.; Magallón-Cacho, L.; Ramírez-Aparicio, J.; Ortega-Guzmán, J.; Borja-Arco, E. Methanol oxidation reaction in alkaline media using gold nanoparticles recovered from electronic waste. *Materials* **2024**, *17*, 1267. [CrossRef] [PubMed]

Karuppasamy, L.; Chen, C.Y.; Anandan, S.; Wu, J.J. Sonochemical fabrication of reduced graphene oxide supported Au nano dendrites for ethanol electrooxidation in alkaline medium. *Catal. Today* **2018**, *307*, 308–317. [CrossRef]

Reddy, G.V.; Sekhar, Y.C.; Raghavendra, P.; Reddy, M.N.; Chandana, P.S.; Sarma, L.S. Controlled synthesis of reduced graphene oxide-supported bimetallic Pt-Au nanoparticles for enhanced electrooxidation of methanol. *Solid State Sci.* **2024**, *149*, 107469. [CrossRef]

Zhang, Z.; Wang, Y.; Wang, X. Nanoporous bimetallic Pt-Au alloy nanocomposites with superior catalytic activity towards electro-oxidation of methanol and formic acid. *Nanoscale* **2011**, *3*, 1663–1674. [CrossRef]

Luo, J.; Njoki, P.N.; Lin, Y.; Mott, D.; Wang; Zhong, C. Characterization of carbon-supported AuPt nanoparticles for electrocatalytic methanol oxidation reaction. *Langmuir* **2006**, *22*, 2892–2898. [CrossRef]

Wang, X.; Chen, S.; Reggiano, G.; Wang, X.; Chen, S.; Reggiano, G.; Thota, S.; Wang, Y.; Kerns, P.; Suib, S.L.; et al. Au-Cu-M (M = Pt, Pd, Ag) nanorods with enhanced catalytic efficiency by galvanic replacement reaction. *Chem. Commun.* **2019**, *55*, 1249–1252. [CrossRef]

Yoshii, T.; Nakatsuka, K.; Kuwahara, Y.; Mori, K.; Yamashita, H. Synthesis of carbon-supported Pd-Co bimetallic catalysts templated by Co nanoparticles using the galvanic replacement method for selective hydrogenation. *RSC Adv.* **2017**, *7*, 22294–22300. [CrossRef]

Zhang, L.; Zhang, J.; Kuang, Q.; Xie, S.; Jiang, Z.; Xie, Z.; Zheng, L. Cu^{2+}-assisted synthesis of hexoctahedral Au-Pd alloy nanocrystals with high-index facets. *J. Am. Chem. Soc.* **2011**, *133*, 17114–17117. [CrossRef]

Grgur, B.; Marković, N.; Ross, P. Underpotential deposition of lead on Pt(111) in perchloric acid solution: $RPD_{Pt(111)}E$ measurements. *Langmuir* **1997**, *13*, 6370–6374. [CrossRef]

Stafford, G.; Bertocci, U. In situ stress and nanogravimetric measurements during underpotential deposition of Pb on (111)-textured Au. *J. Phys. Chem. C* **2007**, *111*, 17580–17586. [CrossRef]

Guo, S.; Dong, S.; Wang, E. Pt/Pd bimetallic nanotubes with petal-like surfaces for enhanced catalytic activity and stability towards ethanol electrooxidation. *Energy Environ. Sci.* **2010**, *3*, 1307–1310. [CrossRef]

Tian, N.; Zhou, Z.; Sun, S.; Ding, Y.; Wang, Z. Synthesis of tetrahexahedral platinum nanocrystals with high-index facets and high electro-oxidation activity. *Science* **2007**, *316*, 732–735. [CrossRef] [PubMed]

Jiang, Q.; Jiang, L.; Qi, J.; Sun, G. Experimental and density functional theory studies on PtPb/C bimetallic electrocatalysts for methanol electrooxidation reaction in alkaline media. *Electrochim. Acta* **2011**, *56*, 6431–6440. [CrossRef]

Burda, C.; Chen, X.; Narayanan, R.; Ei-Sayed, M. Chemistry and properties of nanocrystals of different shapes. *Chem. Rev.* **2005**, *105*, 1025–1102. [CrossRef]

Du, B.; Tong, Y. A coverage-dependent study of Pt spontaneously deposited on to Au and Ru surfaces: Direct experimental evidence of the ensemble effect for methanol electro-oxidation on Pt. *J. Phys. Chem. B* **2005**, *109*, 17775–17780. [CrossRef]

Neurock, M.; Janik, M.; Wieckowski, A. A first principles comparison of the mechanism and site requirements for the electrocatalytic oxidation of methanol and formic acid over Pt. *Faraday Discuss.* **2009**, *140*, 363–378. [CrossRef]

29. Tan, Y.; Xie, Q.; Huang, J.; Duan, W.; Ma, M.; Yao, S. Study on glucose biofuel cells using an electrochemical noise de
Electroanalysis **2008**, *20*, 1599–1606. [CrossRef]
30. Huang, Z.; Tang, Z.; Chao, L. Double-cabin galvanic cell-synthesizing nanoporous, flower-like, Pb-containing Pd-Au nanopart
for nonenzymatic formaldehyde sensor. *Molecules* **2024**, *29*, 2772. [CrossRef]

molecules

MDPI

cle

Strategy for Anode Recovery and Upgrading by In Situ rowth of Iron-Based Oxides on Microwave-Puffed Graphite

axin Chen, Jing Sun *, Pingshan Jia, Wenlong Wang, Zhanlong Song, Ziliang Wang, Xiqiang Zhao Yanpeng Mao

National Engineering Laboratory for Reducing Emissions from Coal Combustion, Engineering Research Center of Environmental Thermal Technology of Ministry of Education, Shandong Key Laboratory of Energy Carbon Reduction and Resource Utilization, School of Energy and Power Engineering, Shandong University, Jinan 250061, China

* Correspondence: sunjing0108@163.com; Tel.: +86-18806409150

Abstract: Faced with the increasing volume of retired lithium-ion batteries (LIBs), recycling and reusing the spent graphite (SG) is of great significance for resource sustainability. Here, a facile method for transforming the SG into a carbon framework as well as loading Fe_2O_3 to form a composite anode with a sandwich structure is proposed. Taking advantage of the fact that the layer spacing of the spent graphite naturally expands, impurities and intercalants are eliminated through microwave thermal shock to produce microwave-puffed graphite (MPG) with a distinct three-dimensional structure. Based on the mechanism of microwave-induced gasification intercalation, a Fe_2O_3-MPG intercalation compound (Fe_2O_3-MPGIC) anode material was constructed by introducing iron precursors between the framework layers and subsequently converting them into Fe_2O_3 through annealing. The Fe_2O_3-MPGIC anode exhibits a high reversible capacity of 1000.6 mAh g^{-1} at 200 mA g^{-1} after 100 cycles and a good cycling stability of 504.4 mAh g^{-1} at 2000 mA g^{-1} after 500 cycles. This work can provide a reference for the feasible recycling of SG and development of high-performance anode materials for LIBs.

Keywords: lithium-ion batteries; spent graphite; transition metal oxides; graphite intercalation compounds; composite anode

1. Introduction

As efficient energy storage devices in the burgeoning renewable energy sector, lithium-ion batteries (LIBs), have become the main power source for electric vehicles and hybrid electric vehicles, contributing significantly to the reduction in CO_2 emissions. However, the expansion of LIBs manufacturing is accompanied by a growing demand for raw materials and a subsequent increase in retired batteries due to their limited lifespan. It is estimated that at least 650 GWh of LIBs will be scrapped in 2025 [1]. Improper disposal of spent batteries presents potential hazards such as explosions and pollutant emissions, and results in the squandering of valuable raw materials. Consequently, the recycling of spent batteries is imperative to safeguard the environment and ensure the sustainability of resources.

The recycling of cathode materials has received considerable attention due to their high metal content [2,3]. However, anode materials (e.g., graphite, hard carbon) are often overlooked or simply used as reducing agents [4]. Graphite, the most commonly used anode material, constitutes approximately 20% of the total mass of LIBs. Proper recycling of graphite is crucial for alleviating the graphite resource shortage and reducing the energy and costs associated with graphite mining and production.

Compared with natural graphite, spent graphite (SG) includes various metal and organic impurities that need to be removed. While SG can be regenerated through acid leaching [5] and pyrometallurgy [6] for reuse as anode material, this process is not attractive due to its chemical and energy consumption [7]. More importantly, the inherent low capacity of graphite cannot meet the demand for high-energy-density LIBs, and high-capacity anodes such as silicon [8] and metal oxides [9] have become the main focus of research.

Ferric oxide (Fe_2O_3), as one of the most promising alternatives to graphite anodes, offers advantages such as wide availability, low cost, high theoretical reversible specific capacity, and good safety. However, they suffer from volume expansion and low conductivity throughout the cycling process, which markedly affect cycling stability and rate performance. Compounding with carbon materials is one of the most common methods for enhancing Fe_2O_3 anodes. Guo et al. [10] synthesized uniform α-Fe_2O_3 nanoparticles with a narrow gap (~1.4 nm), which were immobilized on CNTs through N-doped carbon (α-Fe_2O_3/CNTs-NC). Chen et al. [11] chose to embed Fe_2O_3 nanoparticles within the shells of hollow, highly graphitized carbon fibers, which are coated by N-doped graphitized carbon (GF-FeO@NGC). Numerous studies have demonstrated the favorable specific capacity and cycling stability of these composite electrodes, highlighting their promising potential for practical applications.

The interlayer space of graphite presents a novel option for loading Fe_2O_3. Notably, the repeated insertion of Li^+ causes an expansion in the layer spacing of SG, weakening the Van der Waals force between layers, which makes the intercalation easier. Moreover, due to the excellent microwave absorbing and conductive properties of graphite, the efficient coupling of microwave radiation with graphite facilitates the rapid removal of impurities and intercalants. Microwave radiation on SG induces a Joule heat–discharge–plasma coupled effect, resulting in rapid heating characterized by thermal shock. This causes the impurities and intercalants to be quickly released, leading to further expansion of the layer spacing and forming microwave-puffed graphite (MPG). Fe_2O_3-MPG intercalation compound (Fe_2O_3-MPGIC) can be fabricated by integrating Fe_2O_3 into the layers of MPG, which can significantly improve the electrical conductivity of Fe_2O_3 and mitigate issues arising from volume expansion [12], such as inadequate mechanical stability and detachment of copper foil.

However, the strong agglomeration tendency and thermal stability of Fe_2O_3 particles present challenges when attempting to embed it directly into graphite layers. The potential of graphite frameworks can be fully realized by embedding appropriate precursors into graphite interlayers and subsequently facilitating their growth into Fe_2O_3 particles. The embedding of metal chlorides, especially $FeCl_3$, into graphite interlayers as precursors has attracted considerable interest [12–14]. Notably, the graphite framework can be rapidly intercalated with $FeCl_3$ to form GICs since the microwave-induced Joule heat can enhance the internal energy of gaseous reactant molecules and strengthen the kinetics of intercalation reaction [15].

In this paper, a microwave-assisted method for recovering spent graphite to construct a high-performance Fe_2O_3-MPGIC anode with a sandwich structure is proposed. The method utilizes microwave-induced thermal shock to instantaneously release interlayer impurities and intercalating agents, leading to an increase in interlayer spacing and the formation of a three-dimensional framework. $FeCl_3$ infiltrates the interlayer of the MPG through microwave-induced gasification intercalation and is converted in situ to Fe_2O_3 particles during heat treatment. The Fe_2O_3-MPGIC with a sandwich structure not only maintains the stability and electrical conductivity of graphite but also exhibits enhanced electrochemical properties. This innovative approach offers an efficient way to process spent graphite and fabricate anodes for high-capacity lithium storage.

2. Results and Discussion

2.1. Characterization

The microstructure of the samples was clearly visualized using SEM. As illustrated in Figure 1a, the SG exhibits clumping due to the presence of residual polymer binders. The surface of the SG is characterized by numerous aggregated particles, predominantly composed of residual binders, derived products of SEI, and decomposition products of the electrolyte [16,17]. MPG exhibits an obvious layered structure and expanded layer spacing in Figure 1b, which is attributed to the instant release of impurities and intercalants in the interlayer by microwave thermal shock. As shown in Figure 1c, the $FeCl_3$-MPGIC formed by microwave irradiation after mixing MPG with $FeCl_3$ has a larger layer spacing which is the result of the diffusion of iron chlorides, such as $[FeCl_4]^-$, into the interlayer. In addition, numerous small particles are uniformly distributed in the interlayer of the MPG framework, which are the iron chlorides deposited on the MPG framework after cooling. After heat treatment, these iron chlorides gradually convert to Fe_2O_3 and agglomerate into particles that are larger in size compared to the iron chlorides.

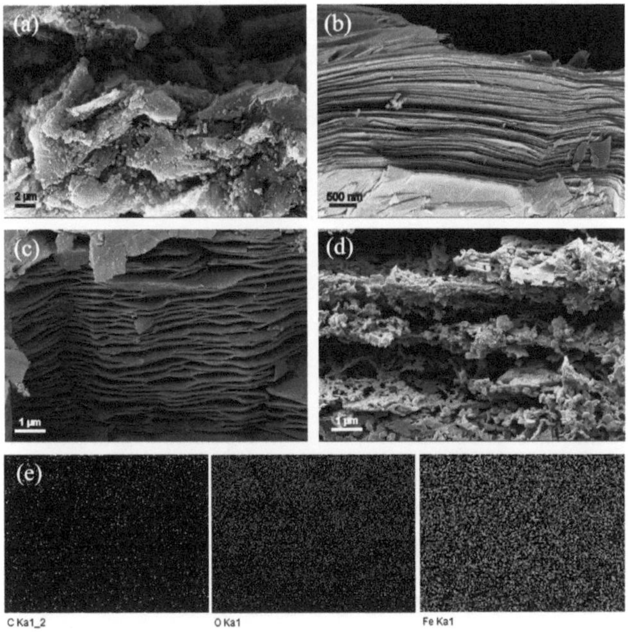

Figure 1. SEM of (**a**) SG, (**b**) MPG, (**c**) $FeCl_3$-MPGIC, and (**d**) Fe_2O_3-MPGIC as well as (**e**) EDS of Fe_2O_3-MPGIC.

The significant increase in particle size of Fe_2O_3 causes a further expansion of the layer spacing of MPG, so that the Fe_2O_3-MPGIC exhibits a distinct sandwich structure, as shown in Figure 1d. Fe_2O_3 particles in Fe_2O_3-MPGIC are widely and uniformly distributed within the interlayers of the MPG framework, which can restrict the volume expansion of Fe_2O_3 particles during the lithiation process, thereby ensuring the stability of the interface between Fe_2O_3 particles and the electrolyte. In addition, a large number of pores were etched by Fe^{3+} on the MPG framework during the heat treatment. The enhanced pore structure facilitates the infiltration and permeation of the electrolyte and promotes charge transfer on the Fe_2O_3 particles in the MPG. Figure 1e presents an energy dispersive spectroscopy (EDS) elemental map that vividly illustrates the distribution of carbon (C), oxygen (O), and iron (Fe) across the sample. The uniform dispersion of these elements is indicative of the Fe_2O_3 particles being homogeneously integrated within the MPG framework., confirming the successful creation of Fe_2O_3-MPGIC composite materials.

TEM was conducted for the in-depth research on the pore structure and surface defects in Fe_2O_3-MPGIC. As depicted in Figure 2, the Fe_2O_3-MPGIC structure is characterized by uniformly sized Fe_2O_3 particles that are firmly anchored to graphite sheets. This observation suggests that ferric chloride, during the intercalation process, is evenly dispersed within the interlayers of $FeCl_3$-MPGIC and subsequently aggregates and converts into

Fe_2O_3 particles upon heat treatment, resulting in a consistent particle distribution. In addition, visible surface defects (Figure 2a) and pore structure (Figure 2b) can be observed on the graphite sheet. The TEM results further demonstrate that the MPG framework is strongly etched by Fe^{3+} during the thermal treatment, resulting in the formation of abundant pore structures and surface defects. This will help increase the contact between the Fe_2O_3-MPGIC electrode and the electrolyte, thereby improving the Li ions and electrons transport efficiency.

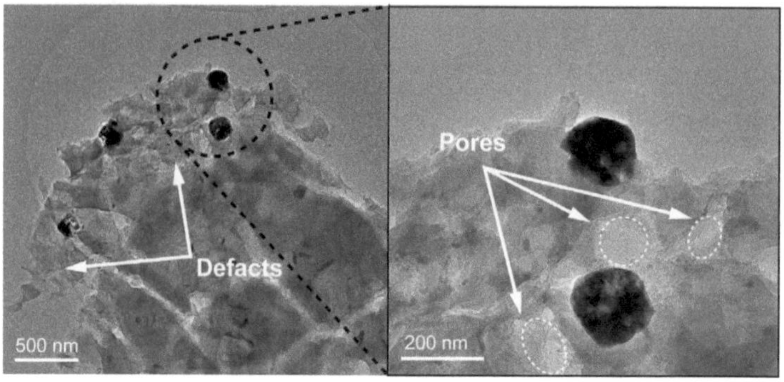

(**a**) (**b**)

Figure 2. TEM of (**a**) defects and (**b**) pore structure of Fe_2O_3-MPGIC.

The crystal structure of the obtained samples was studied through XRD. In general, GICs with a layer of intercalators every *n* layers of graphite are called stage-*n* GICs. According to the diffraction peaks, it can be concluded that $FeCl_3$ has been embedded into the graphite layers after microwave irradiation, and the $FeCl_3$-MPGIC exhibits a typical GIC structure. As shown in Figure 3a, obvious diffraction peaks are observed at 9.18°, 18.6°, 28.24°, and 50.58° in the $FeCl_3$-MPGIC sample, corresponding to the stage-1 intercalation structure of $FeCl_3$ with graphite [18]. And the diffraction peaks at 14.1° and 21° correspond to stage-2, while the diffraction peak at 16.9° corresponds to the stage-3 intercalation structures [19].

In addition, the effect of microwave irradiation on the sample structure was analyzed by comparing the XRD patterns of samples obtained at different irradiation times. Compared to $FeCl_3$-MPGIC irradiated for 3 min, the diffraction peaks located at 9.18°, 14.1°, and 18.6° are more pronounced in samples irradiated for more than 5 min. This indicates that after 5 min of irradiation, the samples obtained have a higher degree of intercalation structure. For the $FeCl_3$-MPGIC irradiated for 7 min, the intensity of the diffraction peak located at 14.1° diminishes, which may be attributed to the high temperature caused by prolonged irradiation inhibiting the kinetics of $FeCl_3$ intercalation into the graphite layers, thus inhibiting the formation of second-order intercalation structures. Therefore, 5 min of irradiation achieves the optimal intercalation structure of $FeCl_3$-MPGIC with the shortest irradiation time and minimum energy consumption.

Further analysis was conducted on the crystal structure of Fe_2O_3-MPGIC to reveal the effect of the mass ratio between $FeCl_3$ and MPG. As illustrated in Figure 3b, the Fe_2O_3-MPGIC samples exhibit obvious diffraction peaks at 24.15°, 33.16°, 35.63°, 40.86°, 49.46°, 54.07°, 62.44°, and 64°, respectively, corresponding to the (0 1 2), (1 0 4), (1 1 0), (1 1 3), (0 2 4), (1 1 6), and (2 1 4) crystal planes of Fe_2O_3, which are consistent with the standard diffraction peaks in PDF # 87-1166. This indicates that the iron chloride such as $[FeCl_4]^-$ in the interlayer of $FeCl_3$-MPGIC can be converted to Fe_2O_3 after heat treatment. The intensities of the Fe_2O_3 characteristic peaks in the XRD pattern of Fe_2O_3-MPGIC increased with the rise in the proportion of ferric chloride. This increase demonstrated that Fe_2O_3-MPGIC with various loadings was synthesized by mixing MPG with ferric chloride in different ratios. In addition, compared to the diffraction peak of $FeCl_3$-MPGIC, Fe_2O_3-MPGIC has almost no corresponding peaks of intercalation structure, which is mainly attributed to the significant increase in layer spacing caused by the aggregation of iron chloride and the growth of Fe_2O_3 during the conversion process, destroying the original form of crystal planes.

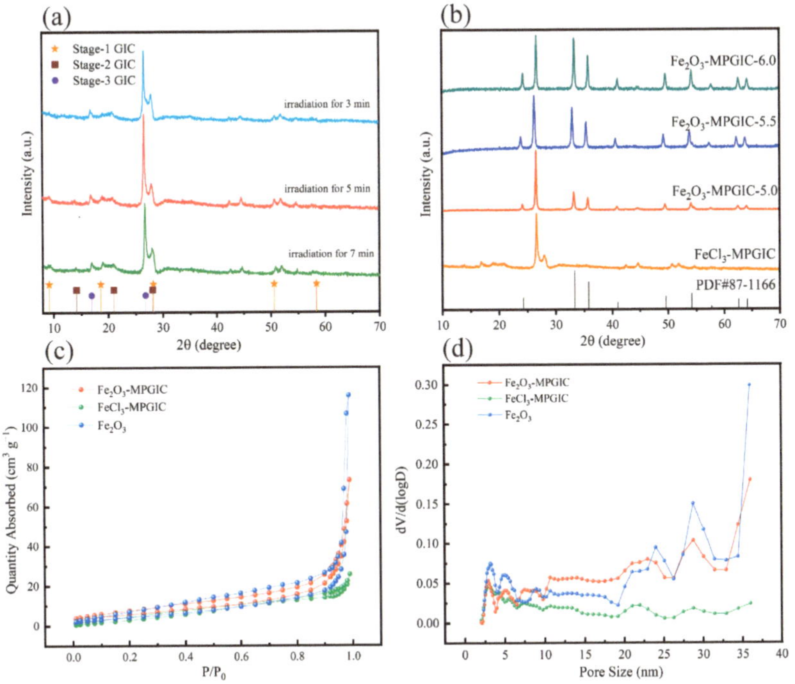

Figure 3. XRD patterns of (**a**) $FeCl_3$-MPGIC and (**b**) Fe_2O_3-MPGIC, and BET results (**c**,**d**) of Fe_2O_3, $FeCl_3$-MPGIC and Fe_2O_3-MPGIC.

N_2-adsorption desorption testing was performed to characterize the changes in the pore structure of Fe_2O_3-MPGIC and further investigate its surface characteristics. Figure 3c shows the isothermal curves of Fe_2O_3, $FeCl_3$-MPGIC and Fe_2O_3-MPGIC. Fe_2O_3-MPGIC has a higher adsorption capacity compared to Fe_2O_3 and $FeCl_3$-MPGIC ($0.1 < P/P_0 < 0.3$). This can be attributed to the expansion of the layer spacing of the MPG framework during the formation of Fe_2O_3 particles and the pore structure etched by Fe^{3+} during thermal treatment. The increased adsorption capacity also results in an increase in the specific surface area. As depicted in Table 1, the specific surface area of Fe_2O_3-MPGIC has been increased due to the growth and accumulation of Fe_2O_3 particle, coupled with the additional expansion of the MPG interlayers during the synthesis process. In addition, the changes in the specific surface area of Fe_2O_3-MPGIC under different material ratios are further analyzed. With an increase in the mass ratio of $FeCl_3$, the specific surface area of Fe_2O_3-MPGIC significantly decreases. This is primarily due to the increased presence of Fe^{3+}, which enhances the etching effect on the MPG framework surface during thermal treatment. Therefore, Fe_2O_3-MPGIC-6.0, with a lower specific surface area, exhibits a more abundant pore structure and more prominent surface defects.

The pore size distribution of the Fe_2O_3-MPGIC surface is analyzed to further study its pore structure. As depicted in Figure 3d, the pore structure of Fe_2O_3-MPGIC remains relatively consistent at small pore sizes ($\leq$3 nm) in comparison to $FeCl_3$-MPGIC. However, the amount of pore structure increases significantly for pores larger than 7 nm, particularly for pores larger than 20 nm, which is the predominant form of the Fe_2O_3-MPGIC pore structure. This suggests that the inclusion of Fe_2O_3-MPGIC enriches the pore structure of the samples, resulting in a broader distribution of pore sizes in the composite material. Rich mesopores (2~50 nm) and the larger surface area can provide ion transport channels, and reduce diffusion resistance [20]. Therefore, Fe_2O_3-MPGIC demonstrates the potential as a lithium storage material due to its excellent structural characteristics.

Table 1. Specific surface area of samples.

Samples	Specific Surface Area ($m^2\ g^{-1}$)
Fe_2O_3	17.408
MPG	13.267
$FeCl_3$-MPGIC-6.0	17.335
Fe_2O_3-MPGIC-5.0	29.435
Fe_2O_3-MPGIC-5.5	24.067
Fe_2O_3-MPGIC-6.0	22.017

The Fe_2O_3-MPGIC not only has a rich pore structure but also exhibits various defects. The Raman spectrum is utilized to investigate the defects and characteristics of the carbon structure in Fe_2O_3-MPGIC. As shown in Figure 4, Fe_2O_3-MPGIC and $FeCl_3$-MPGIC display distinct peaks at 1360 and 1590 cm^{-1}, which corresponds to the D and G bands, respectively, characterizing the carbon structure of the material. The crystallinity of the carbon materials in the samples can be compared by calculating the ID/IG values, which correspond to the disordered carbon structure (D band) and the graphitized carbon structure (G band). The I_D/I_G value of $FeCl_3$-MPGIC is 0.17, indicating a high degree of graphitization and minimal structural defects in the carbon structure of the material. The I_D/I_G values of the Fe_2O_3-MPGIC increase gradually with the increase in mass ratio, reaching 0.32, 0.65, and 0.82, respectively. This indicates that the etching effect of Fe^{3+} on the MPG framework becomes more pronounced as the volume of $FeCl_3$ increases. It indicates that the structural defects of Fe_2O_3-MPGIC significantly increase with the increase in mass ratio, which aligns with the presence of various defects observed in the microstructure of Fe_2O_3-MPGIC in the SEM results.

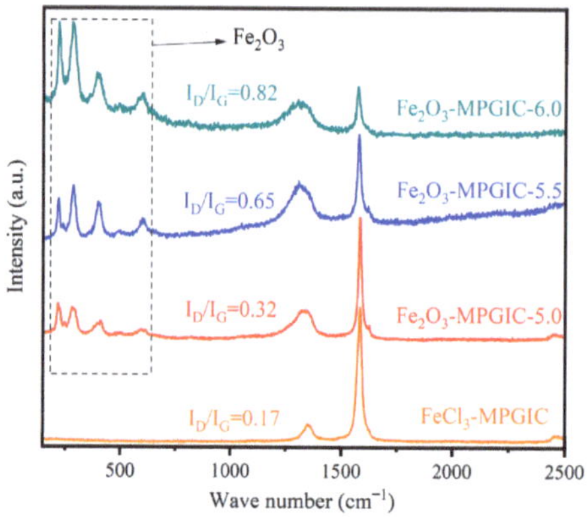

Figure 4. Raman spectra of $FeCl_3$-MPGIC and Fe_2O_3-MPGIC.

Moreover, Fe_2O_3-MPGIC exhibits significant peaks at 215, 275, 390, and 590 cm^{-1}, which correspond to the Fe-O bonds in the Fe_2O_3 structure [21]. As the mass ratio increases, the peak intensity also rises. On one hand, the increase in Fe_2O_3 content leads to an enhancement of the vibration signal of the Fe-O bond, On the other hand, the Fe_2O_3-MPGIC framework contains numerous structural defects, which expose more Fe_2O_3 particles and result in more significant molecular signals. However, few peaks are observed at 200~600 cm^{-1} in $FeCl_3$-MPGIC, mainly due to the substantial amount of $FeCl_3$ within the crystal structure of MPG, which is concealed by the complete graphite layer. As for Fe_2O_3-MPGIC, the increased structural disorder means the formation of numerous structural defects in the composite material. The presence of structural defects exposes more Fe_2O_3 particles and facilitates the infiltration of electrolyte. Therefore, Fe_2O_3-MPGIC exposes more active sites for Li^+ reactions, additionally increasing the lithium storage capacity of the Fe_2O_3-MPGIC electrode.

XPS was performed to further study the elemental composition and chemical properties of Fe_2O_3-MPGIC. As depicted in Figure 5a, the XPS survey spectrum of $FeCl_3$-MPGIC reveals the presence of notable amounts of C, Cl, and Fe elements, confirming the presence of iron chlorides such as $[FeCl_4]^-$ in the MPG framework. The presence of a small amount of the O element is attributed to the sample's affinity for atmospheric oxygen and moisture. The Fe_2O_3-MPGIC clearly contains C, O, and Fe elements, with almost no characteristic peaks of the Cl element, indicating that almost all iron chlorides are converted to Fe_2O_3 after thermal treatment. Furthermore, the more defined O 1s peak observed in the Fe_2O_3-MPGIC spectrum provides compelling evidence that Fe^{3+} ions, having released their chloride components, have combined with oxygen to form the Fe_2O_3 phase.

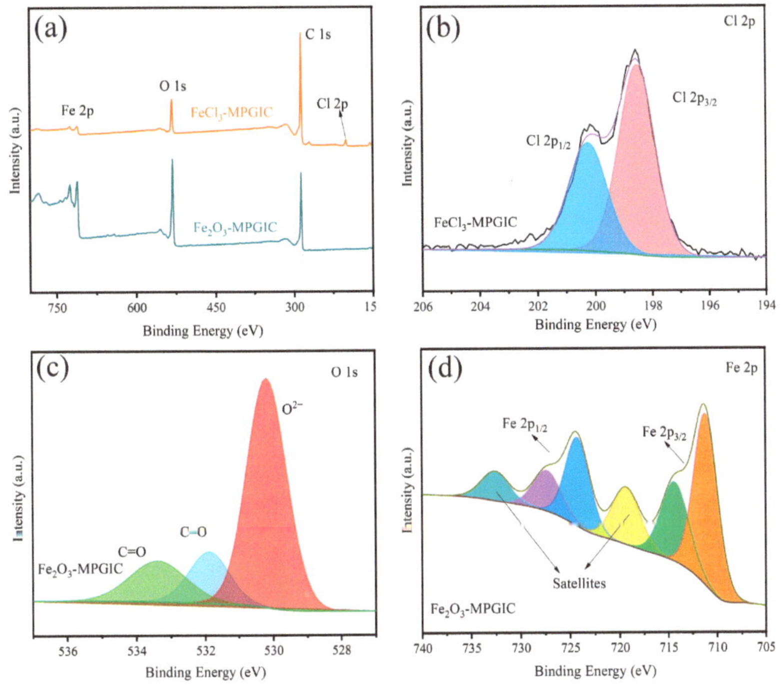

Figure 5. XPS spectra: (**a**) survey of $FeCl_3$-MPGIC and Fe_2O_3-MPGIC, (**b**) Cl 2p of $FeCl_3$-MPGIC, (**c**) O 1s and (**d**) Fe 2p of Fe_2O_3-MPGIC.

Further analysis of the spectra of each element verifies the form of existence of the corresponding components. As shown in Figure 5b, there are two obvious fitting peaks in the Cl 2p map of $FeCl_3$-MPGIC at 198.5 eV and 200.3 eV, which can be identified as Cl $2p_{3/2}$ and Cl $2p_{1/2}$ of chlorides, respectively. It indicates that the iron chlorides embedded in the MPG framework maintain a binding energy close to the crystalline state [14]. An in-depth analysis of the O1s spectrum of Fe_2O_3-MPGIC depicted in Figure 5c is conducted in order to verify the presence of Fe_2O_3. The fitted peaks located at 530.2, 531.9, and 533.4 eV correspond to the presence of three distinct chemical structures of O^{2-}, C-O, and C=O, respectively [22,23]. The significant O^{2-} fitting peak suggests the extensive binding forms of Fe^{3+} and O^{2-}, indicating the presence of Fe_2O_3.

As depicted in Figure 5d, the Fe 2p spectrum of Fe_2O_3-MPGIC exhibits two distinct peaks at 711.3 eV and 724.5 eV, corresponding to Fe 2p3/2 and Fe 2p1/2, respectively, which proves that an iron state exists in the form of trivalent Fe^{3+}. Additionally, two satellite peaks are observed at 719.5 eV and 732.7 eV, confirming the widespread presence of Fe_2O_3 [10]. Therefore, through XPS analysis, it was found that the iron chlorides in $FeCl_3$-MPGIC are widely present in the interlayer of the MPG framework. After thermal treatment, almost all the iron chlorides in the interlayer are converted into Fe_2O_3, thus constructing a sandwich structure of Fe_2O_3-MPGIC.

2.2. Electrochemical Performance

The electrochemical performance of the $FeCl_3$-MPGIC and Fe_2O_3-MPGIC electrodes for LIBs was evaluated by assembling coin-type half cells. The cyclic voltammetry (CV) tests were initially conducted at a scan rate of 0.1 mV s^{-1} within the voltage range of 0.01~3.0 V. Figure 6a shows the CV curve of $FeCl_3$-MPGIC electrode, with prominent peaks observed at 1.19, 0.45, and 0.01 V during the cathodic scanning of the first cycle. Among them, the first two peaks correspond to a multi-step lithiation process in which Li^+ are intercalated into the graphite interlayer and react with iron chlorides such as $[FeCl_4]^-$ within the interlayer. This process is highly irreversible, resulting in the cathodic peaks shifting from 1.19 V and 0.45 V to 1.39 V and 0.71 V, respectively, in subsequent cycles [14]. The cathodic peak located at 0.01 V corresponds to the process of embedding Li^+ into the graphite layer to form LiC_6, indicating that the MPG framework can contribute additional capacity for lithium storage. During the anode scanning phase, the peak located at 0.27 V corresponds to the decomposition of LiC_6 embedded in the interlayer and the removal of Li^+. Subsequently, the reverse process of lithium intercalation occurs at 1.52 V and 2.39 V, corresponding to 1.19 V and 0.45 V, respectively. After the highly irreversible reaction of lithium intercalation during the first cycle, the anodic scanning stage exhibits good reversibility. The analysis of the current intensity in the CV curve indicates that the $FeCl_3$-MPGIC electrode, which contains iron chlorides in the interlayer, has a lithium storage capacity comparable to that of the MPG framework.

After thermal treatment, the iron chlorides in the interlayer transformed into Fe_2O_3, resulting in significant variations in their capacity for lithium storage. As shown in Figure 6b, the CV curve of the Fe_2O_3-MPGIC electrode exhibits notable changes in both peak voltage and peak current. The cathodic peak corresponding to the reaction between Fe_2O_3 and Li^+ to form Fe^0 is located at 0.64 V, which is significantly different from the lithium storage potential of iron chlorides. During the anodic scan, the peak located at 1.69 V corresponds to the oxidation process of Fe^0 converting to Fe^{3+}. Similar to the $FeCl_3$-MPGIC electrode, the MPG framework also contributes a portion of lithium storage capacity to Fe_2O_3-MPGIC, as indicated by the peaks located at 0.01 V and 0.19 V in the CV curve. Comparing the CV curves of $FeCl_3$-MPGIC and Fe_2O_3-MPGIC, it is evident that the Fe_2O_3-MPGIC electrode exhibits a higher peak current. This suggests that Fe_2O_3 plays a significant role in the lithium storage capacity of the composite electrode, indicating that Fe_2O_3 dominates the overall lithium storage capacity.

The cycling performance of Fe_2O_3-MPGIC electrodes was evaluated through galvanostatic charge/discharge testing at a current density of 200 mA g^{-1}. As depicted in Figure 6c, the Fe_2O_3-MPGIC electrode exhibited superior cycling performance and lithium storage capacity compared to the $FeCl_3$-MPGIC electrode. With an increase in the mass ratio, the reversible capacity of the electrode significantly improved, indicating that a higher mass results in a larger volume of iron chlorides in the interlayer, thereby allowing more Fe_2O_3 particles to be loaded into the interlayer of the MPG framework after thermal treatment. Specifically, the initial discharge capacity of Fe_2O_3-MPGIC-6.0 is 1265.6 mAh g^{-1}, with an initial coulombic efficiency is 76.2%. The significant irreversibility is mainly attributed to the irreversible loss of active Li^+ caused by the formation of SEI. The capacity of the electrode slowly increases during the cycling process, indicating that the electrode becomes more thoroughly wetted in the electrolyte and is further activated as the cycling continues [24].

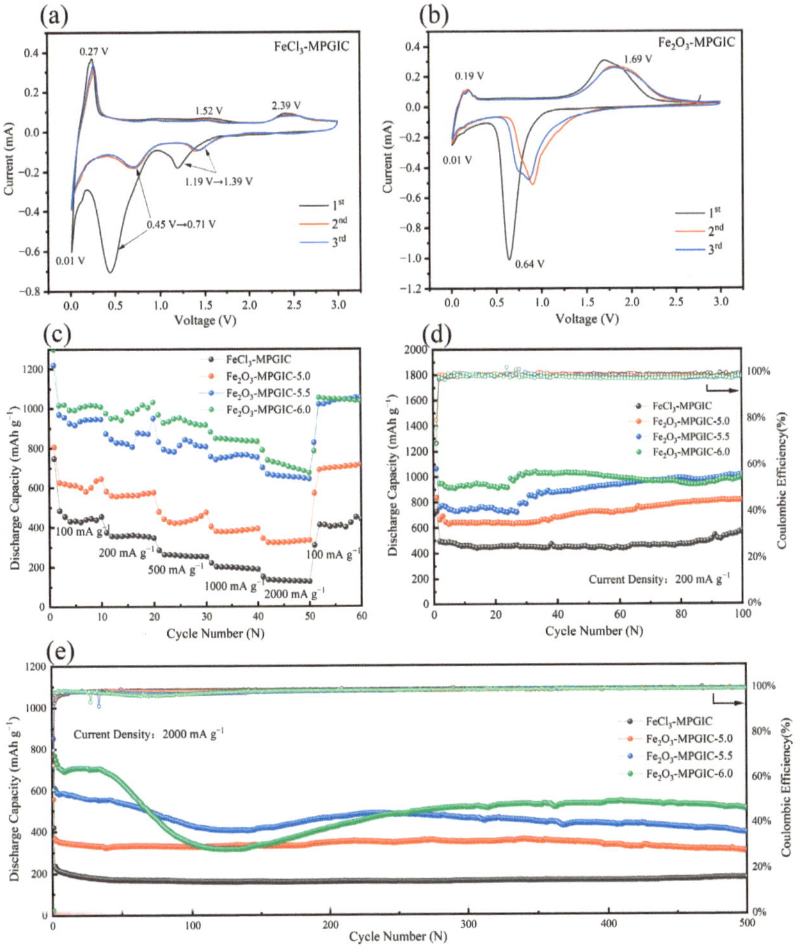

Figure 6. CV curves of (**a**) $FeCl_3$-MPGIC and (**b**) Fe_2O_3-MPGIC, (**c**) cycling performance at 200 mA g^{-1}, (**d**) rate performance, and (**e**) long-cycle performance at 2000 mA g^{-1}.

After 100 cycles, the Fe_2O_3-MPGIC-6.0 electrode maintains a high reversible capacity of 1000.6 mAh g^{-1}, significantly surpassing the 578.3 mAh g^{-1} capacity of $FeCl_3$-MPGIC, which demonstrates the excellent lithium storage capacity and cycling stability of the Fe_2O_3-MPGIC-6.0 electrode. The outstanding cycling performance is primarily attributed to the excellent electrical conductivity provided by the MPG framework and the mitigated volume expansion of Fe_2O_3 through the sandwich structure. In addition, the MPG framework etched by Fe^{3+} possesses a rich pore structure and surface defects, which can promote the electrolyte infiltration and Li^+ diffusion. Therefore, the sandwich structure of the Fe_2O_3-MPGIC electrode improves electron and Li^+ transport, enhancing the electrical contact and resulting in higher reversible capacity and excellent cycling stability.

Furthermore, the rate performance of the electrodes was compared within the current density range of 100 mA g^{-1} to 2000 mA g^{-1}. As shown in Figure 6d, Fe_2O_3-MPGIC-6.0 exhibits excellent rate performance, maintaining high reversible capacities of 1007.7, 1031.2, 915.9, 833, and 673.5 mAh g^{-1} at 100, 200, 500, 1000, and 2000 mA g^{-1}, respectively. The outstanding rate performance is not only 4–5 times higher than the capacity of $FeCl_3$-MPGIC electrode at high current densities, but also up to 10 times higher than that of commercial graphite (62.5 mAh g^{-1} at 2000 mA g^{-1}). When the current density returns to 100 mA g^{-1}, the capacity of Fe_2O_3-MPGIC-6.0 recovers to 1034 mAh g^{-1}, significantly higher than the capacity of $FeCl_3$-MPGIC electrodes (440 mAh g^{-1}) and commercial graphite (369 mAh g^{-1}).

Notably, the excellent rate performance of Fe_2O_3-MPGIC-6.0 draws support from the sandwich structure constructed by the MPG framework to provide reliable mechanical strength for the electrode, so as to ensure that excellent structural stability can be maintained under repeated lithiation and de-lithiation, especially at high current densities. Furthermore, the etching of Fe^{3+} on the MPG framework during the thermal treatment greatly enriches the pore structure and surface defects of the electrode, providing a shorter diffusion path for Li^+ and ensuring efficient charge transfer at high rates. As a result, the electrode exhibits a high reversible capacity, even at high current density. In addition, the excellent rate performance may also be contributed by the pseudo-capacitance on the electrode surface, which can promote rapid electrochemical reactions on the electrode surface [25].

Considering the excellent cycling and rate performance of the Fe_2O_3-MPGIC electrode, the long-term cycling performance of the electrode was further tested at a current density of 2000 mA g^{-1} to study the cycling stability. As shown in Figure 6e, at 2000 mA g^{-1}, Fe_2O_3-MPGIC-6.0 still exhibited excellent cycling stability after 500 cycles, maintaining a high reversible capacity of 504.4 mAh g^{-1}, and the coulombic efficiency remains close to 100% throughout the cycling period. In contrast, the $FeCl_3$-MPGIC electrode exhibited a reversible capacity of only 173.1 mAh g^{-1}. This demonstrates that the sandwich structure can effectively alleviate the volume expansion of Fe_2O_3 particles, thus ensuring the structural stability of the electrode. The rich pore structure and surface defects on the MPG framework also enhance the diffusion of Li^+ and facilitate charge transfer, allowing the electrode to maintain a high reversible capacity while exhibiting excellent cycling stability. However, the long cycle performance curve of Fe_2O_3-MPGIC-6.0 shows a decreasing and then an increasing trend. Upon careful comparison, a similar trend is observed for Fe_2O_3-MPGIC-5.5. This phenomenon may be attributed to the increased loading of Fe_2O_3, which stresses the structural stability of the MPG framework during high rate cycling without activation. While increased loads enhances specific capacity, it may simultaneously diminish the strength of composite materials [26]. Remarkably, as the charge/discharge cycling continues, the material appears to gradually adapt to this structural change, thereby recovering some of its electrochemical properties. This adaptation is corroborated in Figure 7, where the MPG framework still confines most of the Fe_2O_3 particles even after 400 cycles at 2000 mA g^{-1}.

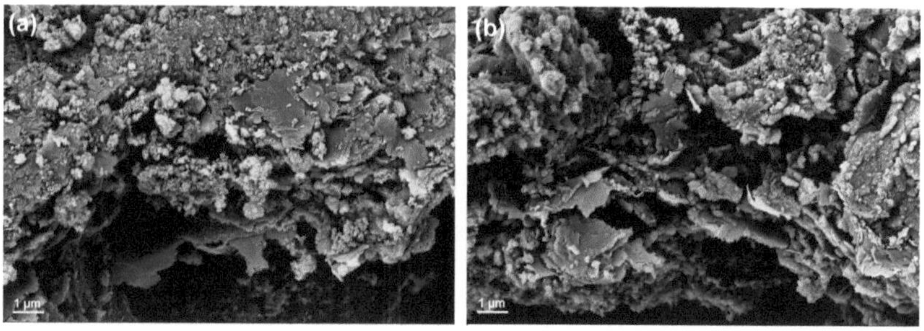

Figure 7. SEM of the anode (**a**) after 150 cycles and (**b**) after 400 cycles at 2000 mA g^{-1}.

EIS was conducted to analyze the electrochemical kinetic behavior of Fe_2O_3-MPGIC electrode. As shown in Figure 8a, the spectrum mainly includes a semicircle at medium and high frequency and a linear track at low frequency. The intercept between the line and the horizontal axis in the high-frequency region corresponds to the electrolyte related impedance (R_S), the semicircle diameter corresponds to the charge transfer impedance (R_{CT}), and the linear trajectory in the low-frequency region corresponds to the Weber impedance (Z_W), which can reflect the diffusion ability of Li^+ of the electrode [27]. Z-view software is employed to fit and analyze the curves. A solution resistance (R_S), charge transfer resistance (R_{CT}), and Weber impedance (W_O) constitute the equivalent circuit together which is connected in parallel with a constant phase element (CPE). Table 2 shows the fitted R_S and R_{CT} values. The R_S of Fe_2O_3-MPGIC is basically the same, indicating that the impedance of the electrode to the electrolyte is basically the same. The R_{CT} value of the electrode shows a trend of increasing and then decreasing with the increase in mass ratio. The increase in R_{CT} may be attributed to the aggregation of Fe_2O_3 particles which disrupts the layered structure of the MPG framework, thereby weakening the efficiency of charge transfer. The subsequent decrease in impedance is caused by the increase in pore structure and surface defects on the MPG framework, which facilitate the infiltration of electrolyte and enhance the charge transfer. The above results indicate that the rich pore structure and surface defects play an important role in reducing the internal resistance of batteries and promoting charge transfer.

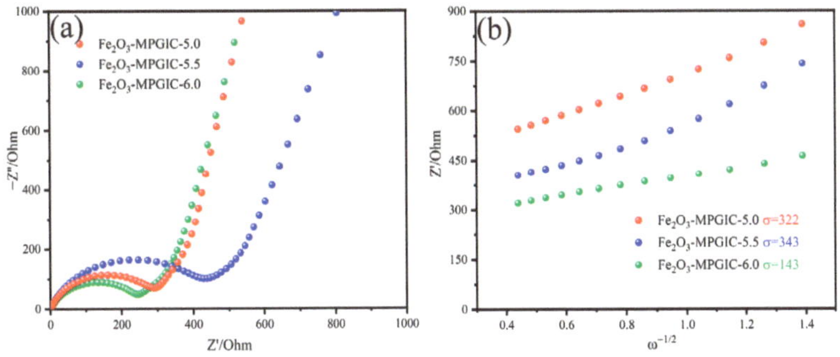

Figure 8. (**a**) EIS spectra and (**b**) the linear relation between Z′ and $\omega^{-1/2}$ of $FeCl_3$-MPGIC and Fe_2O_3-MPGIC.

Table 2. R_S and R_{CT} of Fe_2O_3-MPGIC electrode.

Samples	R_S	R_{CT}
Fe_2O_3-MPGIC-5.0	1.626	300
Fe_2O_3-MPGIC-5.5	1.764	442.3
Fe_2O_3-MPGIC-6.0	1.749	246.3

Furthermore, the following equation is used to fit the curve in the low-frequency region to characterize the Li^+ diffusion ability of the electrode:

$$D = R^2T^2/2A^2n^4F^4C^2\sigma^2 \quad (1)$$

$$Z' = R_S + R_{CT} + \sigma\omega^{-1/2} \quad (2)$$

where R is the gas constant, T represents the Kelvin temperature, usually taken as 298 K, A indicates the surface area of the electrode, n is the number of charge transfers during the redox process, F denotes the Faraday constant, C depicts the concentration of lithium ions, and σ symbolizes the Weber coefficient. By fitting the linear relationship between Z' and $\omega^{-1/2}$ (ω is the angular frequency), the value of σ can be obtained [28].

The value of σ^2 is inversely proportional to the diffusion coefficient D, where lower values σ corresponding to a larger Li^+ diffusion coefficient, reflecting a stronger Li^+ diffusion ability. As shown in Figure 8b, similar to the impedance results fitting from the medium and high frequency, the Li^+ diffusion ability of Fe_2O_3-MPGIC electrode also first decreases and then increases. This is mainly attributed to the increase in the mass ratio weakening the promotion of graphite on Li^+ diffusion, resulting in a reduced Li^+ diffusion ability. As the mass ratio increases, the etching effect of Fe^{3+} on the MPG framework is enhanced. The rich pore structure and surface defects promote electrolyte infiltration, which shortens the transfer path of Li^+ and enhances its diffusion ability. The EIS results indicate that the impedance of the Fe_2O_3-MPGIC electrode is generally at a lower level due to the excellent conductivity provided by the MPG framework. Moreover, the rich pore structure and surface defects on MPG facilitate the infiltration of electrolytes, further enhancing the diffusion of Li^+ and charge transfer efficiency.

In order to study the pseudo-capacitance effect of Fe_2O_3-MPGIC-6.0 electrode, CV tests were conducted at scan rates ranging from 0.1 to 2.0 mV s^{-1}. Figure 9a shows the CV curves of the electrodes at 0.1, 0.2, 0.5, 1.0, and 2.0 mV s^{-1}. The relationship between peak current (i) and scanning rate (v) follows the equation

$$i = av^b \tag{3}$$

where i and v refer to the peak current and scanning rate, respectively, and a and b are constants. The value of b, determined by the linear fitting slope of log v-log I, ranges from 0.5 to 1.0, where 0.5 indicates the battery process controlled by diffusion effect and 1.0 represents the capacitive process controlled by surface effect [29]. As shown in Figure 9b, the b values of the cathode peak (A) and anode peak (B) of Fe_2O_3-MPGIC-6.0 are 0.77 and 0.86, respectively, indicating that the reaction current is an interaction between battery process and capacitive process. This suggests that the Fe_2O_3-MPGIC-6.0 electrode is affected by the pseudo-capacitance effect during the cycling process. Based on the following equation, the contribution of pseudo-capacitance is further quantitatively analyzed:

$$i = k_1v + k_2v^{1/2} \tag{4}$$

where k_1, k_2 are the constant [30]. The reaction current is divided into pseudo-capacitance contribution (k_1v) and diffusion process contribution ($k_2v^{1/2}$), where the value of k_1 can be calculated through the linear relationship between $iv^{-1/2}$ and $v^{1/2}$. As shown in Figure 9c, when the scanning rate is 1.0 mV s^{-1}, the pseudo-capacitance contribution of the Fe_2O_3-MPGIC-6.0 electrode is 75.2%. As the scanning rate increased from 0.1 to 2 mV s^{-1} (Figure 9d), the pseudo-capacitance contribution of Fe_2O_3-MPGIC-6.0 electrode increased from 65.4% to 91.9%. This indicates that the pseudo-capacitance effect has a significant contribution to the lithium storage capacity, especially promoting the storage process of Li^+ during high-rate cycling. Combined with the excellent rate performance of Fe_2O_3-MPGIC-6.0 electrode, the sandwich structure provides structural stability while realizing enhanced charge transfer, and the remarkable pseudo-capacitance effect also effectively enhanced the lithium storage capacity of the electrode at a high current density.

Compared with the relevant studies on graphite-based Fe_2O_3 composite electrodes (Figure 10), the composite electrode in this study demonstrates superior rate capability while achieving a higher specific capacity. In summary, the obvious enhancement in the Li storage and cycling performance of Fe_2O_3-MPGIC electrodes can be attributed to the following reasons. (1) In Fe_2O_3-MPGIC electrodes, the MPG framework serves as a conductive network to accelerate electron transport, and the expanded interlayer spacing shortens the diffusion path of Li^+, improving charge transfer on the composite electrode, thereby ensuring high reversible capacity and excellent rate performance of the electrode. (2) Owing to the support and restriction provided by the MPG framework on Fe_2O_3 particles, the volume expansion of Fe_2O_3 particles during the cycle can be alleviated, while ensuring the structural stability of the electrode. (3) The etching of the MPG by Fe^{3+} during thermal treatment significantly enriches the surface defects and pores of Fe_2O_3-MPGIC electrode. This leads to a shorter diffusion path for Li^+, faster charge transfer, and an increase in lithium storage sites within the electrode, resulting in higher reversible capacity and excellent cycling performance. (4) The sandwich structure exhibits significant pseudo-capacitance, which ensures efficient charge transfer between electrodes even at high current densities, thereby improving the rate performance of Fe_2O_3-MPGIC electrodes.

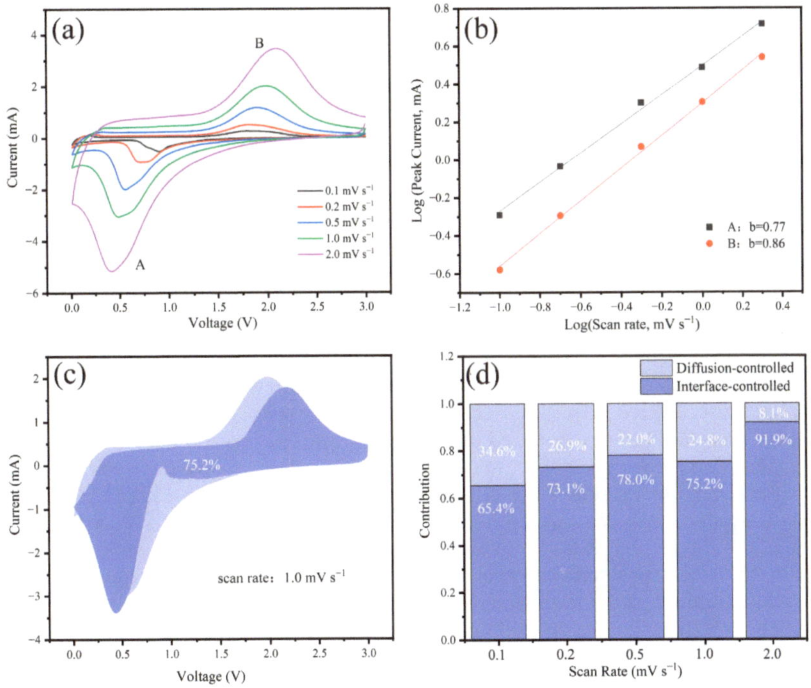

Figure 9. (**a**) CV curves of Fe_2O_3-MPGIC at scan rates ranging from 0.1 to 2.0 mV s^{-1}. (**b**) The fitted linear relationship between log i and log v. (**c**) Pseudo-capacitance contribution at a scan rate of 1.0 mV s^{-1}. (**d**) The ratio of pseudo-capacitance contribution of the Fe_2O_3-MPGIC electrode at various scan rates.

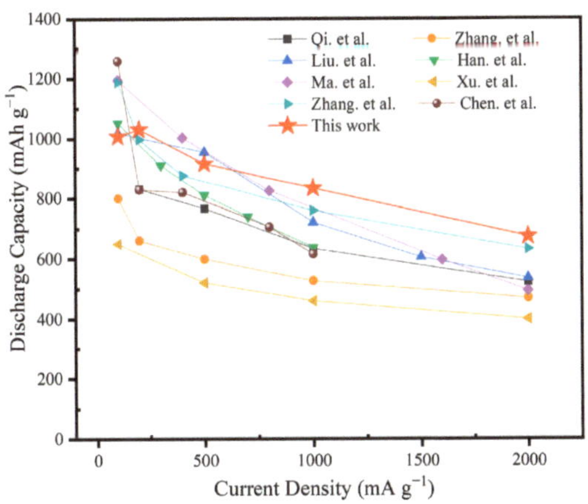

Figure 10. Comparison on lithium storage capacity between Fe_2O_3-MPGIC-6.0 electrode and other relevant studies [31–38].

3. Materials and Methods

3.1. Materials

Spent lithium-ion batteries were obtained from Shandong Jiuli Electronic Technology Co., Ltd., Zaozhuang, China. Potassium persulfate ($K_2S_2O_8$) was obtained from Shanghai McLean Biochemical Co., Ltd., Shanghai, China. Phosphoric acid (H_3PO_4) was obtained from Sinopharm Chemical Reagent Co., Ltd., Shanghai, China. Iron trichloride ($FeCl_3$) was purchased from Shanghai Aladdin Biochemical Technology Co., Ltd., Shanghai, China.

3.2. Acquisition of SG

The spent lithium batteries must be fully discharged prior to manual disassembly to prevent safety hazards. The disassembled cathode material was recycled and used for other research, while the anode material was immersed in a 1 mol L^{-1} KOH solution to separate it from the copper foil. Subsequently, the anode powder was rinsed with acid and ample deionized water to neutralize the pH and eliminate surface impurities. After drying and sieving, the SG was obtained with the copper foil nearly intact.

3.3. Preparation of MPG Framework

The synthesis of MPG has been implemented in previous work [39]. Firstly, microwave treatment was utilized to quickly and effectively remove residual impurities from the SG. The SG was placed in a customized industrial microwave oven and treated for 15 s per 15 s interval at a power of 1000 W for 5 min. Subsequently, oxidative intercalation of the SG after removing impurities was conducted under the synergistic effect of $K_2S_2O_8$ and H_3PO_4. After centrifugation, washing, and drying, the successfully intercalated SG was obtained and then treated for 5 min using the microwave process mentioned earlier. Due to the rapid decomposition and vaporization of the intercalant between graphite layers at high temperatures, part of the interlayer spacing of the SG was expanded, thus successfully obtaining an MPG framework with a clear and complete sandwich structure.

3.4. Preparation of Fe_2O_3-MPGIC

Due to the deliquescence of $FeCl_3$ in the air atmosphere, the sample preparation process needs to be carried out in a glove box filled with argon gas. Firstly, the MPG framework and $FeCl_3$ were mixed by grinding them together in a mass ratio of 1:5.0, 1:5.5, and 1:6.0 in the glove box. Subsequently, the mixture is loaded into a custom quartz reactor. The quartz reactor was placed in a microwave oven and purged with Ar gas at a flow rate of 200 mL min^{-1} to ensure an anhydrous and inert atmosphere.

Intermittent microwave irradiation was used for the preparation of $FeCl_3$-MPGIC. The detailed procedure is as follows. Firstly, the quartz reactor was irradiated at 900 W for 30 s, followed by intermittent irradiation (on: 9 s, off: 21 s) at 900 W for 3 to 7 min. Severe discharge phenomena can be observed in the reactor during this process. $FeCl_3$ was excited into a gaseous mixture of Cl_2 and ferric chloride, migrating toward the interlayer of the MPG framework. Intermittent irradiation can supply enough energy for the migration through excitation and discharge, while also preventing the direct decomposition of $FeCl_3$ due to excessively high temperatures. Secondly, the $FeCl_3$-MPGIC was obtained after multiple washings, vacuum filtration, and overnight drying. Lastly, Fe_2O_3-MPGIC was obtained by heating $FeCl_3$-MPGIC to 550 °C at a rate of 5 °C min^{-1} and then holding it for 3 h in a muffle furnace, wherein the iron chloride embedded in the interlayer was in situ converted into Fe_2O_3. According to the difference in mass ratio, the obtained samples were recorded as Fe_2O_3-MPGIC-5.0, Fe_2O_3-MPGIC-5.5, and Fe_2O_3-MPGIC-6.0. The mechanism of synthesizing Fe_2O_3-MPGIC is primarily illustrated in Figure 11.

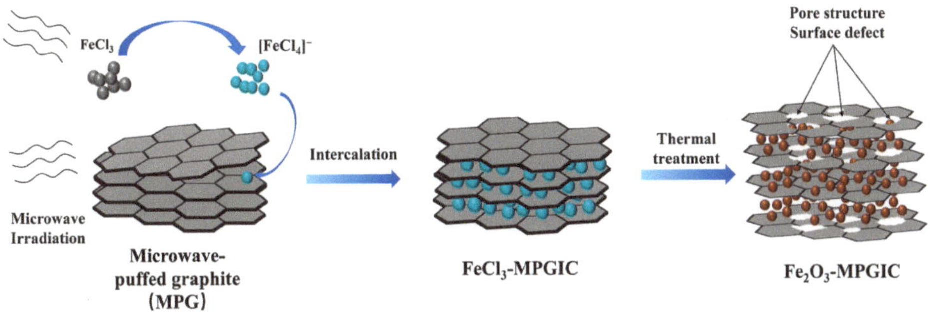

Figure 11. Schematic of synthesizing Fe_2O_3-MPGIC.

3.5. Characterization and Electrochemical Test

The morphology observation and microstructure measurements of as-prepared samples were carried out using scanning electron microscopy (SEM; SUPRA™55, ZEISS, Oberkochen, Germany) and transmission electron microscope (TEM; FEITalos F200x, FEI Company, Hillsboro, OR, USA). Moreover, the crystal structure was examined by X-ray diffraction (XRD; AERIS, PANalytical B.V., Almelo, The Netherlands) using Cu-K radiation, with 2θ values ranging from 8 to 70°. Furthermore, the carbon structure was investigated using Raman spectroscopy (HORIBA LabRAM HR Evolution, HORIBA Scientific, Paris, France) and the laser wavelength used in this experiment is 532 nm. X-ray photoelectron spectroscopy (XPS; Thermo Scientific K-Alpha, Thermo Fisher Scientific, Waltham, MA, USA) was employed to further analyze the chemical characteristics. In addition, the surface area and pore distribution of the sample were modeled by the N_2 adsorption–desorption isotherm (Quanta Chrome Instruments Co., Ltd., Boyertown, PA, USA).

The electrochemical properties of the samples were analyzed by testing assembled CR2025 half-cells. At first, the active substance, super P carbon and sodium carboxymethyl cellulose (CMC) were mixed and ground in the mass ratio of 7:2:1 with deionized water as solvent. Then, the working electrode sheet was uniformly coated on the copper foil and vacuum dried at 60 °C for 12 h. The slurry thickness of the obtained electrode sheet was approximately 20 μm, and the mass loading of the active substance was calculated to be about 1.0 to 1.2 mg cm^{-2}. When assembling the battery, the lithium metal sheet was used as the counter electrode, and the polyethylene porous composite film (Celgard 2400) was placed between it and the working electrode as the diaphragm. The electrolyte contained 1.0 mol L^{-1} $LiPF_6$ in a solution of ethylene carbonate (EC) and dimethyl carbonate (DMC) (1:1 Vol%). The entire assembly process should be completed in a glove box under high purity Ar atmosphere (water and oxygen values not exceeding 0.1 ppm).

The galvanostatic charge/discharge performance test from 0.01 to 3 V at room temperature were performed on a LAND battery testing system (CT3001A, Wuhan Lanhe, Wuhan, China). Moreover, cyclic voltammetry (CV) was conducted on an electrochemical workstation (CHI 660e, Shanghai Chenhua, Shanghai, China), in which the test voltage was in the range of 0.01~3 V (versus Li/Li$^+$), and the scan rate ranged from 0.1 mV s^{-1} to 2 mV s^{-1}. In addition, electrochemical impedance spectroscopy (EIS) was carried out on the CHI 660e electrochemical workstation with a voltage amplitude of 10 mV and a frequency range of 100 kHz to 0.01 Hz. Z-view software (Zview2, Scribner Associates, Inc., Southern Pines, NC, USA) is employed to fit and analyze the curves.

4. Conclusions

This work not only develops green, economic, and high-performance anodes for LIBs, which have shown great potential in promoting the practical applications, but also provides a valuable reference for the resource disposal of SG, which is of great significance in the resource closed-loop management in the LIB industry. Fe_2O_3-MPGIC anode leverages the high theoretical specific capacity of Fe_2O_3 while maintaining the high electrical conductivity and structural stability of graphite. Additionally, the etching effect enriches the graphite layers with a porous structure, which provides more lithium-ion transport channels, enhancing the long cycles and rate performance of Fe_2O_3-MPGIC. In electrochemical tests, Fe_2O_3-MPGIC delivered a high reversible capacity of 1000.6 mAh g^{-1} at 200 mA g^{-1} after 100 cycles and exhibited excellent rate performance, retaining a capacity of 673 mAh g^{-1} when the current density was increased to 2000 mA g^{-1}. It also presented obvious cycling stability, maintaining a capacity of 504.4 mAh g^{-1} at 2000 mA g^{-1} after 500 cycles. However, while increased Fe_2O_3 content can enhance the capacity, it may concurrently introduce challenges that could compromise the electrode's long-term stability. Achieving an optimal balance between the mass fraction of Fe_2O_3 and the preservation of structural stability stands as a pivotal issue that must be addressed in the development of future electrodes of this category.

Author Contributions: Conceptualization, J.S.; investigation, P.J. and W.C.; writing—original draft preparation, W.C., P.J. and J.S.; writing—review and editing, W.C. and J.S.; project administration, W.W., Z.S., Z.W., X.Z. and Y.M.; funding acquisition, J.S. All authors have read and agreed to the published version of the manuscript.

Funding: This research was funded by National Natural Science Foundation of China (No. 52176203 and 52370140), the Young Scholars Program of Shandong University (2018WLJH75), and Natural Science Foundation of Shandong Province (ZR2019MEE035).

Institutional Review Board Statement: Not applicable.

Informed Consent Statement: Not applicable.

Data Availability Statement: Data are contained within the article.

Conflicts of Interest: The authors declare no conflicts of interest.

References

1. Yi, C.; Zhou, L.; Wu, X.; Sun, W.; Yi, L.; Yang, Y. Technology for Recycling and Regenerating Graphite from Spent Lithium Batteries. *Chin. J. Chem. Eng.* **2021**, *39*, 37–50. [CrossRef]
2. Lie, J.; Liu, J.-C. Closed-Vessel Microwave Leaching of Valuable Metals from Spent Lithium-Ion Batteries (LIBs) Using D Function Leaching Agent: Ascorbic Acid. *Sep. Purif. Technol.* **2021**, *266*, 118458. [CrossRef]
3. Qi, Y.; Meng, F.; Yi, X.; Shu, J.; Chen, M.; Sun, Z.; Sun, S.; Xiu, F.-R. A Novel and Efficient Ammonia Leaching Metho Recycling Waste Lithium Ion Batteries. *J. Clean. Prod.* **2020**, *251*, 119665. [CrossRef]
4. Hu, Q.; Luo, Z.; Zhou, H.; Cao, Z. High-Efficiency Selective Leaching of Valuable Metals from Spent Lithium-Ion Batteries: Ef of Na2S2O8 on the Leaching of Metals. *Waste Manag.* **2023**, *167*, 204–212. [CrossRef] [PubMed]
5. Yang, J.; Fan, E.; Lin, J.; Arshad, F.; Zhang, X.; Wang, H.; Wu, F.; Chen, R.; Li, L. Recovery and Reuse of Anode Graphite f Spent Lithium-Ion Batteries via Citric Acid Leaching. *ACS Appl. Energy Mater.* **2021**, *4*, 6261–6268. [CrossRef]
6. Natarajan, S.; Aravindan, V. An Urgent Call to Spent LIB Recycling: Whys and Wherefores for Graphite Recovery. *Adv. En Mater.* **2020**, *10*, 2002238. [CrossRef]
7. Gao, Y.; Zhang, J.; Jin, H.; Liang, G.; Ma, L.; Chen, Y.; Wang, C. Regenerating Spent Graphite from Scrapped Lithium-Ion Ba by High-Temperature Treatment. *Carbon* **2022**, *189*, 493–502. [CrossRef]
8. Jana, S.S.; Kumari, S.; Maiti, T. Synergistic Improvement in Thermoelectric Performance by Forming Environmentally Be Transition Metal Oxide Composites with Graphite. *J. Phys. Chem. C* **2023**, *127*, 12849–12856. [CrossRef]
9. Bracamonte, M.V.; Primo, E.N.; Luque, G.L.; Venosta, L.; Bercoff, P.G.; Barraco, D.E. Lithium Dual Uptake Anode Mate Crystalline Fe_3O_4 Nanoparticles Supported over Graphite for Lithium-Ion Batteries. *Electrochim. Acta* **2017**, *258*, 192- [CrossRef]
10. Guo, J.; Wang, S.; Sun, X.; Guo, R.; Xiong, K.; Ling, R.; Shen, K.; Cui, J.; Ma, H.; Zhou, J.; et al. Uniform α-Fe_2O_3 Nanopart with Narrow Gap Immobilized on CNTs through N-Doped Carbon as High-Performance Lithium-Ion Batteries Anode. *Ce Int.* **2021**, *47*, 15743–15749. [CrossRef]
11. Chen, S.; Wu, F.; Wang, H.; Gao, S.; Chen, J.; Chen, Z.; Fu, J. N-Doped Graphitized Carbon-Coated Fe_2O_3 Nanoparticles in Hi Graphitized Carbon Hollow Fibers for Advanced Lithium-Ion Batteries Anodes. *Electrochim. Acta* **2023**, *467*, 143032. [Cross

Wang, F.; Yi, J.; Wang, Y.; Wang, C.; Wang, J.; Xia, Y. Graphite Intercalation Compounds (GICs): A New Type of Promising Anode Material for Lithium-Ion Batteries. *Adv. Energy Mater.* **2014**, *4*, 1300600. [CrossRef]
Seiler, S.; Halbig, C.E.; Grote, F.; Rietsch, P.; Börrnert, F.; Kaiser, U.; Meyer, B.; Eigler, S. Effect of Friction on Oxidative Graphite Intercalation and High-Quality Graphene Formation. *Nat. Commun.* **2018**, *9*, 836. [CrossRef]
Li, D.; Zhu, M.; Chen, L.; Chen, L.; Zhai, W.; Ai, Q.; Hou, G.; Sun, Q.; Liu, Y.; Liang, Z.; et al. Sandwich-Like $FeCl_3$ @C as High-Performance Anode Materials for Potassium-Ion Batteries. *Adv. Mater. Inter.* **2018**, *5*, 1800606. [CrossRef]
Wang, Z.; Yu, C.; Huang, H.; Guo, W.; Zhao, C.; Ren, W.; Xie, Y.; Qiu, J. Energy Accumulation Enabling Fast Synthesis of Intercalated Graphite and Operando Decoupling for Lithium Storage. *Adv. Funct. Mater.* **2021**, *31*, 2009801. [CrossRef]
Fan, W.; Zhang, J.; Ma, R.; Chen, Y.; Wang, C. Regeneration of Graphite Anode from Spent Lithium-Ion Batteries via Microwave Calcination. *J. Electroanal. Chem.* **2022**, *908*, 116087. [CrossRef]
Yi, C.; Yang, Y.; Zhang, T.; Wu, X.; Sun, W.; Yi, L. A Green and Facile Approach for Regeneration of Graphite from Spent Lithium Ion Battery. *J. Clean. Prod.* **2020**, *277*, 123585. [CrossRef]
Sun, Y.; Han, F.; Zhang, C.; Zhang, F.; Zhou, D.; Liu, H.; Fan, C.; Li, X.; Liu, J. $FeCl_3$ Intercalated Microcrystalline Graphite Enables High Volumetric Capacity and Good Cycle Stability for Lithium-Ion Batteries. *Energy Technol.* **2019**, *7*, 1801091. [CrossRef]
Zhang, C.; Ma, J.; Han, F.; Liu, H.; Zhang, F.; Fan, C.; Liu, J.; Li, X. Strong Anchoring Effect of Ferric Chloride-Graphite Intercalation Compounds ($FeCl_3$-GICs) with Tailored Epoxy Groups for High-Capacity and Stable Lithium Storage. *J. Mater. Chem. A* **2018**, *6*, 17982–17993. [CrossRef]
Yang, K.; Fan, Q.; Song, C.; Zhang, Y.; Sun, Y.; Jiang, W.; Fu, P. Enhanced Functional Properties of Porous Carbon Materials as High-Performance Electrode Materials for Supercapacitors. *Green Energy Resour.* **2023**, *1*, 100030. [CrossRef]
Reddy, M.V.; Yu, T.; Sow, C.H.; Shen, Z.X.; Lim, C.T.; Subba Rao, G.V.; Chowdari, B.V.R. α-Fe_2O_3 Nanoflakes as an Anode Material for Li-Ion Batteries. *Adv. Funct. Mater.* **2007**, *17*, 2792–2799. [CrossRef]
Song, J.; Maulana, A.Y.; Kim, H.; Yun, B.; Gim, H.; Jeong, Y.; An, N.; Futalan, C.M.; Kim, J. N-Doped Graphitic Carbon Coated Fe_2O_3 Using Dopamine as an Anode Material for Sodium-Ion Batteries. *J. Alloys Compd.* **2022**, *921*, 166082. [CrossRef]
Zhong, M.; Guo, D.; Meng, X.; Bian, L.; Song, Y.; Sun, X.; Liu, X. Heterostructured Polypyrrole/Hybrid Iron Oxide Composite Film as Highly Stable Anode for Pseudocapacitors. *J. Power Sources* **2021**, *513*, 230550. [CrossRef]
Kong, H.; Wu, Y.; Hong, W.; Yan, C.; Zhao, Y.; Chen, G. Structure-Designed Synthesis of Cu-Doped Co_3O_4@N-Doped Carbon with Interior Void Space for Optimizing Alkali-Ion Storage. *Energy Storage Mater.* **2020**, *24*, 610–617. [CrossRef]
Zhao, B.; Liu, Q.; Chen, Y.; Liu, Q.; Yu, Q.; Wu, H.B. Interface-Induced Pseudocapacitance in Nonporous Heterogeneous Particles for High Volumetric Sodium Storage. *Adv. Funct. Mater.* **2020**, *30*, 2002019. [CrossRef]
Luo, Z.; Quan, J.; Ding, T.; Xu, B.; Li, W.; Mao, Q.; Ma, W.; Li, M.; Xiang, H.; Zhu, M. Recent Advances in Iron Oxide/Graphene Composites for Flexible Supercapacitors. *J. Alloys Compd.* **2024**, *980*, 173614. [CrossRef]
Liu, Y.; Miao, X.; Zhang, X.; Chen, S.; Chen, Y.; Cheng, J.; Wang, W.; Zhang, Y. Targeted Interfacial Anchoring and Wrapping of Fe3O4 Nanoparticles onto Graphene by PPy-Derived-Carbon for Stable Lithium-Ion Battery Anodes. *Mater. Res. Bull.* **2019**, *111*, 170–176. [CrossRef]
Cong, B.; Hu, Y.; Sun, S.; Wang, Y.; Wang, B.; Kong, H.; Chen, G. Metal–Organic Framework Derived Amorphous VO_x Coated Fe_3O_4/C Hierarchical Nanospindle as Anode Material for Superior Lithium-Ion Batteries. *Nanoscale* **2020**, *12*, 16901–16909. [CrossRef]
Hou, T.; Sun, X.; Xie, D.; Wang, M.; Fan, A.; Chen, Y.; Cai, S.; Zheng, C.; Hu, W. Mesoporous Graphitic Carbon-Encapsulated Fe_2O_3 Nanocomposite as High-Rate Anode Material for Sodium-Ion Batteries. *Chem. A Eur. J* **2018**, *24*, 14786–14793. [CrossRef]
Wang, Q.; Wang, B.; Zhang, Z.; Zhang, Y.; Peng, J.; Zhang, Y.; Wu, H. Tailoring Yolk–Shell FeP@carbon Nanoboxes with Engineered Void Space for Pseudocapacitance-Boosted Lithium Storage. *Inorg. Chem. Front.* **2018**, *5*, 2605–2614. [CrossRef]
Qi, X.; Zhang, H.-B.; Xu, J.; Wu, X.; Yang, D.; Qu, J.; Yu, Z.-Z. Highly Efficient High-Pressure Homogenization Approach for Scalable Production of High-Quality Graphene Sheets and Sandwich-Structured α-Fe_2O_3/Graphene Hybrids for High-Performance Lithium-Ion Batteries. *ACS Appl. Mater. Interfaces* **2017**, *9*, 11025–11034. [CrossRef] [PubMed]
Zhang, M.; Liu, E.; Cao, T.; Wang, H.; Shi, C.; Li, J.; He, C.; He, F.; Ma, L.; Zhao, N. Sandwiched Graphene Inserted with Graphene-Encapsulated Yolk–Shell γ-Fe $_2O_3$ Nanoparticles for Efficient Lithium Ion Storage. *J. Mater. Chem. A* **2017**, *5*, 7035–7042. [CrossRef]
Liu, S.; Wang, R.; Liu, M.; Luo, J.; Jin, X.; Sun, J.; Gao, L. Fe_2O_3 @SnO_2 Nanoparticle Decorated Graphene Flexible Films as High-Performance Anode Materials for Lithium-Ion Batteries. *J. Mater. Chem. A* **2014**, *2*, 4598–4604. [CrossRef]
Han, T.; Wei, Y.; Jin, X.; Jiu, H.; Zhang, L.; Sun, Y.; Tian, J.; Shang, R.; Hang, D.; Zhao, R. Hydrothermal Self-Assembly of α-Fe2O3 Nanorings@graphene Aerogel Composites for Enhanced Li Storage Performance. *J. Mater. Sci.* **2019**, *54*, 7119–7130. [CrossRef]
Ma, J.; Kong, Y.; Liu, S.; Li, Y.; Jiang, J.; Zhou, Q.; Huang, Y.; Han, S. Flexible Phosphorus-Doped Graphene/Metal–Organic Framework-Derived Porous Fe_2O_3 Anode for Lithium-Ion Battery. *ACS Appl. Energy Mater.* **2020**, *3*, 11900–11906. [CrossRef]
Xu, B.; Guan, X.; Zhang, L.Y.; Liu, X.; Jiao, Z.; Liu, X.; Hu, X.; Zhao, X.S. A Simple Route to Preparing γ-Fe_2O_3/RGO Composite Electrode Materials for Lithium Ion Batteries. *J. Mater. Chem. A* **2018**, *6*, 4048–4054. [CrossRef]
Zhang, G.; Shi, Y.; Wang, H.; Jiang, L.; Yu, X.; Jing, S.; Xing, S.; Tsiakaras, P. A Facile Route to Achieve Ultrafine Fe_2O_3 Nanorods Anchored on Graphene Oxide for Application in Lithium-Ion Battery. *J. Power Sources* **2019**, *416*, 118–124. [CrossRef]

38. Chen, Y.; Zhang, S.; Feng, Y.; Yang, G.; Ji, H.; Miao, X. Characterization of Fe_2O_3/Graphene Composites Synthesized Using Situ Reaction of Inexpensive Graphite Oxide and $FeCl_3$. *ChemElectroChem* **2020**, *7*, 5013–5020. [CrossRef]
39. Jia, P.; Sun, J.; Li, S.; Wang, W.; Song, Z.; Zhao, X.; Mao, Y. A Feasible Recycling Route for Spent Graphite: Microwave-Ass Puffing with Fe_2O_3 Loading to Construct High-Performance Anode for Lithium-Ion Batteries. *Mater. Today Sustain.* 25, 100620. [CrossRef]

molecules

MDPI

cle

ı Electrochemical Immunosensor Based on ıitosan–Graphene Nanosheets for Aflatoxin B1 Detection Corn

ai Zhang [1,2,3], Caizhang Wu [3,*], Zhike Zhao [3] and Kun Xu [1,2,3,*]

[1] Key Laboratory of Grain Information Processing and Control (Henan University of Technology), Ministry of Education, Zhengzhou 450001, China; z.shuai69@stu.haut.edu.cn
[2] Henan Key Laboratory of Grain Photoelectric Detection and Control, Henan University of Technology, Zhengzhou 450001, China
[3] College of Electrical Engineering, Henan University of Technology, Zhengzhou 450001, China; zhaozhike@haut.edu.cn
* Correspondence: wucaizhang@haut.edu.cn (C.W.); k.xu018@haut.edu.cn (K.X.)

Abstract: We reported a highly efficient electrochemical immunosensor utilizing chitosan–graphene nanosheets (CS-GNs) nanocomposites for the detection of aflatoxin B_1 (AFB_1) in corn samples. The CS-GNs nanocomposites, serving as a modifying layer, provide a significant specific surface area and biocompatibility, thereby enhancing both the electron transfer rate and the efficiency of antibody immobilization. The electrochemical characterization was conducted utilizing both differential pulse voltammetry (DPV) and electrochemical impedance spectroscopy (EIS). Moreover, the antibody concentration, pH, antibody immobilization time, and immunoreaction time, were optimized. The results showed that the current change (ΔI) before and after the immunoreaction demonstrated a strong linear relationship ($R^2 = 0.990$) with the AFB_1 concentration, as well as good specificity and stability. The linear range extended from 0.05 to 25 ng/mL, with a detection limit of 0.021 ng/mL ($S/N = 3$). The immunosensor exhibited a recovery rate ranging from 97.3% to 101.4% in corn samples, showing a promising performance using an efficient method, and indicating a remarkable prospect for the detection of fungal toxins in grains.

Keywords: electrochemical immunosensor; aflatoxin B_1; chitosan; graphene nanosheets; nanocomposites

1. Introduction

Aflatoxin is a highly toxic furanocoumarin derivative produced by Aspergillus flavus and Aspergillus parasiticus. It is frequently encountered in moldy grains such as rice, soybeans, and peanuts [1]. Aflatoxin B_1 (AFB_1) is known for its extreme toxicity and is widely acknowledged as one of the most potent carcinogens to date [2]. The permissible levels of aflatoxin B_1 in various food items that are highly prone to contamination are stipulated by the Chinese food hygiene standards. For corn, peanuts, and peanut oil, the permitted level of aflatoxin B_1 is set at $\leq$20 μg/kg [3]. The regulation of the levels of AFB_1 in grain, peanuts, and their products has been established by the European Union and other nations. For peanuts intended for immediate consumption, the permissible level of AFB_1 must not exceed 2 μg/kg. Additionally, for import purposes, peanuts utilized as food ingredients must not exceed an AFB_1 content of 8 μg/kg [4].

Several techniques have been reported for the detection and analysis of AFB_1, including high-performance liquid chromatography (HPLC) [5], enzyme-linked immunosorbent assay (ELISA) [6], and thin-layer chromatography (TLC) [7]. Although these aforementioned methods offer high sensitivity and accuracy, they also come with distinct limitations. These methods necessitate experimenters to possess proficient operational skills, as well as expensive equipment, materials, and intricate sample preparation procedures [8]. In addition to the aforementioned methods, in recent years, emerging techniques such as electrochemistry [9], fluorescence [10], chemiluminescence [11], optical fibers [12], and surface plasmon resonance (SPR) [13] have also been employed for the detection of AFB_1 concentration.

Researchers have demonstrated significant interest in electrochemical immunosensors as a promising technology for detecting AFB_1 in food. Their attractiveness stems from their ability to provide high sensitivity, cost-effectiveness, and rapid response times. In recent years, carbon nanomaterials have emerged as a prominent focus of research due to their exceptional characteristics of high specific surface area [14] and excellent conductivity [15]. These properties render them widely applicable in the electrochemical detection of AFB_1, leading to significant improvements in both the accuracy and efficiency of the detection process. Shi et al. magnetically stirred a dispersion of PVP, ascorbic acid, and COOH-GO at 90 °C for 10 min. Subsequently, they mixed the dispersion with a HAuCl4 solution for 3 h. Through this procedure, they obtained Au-COOH-GO nanocomposites to fabricate electrochemical immunosensors for AFB_1 detection [16]. Srivastava et al. synthesized graphene oxide (GO) using the modified Hummers method. Initially, graphite powder was pre-oxidized by reacting it with a mixture of H_2SO_4, $K_2S_2O_8$, and P_2O_5 for 4 h at 80 °C. Subsequently, it was stirred in H_2SO_4/H_3PO_4, followed by the addition of KMnO4, and the mixture was stirred for 15 h at 50 °C to obtain GO. The synthesized GO was then utilized in the fabricate electrochemical immunosensors for detecting AFB_1 [17]. Bhardwaj et al. synthesized GO also using the modified Hummers method and subsequently subjected the resulting GO to hydrothermal treatment at 200 °C for 10 h to obtain GQDs. The GQDs were synthesized to fabricate electrochemical immunosensors for the detection of AFB_1 [18]. Although the electrochemical methods developed above demonstrate good performance in detecting AFB_1, the fabrication process of the nanocomposites is intricate and the preparation efficiency is relatively low.

In conclusion, the application of graphene and its modified materials in electrochemical immunosensors has provided a new approach for the detection of AFB_1. These studies have laid the foundation for the development of more sensitive and highly selective methods for detecting AFB_1, holding significant importance in the field of food safety [19,20]. Additionally, graphene has significant advantages due to its large specific surface area and high conductivity for electrochemical biosensors. However, it is naturally hydrophobic and tends to aggregate in hydrophilic solvents [21]. Due to its excellent film-forming properties and effective dispersion effect, chitosan has emerged as a popular dispersant for graphene [22]. Furthermore, chitosan exhibits excellent biocompatibility and the ability to immobilize various functional groups, rendering it a desirable substrate for the immobilization of biosensors [23]. In this work, CS-GNs nanocomposites were synthesized and immobilized on glass carbon electrodes (GCE). Chitosan, known for its biocompatibility, was utilized to immobilize the AFB_1 antibody, thereby enhancing the specificity of the sensor. Through this approach, a straightforward, effective, and exceptionally precise electrochemical immunosensor was constructed and utilized for the detection of AFB_1 concentration in actual corn samples.

2. Results and Discussion

2.1. Characterization of CS-GNs Nanocomposites

Raman spectroscopy is a powerful tool for characterizing the structure and properties of graphene [24]. We conducted an analysis of GN samples using the Raman spectroscopy technique. As shown in Figure 1, three main peaks are observed, namely the G band, 2D band, and D band. The G band is located at approximately 1579 cm^{-1}, representing the E2g vibrational mode within the GNs lattice, corresponding to the in-plane vibrations between carbon atoms. The 2D band is located at around 2717 cm^{-1}, representing the double resonance mode between the layers of GNs. The D band is situated at approximately 1355 cm^{-1}, indicating structural distortions caused by defects and impurities within GNs. It is notable that the intensity of the D band is relatively weak, suggesting that the GNs sample exhibits high crystallinity and fewer defects.

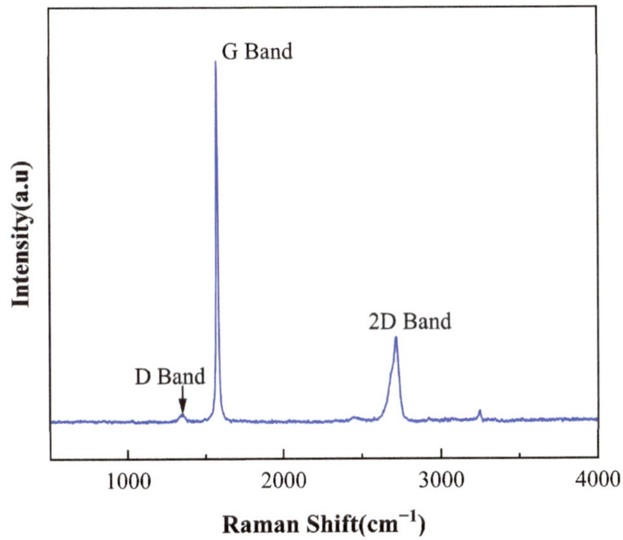

Figure 1. Raman spectra of GNs.

The morphology of the CS-GNs was characterized using SEM and EDS. To study the SEM of CS-GNs nanocomposites, a droplet of CS-GN dispersion was applied onto tin foil for scanning. The SEM image of Figure 2a shows the size dimensions of the GNs around 10 μm. The SEM image in Figure 2b illustrates the structure of the GNs. The image clearly showed the structure of the overlapped graphene sheets, with visible layer edges and folds. The SEM image in Figure 2c shows the dispersion of GNs. It is evident that the graphene was uniformly distributed in the chitosan solution, indicating that graphene is relatively well dispersed in chitosan, with a homogeneous morphology and a substantial biocompatible membrane surface area.

Furthermore, the EDS patterns of CS-GNs nanocomposites are depicted in Figure 2d. From the patterns, it is evident that the main elements detected include C, O, and Al elements. Among them, the weight percentage of element C was 80.90%, with an atomic percentage of 88.06%; the weight percentage of the element O was 8.05% and the atomic percentage was 6.58%; and the weight percentage of element Al was 11.05% and the atomic percentage was 5.35%. Since tin foil was utilized as the substrate for the SEM inspection of CS-GNs nanocomposites, and the predominant element in tin foil paper is aluminum, a significant peak of aluminum is observed in the EDS pattern. These results showed that the CS-GNs nanocomposites were successfully prepared.

The FTIR spectra of the GN dispersion solution, CS solution, and CS-GNs nanocomposites are shown in Figure 3. It can be observed that there are more oxygen-containing functional groups in the GNs dispersion. The vibrational bands observed around 3158 cm^{-1} correspond to the −OH stretching vibration peak. The stretching vibration peak of the skeleton C=C is at 1389 cm^{-1}. The stretching vibration peak at 1672 cm^{-1} corresponds to C=O, and the stretching vibration peak of the epoxy bond C–O–C is at 1095 cm^{-1}. In the CS solution, the N–H stretching vibrations originating from amino and $-NH_2$ groups are observed at 3452 cm^{-1}. The peak at 1638 cm^{-1} corresponds to the stretching vibration of the C=O group in acetylated amino units. Additionally, the peaks observed at 1152 cm^{-1} and 1015 cm^{-1} are attributed to the stretching vibrations of the C_6–OH primary alcohol group and the C_3–OH secondary alcohol group in CS, respectively. The FTIR spectra of CS-GNs is generally similar to that of the dispersed GNs and the CS solution, with no new characteristic peaks observed. This indicates that there is no chemical reaction between GNs and CS. The peak observed at 3431 cm^{-1} in the CS-GNs spectrum is attributed to the interaction between the $-OH$ groups of GNs and the $-NH_2$ groups of CS. Compared to the FTIR spectra of GNs and CS, the intensities of the characteristic peaks in the CS-GNs spectrum are enhanced, indicating the formation of hydrogen bond interactions between GNs and CS.

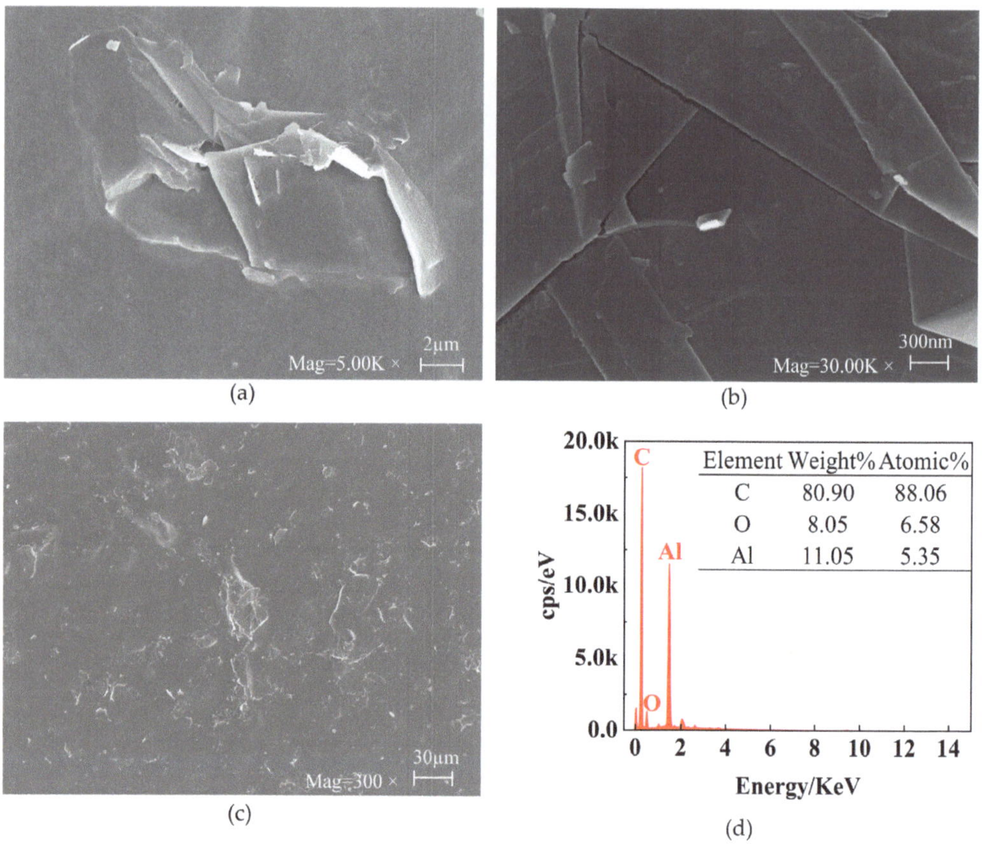

Figure 2. SEM images (**a**–**c**) of CS-GNs nanocomposites at different magnifications. EDS patterns (**d**) of CS-GNs nanocomposites.

Due to its structural characteristics, CS-GNs exhibit a relatively high specific surface area. Firstly, graphene nanosheets, as a component of CS-GNs, possess a two-dimensional structure and a monolayer arrangement of carbon atoms, resulting in a significantly large specific surface area [25]. This characteristic endows graphene nanosheets with excellent performance in adsorption, catalysis, and other fields. Secondly, chitosan is a polysaccharide polymer containing abundant hydroxyl functional groups, enabling it to interact favorably with graphene nanosheets at the molecular level [26]. Through the composite of chitosan with graphene, a greater surface area of nanocomposite materials can be achieved. The advantage of this composite structure not only increases its specific surface area but also enhances its performance in applications such as adsorption, catalysis, sensing, and others, making it a material with promising and wide-ranging application prospects [27].

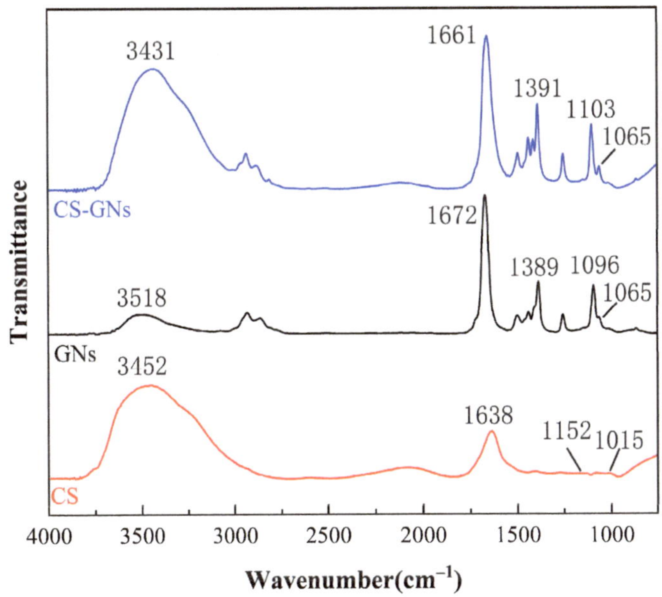

Figure 3. FTIR spectra of GNs, CS, and CS-GNs.

2.2. Characterization of the Immunosensor

To examine the characteristics of the immunosensor interface, experiments were conducted using CV and EIS methods. The base solution consisted of 0.2 M PBS containing 5.0 mM $K_3[Fe(CN)_6]$ and 0.1 M KCl. Figure 4 shows the CV and EIS scan the results of different modified electrodes: bare CS-GNs/GCE (curve a), anti-AFB_1/CS-GNs/GCE (curve b), BSA/anti-AFB_1/CS-GNs/GCE (curve c), and AFB_1/BSA/anti-AFB_1/CS-GNs/GCE (curve d).

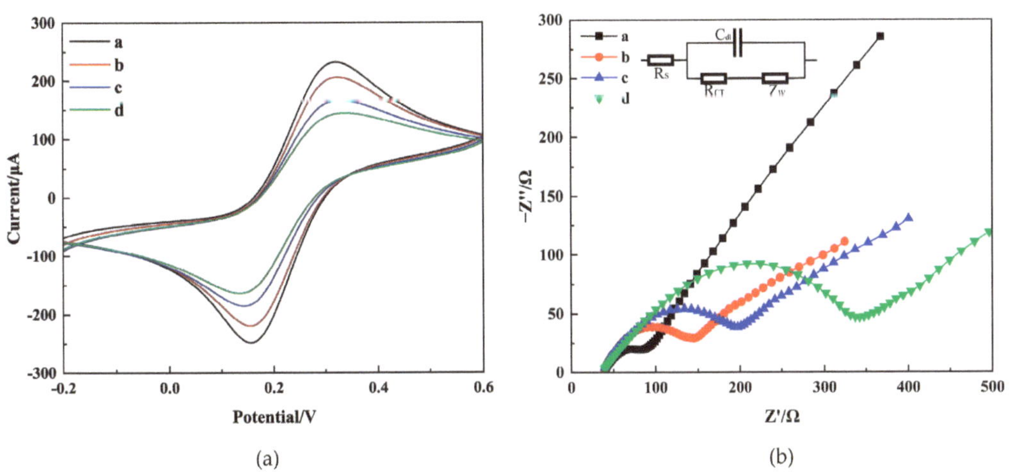

Figure 4. CV (**a**) and EIS (**b**) on CS-GNs/GCE (curve a), anti-AFB_1/CS-GNs/GCE (curve b), BSA/anti-AFB_1/CS-GNs/GCE (curve c), AFB_1/BSA/anti-AFB_1/CS-GNs/GCE (curve d) in 0.2 M pH = 7.2 PBS containing 5.0 mM $K_3[Fe(CN)_6]$ and 0.1 M KCl. The concentration of AFB_1 is 15 ng/mL.

CV scans were performed on the modified electrodes at a rate of 25 mV/S between −0.2 and 0.6 V. In Figure 4a, two separate peaks are displayed by the CS-GNs/GCE with a ΔE_p ($E_{pa} - E_{pc}$) value of 163 mV. The current values for I_{pa} and I_{pc} are 232.7 μA and −248.3 μA, respectively. After the incubation of the antibodies on CS-GNs/GCE, the ΔE_p value exhibited an increase to 168 mV, while the I_{pa} and I_{pc} values displayed a decrease to 206.4 μA and −218.9 μA, respectively. Furthermore, the obstruction of active sites by antibodies caused a hindrance to the electron transfer between $[Fe(CN)_6]^{3-}$ and the electrode. After BSA was immobilized on the electrode surface, the peak current decreased even further. The ΔE_p value increased to 185 mV, and the I_{pa} and I_{pc} values decreased to 166.7 μA and −185.4 μA, respectively. The results indicated that the active sites responsible for nonspecific adsorption were successfully obstructed [28]. When the immunosensor was incubated with AFB_1 (15 ng/mL), a clear reduction in peak current was detected. The ΔE_p value increased to 198 mV, and the I_{pa} and I_{pc} values decreased to 144.7 μA and −163.4 μA, respectively. The results indicated that the immunoreaction occurred, and the AFB_1 captured on the electrode surface hindered the reaction of $[Fe(CN)_6]^{3-}$, indicating the successful formation of the immune complex on the electrode surface [29].

EIS was also a powerful tool for characterizing the step-by-step manufacturing process of the electrode [30]. The electron transfer resistance (R_{CT}) was analyzed by fitting the diameter of the semicircle using the Randles equivalent circuit (inset in Figure 4b). The Nyquist plot presents the R_{CT} behavior of an electrode. A semicircle diameter forms at the higher frequency region, indicating an electron-limiting process. Meanwhile, the low frequency region exhibits a diffusion-controlled process [31]. As depicted in Figure 4b, the diameter of the semicircle observed in the CS-GNs/GCE was significantly smaller compared to the others, indicating a large electrode surface area and superior conductivity. The R_{CT} value of CS-GNs/GCE was measured at 66.72 Ω. Following the incubation of antibodies on the CS-GNs/GCE, the R_{CT} value increased to 129.2 Ω, surpassing that of the CS-GNs/GCE. This may be attributed to the blocking of electron transfer by the antibodies. After BSA was immobilized on the electrode surface, the semicircular domain increased, and the R_{CT} value was found to increase up to 197.2 Ω. This indicates that the active sites causing nonspecific adsorption were successfully blocked by BSA. Finally, the AFB_1 (15 ng/mL) was immobilized on the electrode, a significant increase in the semicircular domain was observed, and the R_{CT} value increased to 320.1 Ω. This result suggested that the immune complex effectively formed on the electrode surface, thereby impeding the electron transfer. As expected, EIS was utilized to assess the precise state of the immunosensor during each stage of assembly. The results indicate that the immunosensor was successfully fabricated. Consequently, data derived from both CV and EIS showed that the successful fabrication of the immunosensor.

2.3. *Optimization of Experimental Conditions*

To examine the immunosensor's optimum sensing capabilities, we investigated the impacts of various factors on its performance. These factors included the concentration of the immobilized antibody, pH levels, incubation duration of the antibody, and immunoreaction time.

The performance of the sensor is highly dependent on the concentration of antibodies immobilized on the electrode surface, as they create binding sites for antigens. We conducted an experiment to investigate the impact of different antibody concentrations (25, 50, 100, 150, and 200 μg/mL) on detecting AFB_1 at a concentration of 15 ng/mL using the immunosensor. Figure 5a illustrates the change in peak current (ΔI) before and after the immunoreaction. It is observed that ΔI increases until reaching 150 μg/mL, after which it begins to decrease. This may be attributed to antibodies' saturation at this concentration, consistent with the findings in the existing literature [32]. Consequently, 150 μg/mL was determined as the optimal antibody concentration.

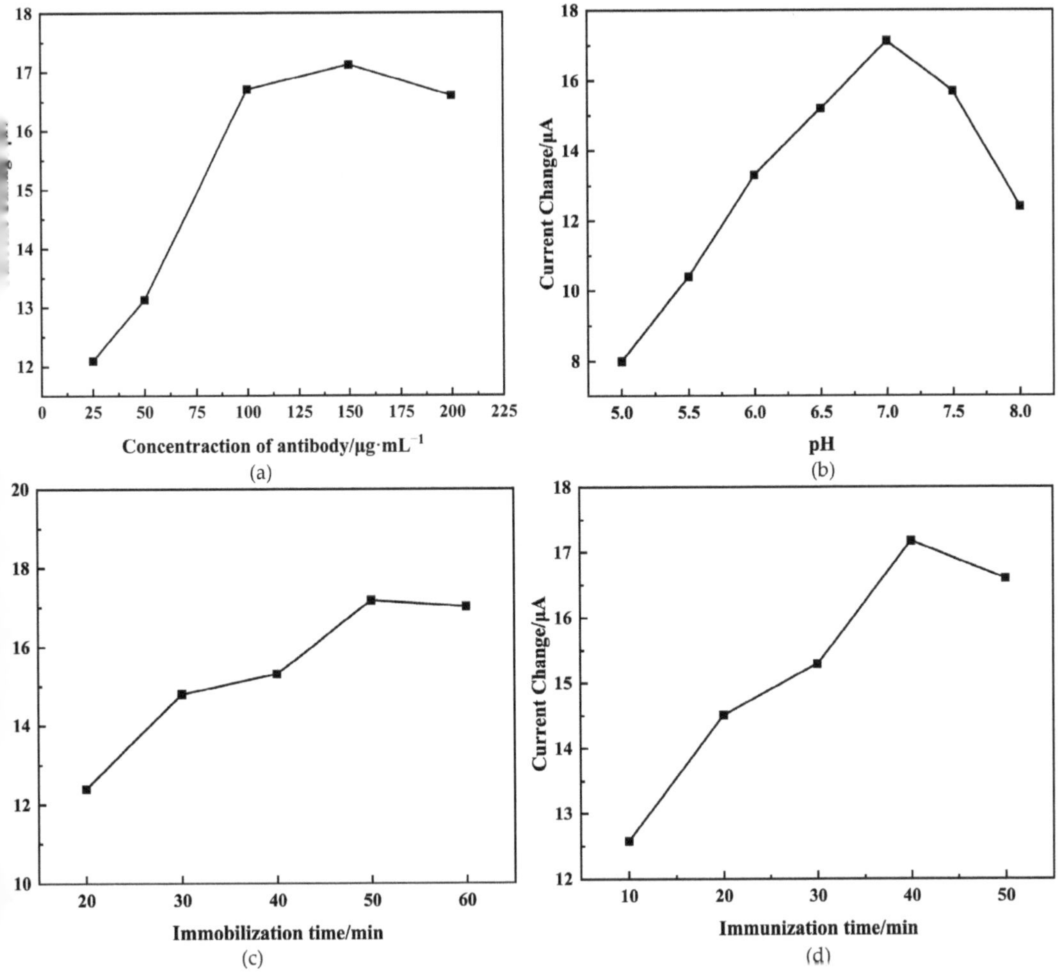

Figure 5. Effects of antibody concentration (**a**), pH (**b**), antibody immobilization time (**c**), and immunoreaction time (**d**) on peak current change of immunosensor. The concentration of antigen is 15 ng/mL.

The pH value of the base solution was a crucial parameter, potentially resulting in protein denaturalization or the instability of the immunosensor [33]. Figure 5b illustrates that the peak current change (ΔI) gradually increased with the pH value of the base solution increased, reaching its peak at 7.0. This happens because extreme acidity or alkalinity can damage the immobilized protein, especially under alkaline conditions [34]. As a result, the pH value of the base solution was adjusted to 7.0 for further investigation.

The performance of the sensor can be affected by the duration of antibody immobilization. As depicted in Figure 5c, the change in peak current (ΔI) exhibited a gradual increase with prolonged antibody immobilization time until it reached a plateau at 50 min. This could be attributed to the antibody reaching its saturation point in terms of activity [35]. Therefore, based on the experiment, 50 min was determined as the optimal duration.

The duration of the immunoreaction between the antigen and antibody significantly affects the performance of the sensor. As depicted in Figure 5d, there was a gradual increase in the peak current change (ΔI), increasing the immunoreaction time, which eventually leveled off at 40 min. This outcome suggests that the immunoreaction between the antigen and antibody reached saturation after 40 min. Thus, 40 min was identified as the optimal duration for the immunoreaction between the antigen and antibody.

2.4. Analytical Performance

Under the optimal conditions, the performance of the prepared immunosensor was evaluated for various concentrations of AFB_1 using the DPV technique.

As depicted in Figure 6a, the DPV peak currents exhibited a notable decrease with an increasing AFB_1 concentration within the range of 0–25 ng/mL. This decrease can be attributed to the heightened hindrance of the immunocomplex to electron transfer. As depicted in Figure 6b, the current change (ΔI) before and after immunization displayed a linear relationship with AFB_1 concentrations from 0.05 ng/mL to 25 ng/mL, with a low detection limit of 0.021 ng/mL ($S/N = 3$). The limit of detection (LOD) was determined using the regression curve parameters, $LOD = 3{\cdot}S_b/s$, where "S_b" represents the ard deviation of the blank sample and "s" represents the slope [36].The calibrated regression equation is:

$$\Delta I = 0.822{\cdot}C + 6.504,$$

with a correlation coefficient of 0.99. The proposed immunosensor was compared with other reported AFB_1 immunosensors reported in the literature. The acceptable linear range and detection limit of the proposed immunosensor are described in Table 1, indicating its excellent performance for AFB_1 detection. The outstanding electrochemical performance of the proposed immunosensor stemmed from the large surface area and exceptional conductivity of CS-GNs nanocomposites.

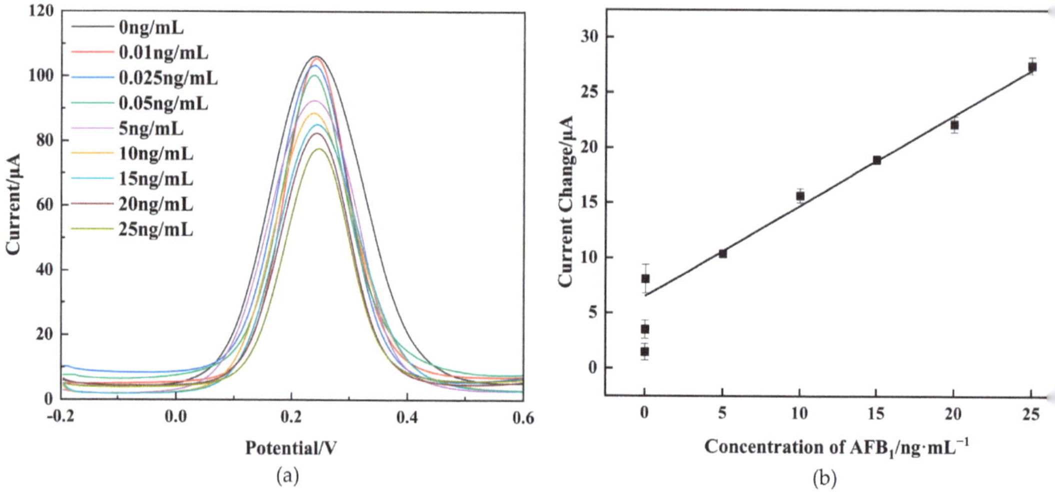

Figure 6. Immunosensor response to DPV at different concentrations of AFB_1 (**a**). ΔI calibration curves of the immunosensor for different concentrations of AFB_1 (**b**).

The LOD of the developed immunosensor surpassed that of most reported electrochemical methods for the detection of AFB_1. The LOD of the immunosensor was comparable to that achieved by Li et al. [37], which developed a biosensor based on aptamers for AFB_1 detection. However, the biosensor developed by Li et al. [37] had a complex structure, was costly, and cumbersome to prepare.

Table 1. Comparison of the proposed immunosensor and other sensors.

Immunosensors	Linear Range (ng/mL)	Detection Limit (ng/mL)	References
AFB_1/Fc-apt/MCH/cDNA/AuNPs/THI-rGO/GCE	0.05–20	0.016	[37]
AFB_1/BSA/anti-AFB_1/AuNPs/Zn/Ni-ZIF-8-800@Graphene/GCE	0.18–100	0.18	[38]
AFB_1/BSA/anti-AFB_1/Au@PEI@CNFs/GCE	0.05–25	0.027	[39]
AFB_1/BSA/anti-AFB_1/Au-COOH-GO/GCE	0.05–25	0.05	[16]
AFB_1/MCH/pept/porous/AuNPs/GCE	10–20,000	0.94	[40]
AFB_1/BSA/anti-AFB_1/CS-GNs/GCE	0.05–25	0.021	This work

2.5. Reproducibility, Stability, and Selectivity

In order to investigate the reproducibility of the immunosensor, five electrodes were tested to detect 15 ng/mL AFB_1 under the same conditions. The results are depicted in Figure 7a, the relative standard deviation (RSD) of the AFB_1 measurements for the five sensors is 2.4%, which proved that the proposed immunosensor has excellent reproducibility.

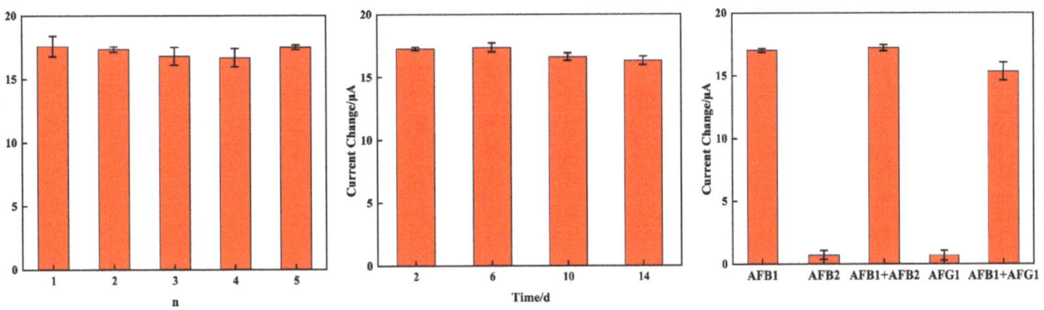

Figure 7. Amperometric change response of the immunosensor to 5 different electrodes treated in the same way (**a**); the time stability study of the immunosensor (**b**); and the current change responses of the immunosensor to AFB_1, AFB_2, AFB_1 + AFB_2, AFG_1, AFB_1 + AFG_1 (**c**).

The stability of the immunosensor was assessed by detecting the electrochemical response after the immunosensors were stored at 4 °C for 2, 6, 10, and 14 days. As depicted in Figure 7b, after 14 days of storage, the electrochemical response retained 94.42% of the initial current for 5 ng/mL AFB_1, which indicated the significant stability of the immunosensor.

The specificity of the prepared immunosensor was also crucial for assessing its performance. The specificity was evaluated using interfering substances, which consisted of 15 ng/mL of AFB_2 and AFG_1. As observed in Figure 7c, the peak current change (ΔI) before and after immunization with pure interfering substances exhibited no noticeable variation. The observed peak current change (ΔI) before and after immunization with the mixture solution showed similarity to that of the 15 ng/mL AFB_1 standard solution. All the above observations demonstrate that the immunosensor exhibited a commendable level of specificity.

2.6. Detection of AFB_1 in Corn Samples

In order to assess the precision of the immunosensor, spiked recoveries were measured in pretreated samples of corn. The standard addition method was employed to assess the application of the proposed immunosensor in corn samples. AFB1 was added to the corn samples at spiked concentrations of 5 ng/mL, 10 ng/mL, and 15 ng/mL, respectively. As shown in Table 2, the range of the recovery was from 97.3 to 101.4%. These results demonstrated the practicality of the immunosensor in effectively analyzing the target AFB1 concentrations in real samples.

Table 2. Recovery for different concentrations of AFB_1 spiked in corn samples.

Samples	Added AFB_1 (ng/mL)	Found AFB_1 (ng/mL)	Recovery (%)
1	5.00	4.86	97.3
2	10.00	10.14	101.4
3	15.00	14.64	97.6

3. Materials and Methods

3.1. Materials and Apparatus

The graphene nanosheets (2 nm, with a diameter of 2~3 μm) were purchased from Nanjing Xianfeng Nanomaterials Technology Co., Ltd., located in Nanjing, China. Chitosan was obtained from China National Pharmaceutical Group Chemical Reagent Co., Ltd. (Shanghai, China). Aflatoxin B_1 (AFB_1), aflatoxin B_2 (AFB_2), and aflatoxin G_1 (AFG_1) standard solution (in acetonitrile, 10 μg/mL) were sourced from Beijing ZhongkeErhuan Technology Co., Ltd. (Beijing, China). The anti-Aflatoxin B_1 antibody was provided by Shanghai Sangong Biological Engineering Co., Ltd. (Shanghai, China). Bovine serum albumin (BSA) and 1-ethyl-(3-dimethylaminopropyl) carbodiimide hydrochloride (EDC.HCl) were obtained from Hefei Genial Biotech Co., Ltd. (Hefei, China). *N*-hydroxysuccinimide (NHS), *N,N*-dimethylformamide (DMF), and phosphate-buffered saline (PBS) at a pH range of 7.2–7.4 were procured from Shanghai Titan Technology Co., Ltd. (Shanghai, China), for experimental use. P-aminobenzoic acid (PABA) was acquired from Hefei Qiansheng Biological Technology Co., Ltd. (Hefei, China).

The electrochemical characterization tests, including differential pulse voltammetry (DPV), cyclic voltammetry (CV), and electrochemical impedance spectroscopy (EIS), were performed using the CHI-760E electrochemical workstation from Shanghai Chenhua Instrument Co., Ltd. (Shanghai, China). The electrochemical experiment utilized a conventional three-electrode system, comprising a saturated KCl Ag/AgCl electrode as the reference electrode, a platinum wire (Pt) electrode as the counter electrode, and a glass carbon electrode (GCE) as the working electrode. The scanning electron microscope (SEM) images were captured using the Sigma 300 hot-field scanning electron microscope (Carl Zeiss, Oberkochen, Germany).

3.2. Methods

3.2.1. Preparation of CS-GNs Nanocomposites

The powder of 5 mg chitosan was dissolved in 5 mL of 1.0% (v/v) acetic acid and stirred with a magnetic stirrer for 1 h. After complete dispersion, the solution was stored at 4 °C for later use. Then, 5 mg of graphene nanosheets was dissolved in 5 mL of anhydrous ethanol and sonicated for 10 h. The supernatant was discarded after centrifugation in a centrifuge at 9000 rpm for 15 min. Then, 5 mL of DMF was added and sonicated for more than 2 h until completely dispersed. Subsequently, 5 mL of the prepared graphene dispersion was taken and mixed with 5 mL of prepared CS solution, followed by ultrasonication for 2 h to obtain a uniform CS-GNs dispersion. The CS-GNs dispersion was stored at 4 °C for further use.

3.2.2. Fabrication of the Immunosensor

Before modification, the GCE was treated by a typical purification method. Initially, the bare GCE was polished with a polishing powder containing 0.05 μm Al_2O_3 particles until achieving a highly reflective, mirror-like surface. Secondly, the electrode was cleaned 5 min in ethanol and distilled water, and then dried at room temperature. To activate the GCE, the dried electrode was subjected to cyclic voltammetry scanning (−0.3–1.5 V,

50 mV/s) in a 0.5 M H_2SO_4 solution for 15 cycles. Subsequently, the GCE was cleaned by ultrasonication in distilled water for 10 min and rinsed thoroughly with abundant distilled water.

After being dried at room temperature, the cleaned GCE was subjected to CV scanning (−1.5~1.0 V, 50 mV/s) in 5 mM p-aminobenzoic acid (PABA) solution for 15 cycles, followed by rinsing with distilled water and dried at room temperature. Finally, 10 μL of CS-GNs nanocomposites dispersion was carefully dropped onto the surface of the GCE.

Before immobilizing the antibodies, the surface of CS-GNs/GCE was initially activated using EDC: NHS coupling chemistry. Here, EDC (0.4 M) served as a coupling agent, while NHS (0.1 M) acted as an activator for the covalent immobilization of biomolecules [41].

After the activation, CS-GNs/GCE was thoroughly washed with PBS. Subsequently, 10 μL of 150 μg/mL anti-AFB_1 was carefully dropped onto the surface of the CS-GNs/GCE and then incubated at 37 °C for 50 min. Following that, the fabricated electrode was rinsed with PBS to remove the physically adsorbed antibodies. Subsequently, the electrode was incubated in 10 μL 3% BSA solution at 37 °C for 1 h, in order to block any unreacted active sites on the surface. Afterwards, the electrode was thoroughly rinsed with PBS once more, resulting in the successful fabrication of the AFB1 electrochemical immunosensor using CS-GNs/GCE, which was then stored at 4 °C for further use. Thereafter, the electrode was dropped with 10 μL of AFB_1 solution with diverse concentrations and incubated at 37 °C for 40 min. The physically adsorbed AFB_1 antigen molecules were washed away by PBS, and then electrochemically tested by the DPV method in the base solution (5 mM $K_3[Fe(CN)_6]$ + 0.1 M KCl + 0.2 M PBS). The DPV peak current change (ΔI) before and after the immunoreaction served as the basis for quantifying the AFB_1 concentration in the samples. Figure 8 shows the preparation of CS-GNs nanocomposites and outlines the process for preparing the electrochemical immunosensor.

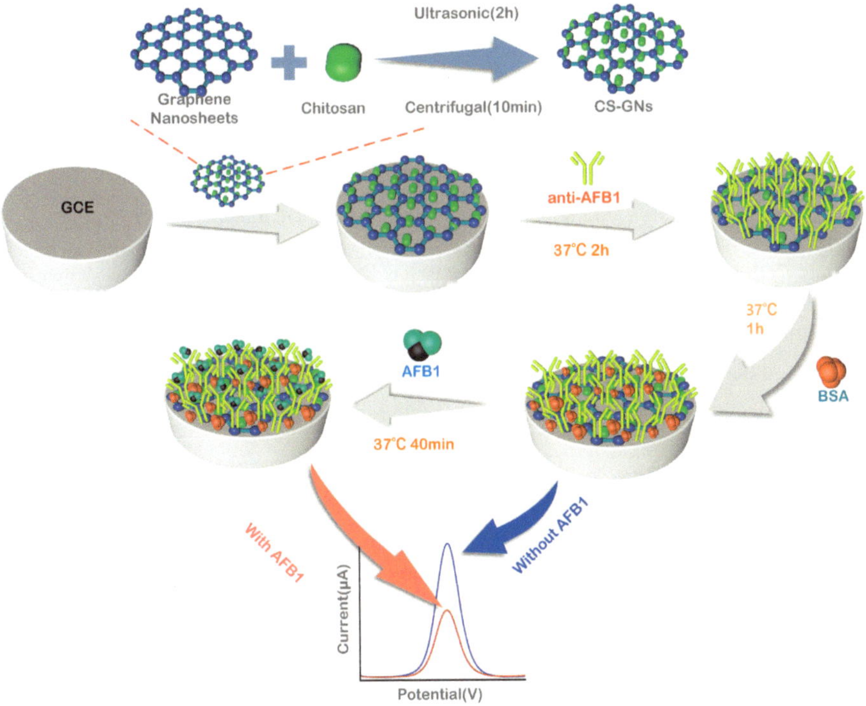

Figure 8. Schematic diagrams for the preparation of CS-GN nanohybrids and the electrochemical immunosensor.

3.2.3. Preparation of Spiked Samples

The spiked sample was prepared using a previously established method [42]. The unaffected corn samples were crushed and ground into powder, weighed 30 mg in a centrifuge tube, added to 30 mL of acetonitrile/water (8:2 v/v) solution, ultrasonicated for 4 h until completely dispersed, and then centrifuged at 9000 r/min for 15 min. The resulting supernatant collected and diluted multiple concentrations with PBS, and the samples were subsequently spiked with varying concentrations of AFB_1 (5.0, 10.0, 15.0 ng/mL), before being stored at 4 °C until use.

4. Conclusions

In this work, we designed an electrochemical immunosensor based on chitosan graphene nanosheets (CS-GNs) for the detection of AFB_1 concentration in corn samples. The CS-GNs nanocomposites exhibited a large specific surface area, excellent biocompatibility, and high electrochemical activity. These properties facilitate the immobilization of antibodies and enhance the rate of electron transfer. The obtained CS-GNs nanocomposites were surface characterized using SEM. The optimization of antibody concentration, pH, antibody incubation time, and immunoreaction time was based on the DPV method. With the best conditions, the change in DPV peak currents before and after immunization was linear over the concentration range of 0.05–25 ng/mL AFB_1, with a detection limit of 0.021 ng/mL ($S/N = 3$). The developed immunosensor exhibited favorable reproducibility, stability, and specificity for detecting the concentration of AFB_1 in corn samples. Additionally, the recovery of AFB_1 detection in corn samples ranged from 97.3% to 101.4%. The results show that the developed AFB_1 immunosensor possesses the advantages of simplicity, sensitivity, and high selectivity. This makes it a valuable reference for the detection and analysis of other biomolecules.

Author Contributions: This full manuscript was obtained from a joint effort of all authors. K.X.: conceptualization, methodology, writing—review and editing. S.Z.: methodology, experiment, writing—original draft preparation, data curation. C.W.: writing, review, editing and funding acquisition. Z.Z.: funding acquisition. All authors have read and agreed to the published version of the manuscript.

Funding: This research was funded by National Natural Science Foundation of China (62303094), the Innovation Funds Plan of Henan University of Technology (2020ZKCJ31), the Open Research Subject of Research Platform of Grain Information Processing Center (Henan University of Technology) (KFJJ2022015), Henan Provincial Science and Technology Research and Development Plan Joint Fund (222103810084), and the Natural Science Project of Zhengzhou Science and Technology Bureau (22ZZRDZX07).

Institutional Review Board Statement: Not applicable.

Informed Consent Statement: Not applicable.

Data Availability Statement: Data are contained within the article.

Conflicts of Interest: The authors declare no conflicts of interest.

References

1. Tian, D.; Wang, J.; Zhuang, Q.; Wu, S.; Yu, Y.; Ding, K. An Electrochemiluminescence Biosensor Based on Graphitic Ca Nitride Luminescence Quenching for Detection of AFB1. *Food Chem.* **2023**, *404*, 134183. [CrossRef]
2. Hassan, H.F.; Kordahi, R.; Dimassi, H.; El Khoury, A.; Daou, R.; Alwan, N.; Merhi, S.; Haddad, J.; Karam, L. Aflatoxin Rice: Effects of Storage Duration, Grain Type and Size, Production Site, and Season. *J. Food Prot.* **2022**, *85*, 938–944. [Cross [PubMed]
3. Zhao, Z.; Wang, H.; Zhai, W.; Feng, X.; Fan, X.; Chen, A.; Wang, M. A Lateral Flow Strip Based on a Truncated Apta Complementary Strand for Detection of Type-B Aflatoxinsofsoft in Nuts and Dried Figs. *Toxins* **2020**, *12*, 136. [CrossRef]
4. Hidalgo-Ruiz, J.L.; Romero-González, R.; Martínez Vidal, J.L.; Garrido Frenich, A. Determination of Mycotoxins in Nuts by High-Performance Liquid Chromatography-Tandem Mass Spectrometry: Looking for a Representative Matrix. *J. Food Co Anal.* **2019**, *82*, 103228. [CrossRef]

Wang, Y.; Hou, C.; Dai, Y.; Chu, L.; Geng, S.; Zheng, S.; Kang, X. Determination of Aflatoxin B1 by Novel Nanofiber-Packed Solid-Phase Extraction Coupled with a High Performance Liquid Chromatography-Fluorescence Detector. *Anal. Methods* **2023**, *15*, 472–481. [CrossRef]

Tang, W.; Qi, Y.; Li, Z. A Portable, Cost-Effective and User-Friendly Instrument for Colorimetric Enzyme-Linked Immunosorbent Assay and Rapid Detection of Aflatoxin B1. *Foods* **2021**, *10*, 2483. [CrossRef]

Xiao, X.; Hu, S.; Lai, X.; Peng, J.; Lai, W. Developmental Trend of Immunoassays for Monitoring Hazards in Food Samples: A Review. *Trends Food Sci. Technol.* **2021**, *111*, 68–88. [CrossRef]

Liu, X.; Wen, Y.; Wang, W.; Zhao, Z.; Han, Y.; Tang, K.; Wang, D. Nanobody-Based Electrochemical Competitive Immunosensor for the Detection of AFB1 through AFB1-HCR as Signal Amplifier. *Microchim. Acta* **2020**, *187*, 352. [CrossRef] [PubMed]

Yadav, N.; Yadav, S.S.; Chhillar, A.K.; Rana, J.S. An Overview of Nanomaterial Based Biosensors for Detection of Aflatoxin B1 Toxicity in Foods. *Food Chem. Toxicol.* **2021**, *152*, 112201. [CrossRef]

Qi, X.; Lv, L.; Wei, D.; Lee, J.J.; Niu, M.; Cui, C.; Guo, Z. Detection of Aflatoxin B1 with a New Label-Free Fluorescence Aptasensor Based on PVP-Coated Single-Walled Carbon Nanohorns and SYBR Gold. *Anal. Bioanal. Chem.* **2022**, *414*, 3087–3094. [CrossRef]

Wang, Z.; Wei, L.; Ruan, S.; Chen, Y. CRISPR/Cas12a-Assisted Chemiluminescence Sensor for Aflatoxin B_1 Detection in Cereal Based on Functional Nucleic Acid and In-Pipet Rolling Circle Amplification. *J. Agric. Food Chem.* **2023**, *71*, 4417–4425. [CrossRef]

Liu, X.; Singh, R.; Li, M.; Li, G.; Min, R.; Marques, C.; Zhang, B.; Kumar, S. Plasmonic Sensor Based on Offset-Splicing and Waist-Expanded Taper Using Multicore Fiber for Detection of Aflatoxins B1 in Critical Sectors. *Opt. Express* **2023**, *31*, 4783. [CrossRef]

Wei, X.; Yin, M.; Zhang, L.; Lin, H.; Wang, J.; Xie, W.; Xu, D. Surface Plasmon Resonance (SPR) Biosensor for Detection of Mycotoxins: A Review. *J. Immunol. Methods* **2022**, *510*, 113349. [CrossRef] [PubMed]

Kosowska, K.; Domalik-Pyzik, P.; Nocuń, M.; Chłopek, J. Chitosan and Graphene Oxide/Reduced Graphene Oxide Hybrid Nanocomposites—Evaluation of Physicochemical Properties. *Mater. Chem. Phys.* **2018**, *216*, 28–36. [CrossRef]

Aghris, S.; Azriouil, M.; Ettadili, F.E.; Loukili, A.; Laghrib, F.; Farahi, A.; Bakasse, M.; Lahrich, S.; El Mhammedi, M.A. An Electrochemical Sensor for Flubendiamide Insecticide Analysis Based on Chitosan/Reduced Graphene Oxide. *Sens. Diagn.* **2023**, *2*, 398–408. [CrossRef]

Shi, L.; Wang, Z.; Wu, N.; Chen, X.; Yang, G.; Liu, W. A Simple Electrochemical Immunosensor for Highly Sensitive Detection of Aflatoxin B1 Based on Gold Nanoparticle Decorated Carboxylated Graphene Oxide. *Int. J. Electrochem. Sci.* **2020**, *15*, 1655–1668. [CrossRef]

Srivastava, S.; Ali, M.A.; Umrao, S.; Parashar, U.K.; Srivastava, A.; Sumana, G.; Malhotra, B.D.; Pandey, S.S.; Hayase, S. Graphene Oxide-Based Biosensor for Food Toxin Detection. *Appl. Biochem. Biotechnol.* **2014**, *174*, 960–970. [CrossRef] [PubMed]

Bhardwaj, H.; Singh, C.; Kotnala, R.K.; Sumana, G. Graphene Quantum Dots-Based Nano-Biointerface Platform for Food Toxin Detection. *Anal. Bioanal. Chem.* **2018**, *410*, 7313–7323. [CrossRef]

Ji, G.; Tian, J.; Xing, F.; Feng, Y. Optical Biosensor Based on Graphene and Its Derivatives for Detecting Biomolecules. *Int. J. Mol. Sci.* **2022**, *23*, 10838. [CrossRef] [PubMed]

Zakharova, O.V.; Mastalygina, E.E.; Golokhvast, K.S.; Gusev, A.A. Graphene Nanoribbons: Prospects of Application in Biomedicine and Toxicity. *Nanomaterials* **2021**, *11*, 2425. [CrossRef]

Reddy, Y.V.M.; Shin, J.H.; Palakollu, V.N.; Sravani, B.; Choi, C.-H.; Park, K.; Kim, S.-K.; Madhavi, G.; Park, J.P.; Shetti, N.P. Strategies, Advances, and Challenges Associated with the Use of Graphene-Based Nanocomposites for Electrochemical Biosensors. *Adv. Colloid Interface Sci.* **2022**, *304*, 102664. [CrossRef]

Liu, X.; Wu, Y.; Zhao, X.; Wang, Z. Fabrication and Applications of Bioactive Chitosan-Based Organic-Inorganic Hybrid Materials: A Review. *Carbohydr. Polym.* **2021**, *267*, 118179. [CrossRef]

Tien, N.D.; Lyngstadaas, S.P.; Mano, J.F.; Blaker, J.J.; Haugen, H.J. Recent Developments in Chitosan-Based Micro/Nanofibers for Sustainable Food Packaging, Smart Textiles, Cosmeceuticals, and Biomedical Applications. *Molecules* **2021**, *26*, 2683. [CrossRef]

Li, Z.; Deng, L.; Kinloch, I.A.; Young, R.J. Raman Spectroscopy of Carbon Materials and Their Composites: Graphene, Nanotubes and Fibres. *Prog. Mater. Sci.* **2023**, *135*, 101089. [CrossRef]

Gao, M.; Wang, L.; Zhao, B.; Gu, X.; Li, T.; Huang, L.; Wu, Q.; Yu, S.; Liu, S. Sandwich Construction of Chitosan/Reduced Graphene Oxide Composite as Additive-Free Electrode Material for High-Performance Supercapacitors. *Carbohydr. Polym.* **2021**, *255*, 117397. [CrossRef] [PubMed]

Gorgieva, S.; Osmić, A.; Hribernik, S.; Božič, M.; Svete, J.; Hacker, V.; Wolf, S.; Genorio, B. Efficient Chitosan/Nitrogen-Doped Reduced Graphene Oxide Composite Membranes for Direct Alkaline Ethanol Fuel Cells. *Int. J. Mol. Sci.* **2021**, *22*, 1740. [CrossRef] [PubMed]

Liu, C.; Hung, C.-W.; Cheng, I.-C.; Hsu, C.-C.; Cheng, I.-C.; Chen, J.-Z. Dielectric Barrier Discharge Plasma Jet (DBDjet) Processed Reduced Graphene Oxide/Polypyrrole/Chitosan Nanocomposite Supercapacitors. *Polymers* **2021**, *13*, 3585. [CrossRef] [PubMed]

Damphathik, C.; Songsiririthigul, C.; Lerdsri, J.; Jakmunee, J.; Wongnongwa, Y.; Jungsuttiwong, S.; Ortner, A.; Kalcher, K.; Samphao, A. A Novel Immunosensor Based on Cobalt Oxide Nanocomposite Modified Single Walled Carbon Nanohorns for the Selective Detection of Aflatoxin B1. *Talanta* **2023**, *258*, 124472. [CrossRef] [PubMed]

Shi, L.; Wang, Z.; Yang, G.; Chen, X.; Gou, G.; Liu, W. Electrochemical Immunosensor for Aflatoxin B1 Based on Polyaniline/Graphene Nanohybrids Decorated with Au Nanoparticle. *Electrochemistry* **2017**, *85*, 384–390. [CrossRef]

30. Dos Santos, D.M.; Migliorini, F.L.; Soares, A.C.; Mattoso, L.H.C.; Oliveira, O.N.; Correa, D.S. Electrochemical Immunose Made with Zein-based Nanofibers for On-site Detection of Aflatoxin B1. *Electroanalysis* **2023**, *35*, e202100672. [CrossRef]
31. Yu, L.; Zhang, Y.; Hu, C.; Wu, H.; Yang, Y.; Huang, C.; Jia, N. Highly Sensitive Electrochemical Impedance Spectros Immunosensor for the Detection of AFB1 in Olive Oil. *Food Chem.* **2015**, *176*, 22–26. [CrossRef]
32. Zhang, S.; Shen, Y.; Shen, G.; Wang, S.; Shen, G.; Yu, R. Electrochemical Immunosensor Based on PdeAu Nanoparticles Supp on Functionalized PDDA-MWCNT Nanocomposites for Aflatoxin B1 Detection. *Anal. Biochem.* **2016**, *494*, 10–15. [CrossRe
33. Wang, Y.; Zhao, G.; Li, X.; Liu, L.; Cao, W.; Wei, Q. Electrochemiluminescent Competitive Immunosensor Base Polyethyleneimine Capped SiO_2 Nanomaterials as Labels to Release Ru(Bpy)32+ Fixed in 3D Cu/Ni Oxalate for the Detecti Aflatoxin B1. *Biosens. Bioelectron.* **2018**, *101*, 290–296. [CrossRef]
34. Ma, H.; Sun, J.; Zhang, Y.; Xia, S. Disposable Amperometric Immunosensor for Simple and Sensitive Determination of Aflato 1 in Wheat. *Biochem. Eng. J.* **2016**, *115*, 38–46. [CrossRef]
35. Chen, Q.; Yu, C.; Gao, R.; Gao, L.; Li, Q.; Yuan, G.; He, J. A Novel Electrochemical Immunosensor Based on the rGO-TEPA-NH2 and AuPt Modified C60 Bimetallic Nanoclusters for the Detection of Vangl1, a Potential Biomarker for Dysontoge *Biosens. Bioelectron.* **2016**, *79*, 364–370. [CrossRef] [PubMed]
36. Sharma, A.; Kumar, A.; Khan, R. Electrochemical Immunosensor Based on Poly (3,4-Ethylenedioxythiophene) Modified Gold Nanoparticle to Detect Aflatoxin B1. *Mater. Sci. Eng. C* **2017**, *76*, 802–809. [CrossRef] [PubMed]
37. Li, Y.; Liu, D.; Zhu, C.; Shen, X.; Liu, Y.; You, T. Sensitivity Programmable Ratiometric Electrochemical Aptasensor Base Signal Engineering for the Detection of Aflatoxin B1 in Peanut. *J. Hazard. Mater.* **2020**, *387*, 122001. [CrossRef] [PubMed]
38. Wang, N.; Liu, Q.; Hu, X.; Wang, F.; Hu, M.; Yu, Q.; Zhang, G. Electrochemical Immunosensor Based on AuNPs/Zn/ Ni-Z 800@graphene for Rapid Detection of Aflatoxin B1 in Peanut Oil. *Anal. Biochem.* **2022**, *650*, 114710. [CrossRef]
39. Huang, Y.; Zhu, F.; Guan, J.; Wei, W.; Zou, L. Label-Free Amperometric Immunosensor Based on Versatile Carbon Nanof Network Coupled with Au Nanoparticles for Aflatoxin B1 Detection. *Biosensors* **2020**, *11*, 5. [CrossRef]
40. Liu, B.; Peng, J.; Wu, Q.; Zhao, Y.; Shang, H.; Wang, S. A Novel Screening on the Specific Peptide by Molecular Simulation Development of the Electrochemical Immunosensor for Aflatoxin B1 in Grains. *Food Chem.* **2022**, *372*, 131322. [CrossRef]
41. Bhardwaj, H.; Singh, C.; Pandey, M.K.; Sumana, G. Star Shaped Zinc Sulphide Quantum Dots Self-Assembled Monola Preparation and Applications in Food Toxin Detection. *Sens. Actuators B Chem.* **2016**, *231*, 624–633. [CrossRef]
42. Ma, H.; Sun, J.; Zhang, Y.; Bian, C.; Xia, S.; Zhen, T. Label-Free Immunosensor Based on One-Step Electrodeposition of Chit Gold Nanoparticles Biocompatible Film on Au Microelectrode for Determination of Aflatoxin B1 in Maize. *Biosens. Bioele* **2016**, *80*, 222–229. [CrossRef] [PubMed]

molecules

MDPI

cle

terfacial Interaction in NiFe LDH/NiS_2/VS_2 for Enhanced ectrocatalytic Water Splitting

;xia Wang *, Xu Zhang, Xiaojiao Yu *, Junpeng Li, Kai Wang and Jinfen Niu

School of Science, Xi'an University of Technology, Xi'an 710054, China; xuzhang0501@126.com (X.Z.); junpengljp@163.com (J.L.); kw225713@163.com (K.W.); niujinfen@xaut.edu.cn (J.N.)
* Correspondence: wangtingxia@xaut.edu.cn (T.W.); yxjw@xaut.edu.cn (X.Y.)

Abstract: A bifunctional electrocatalyst with high efficiency and low costs for overall water splitting is critical to achieving a green hydrogen economy and coping with the energy crisis. However, developing robust electrocatalysts still faces huge challenges, owing to unsatisfactory electron transfer and inherent activity. Herein, NiFe LDH/NiS_2/VS_2 heterojunctions have been designed as freestanding bifunctional electrocatalysts to split water, exhibiting enhanced electron transfer and abundant catalytic sites. The optimum NiFe LDH/NiS_2/VS_2 electrocatalyst exhibits a small overpotential of 380 mV at 10 mA cm^{-2} for overall water splitting and superior electrocatalytic performance in both hydrogen and oxygen evolution reactions (HER/OER). Specifically, the electrocatalyst requires overpotentials of 76 and 286 mV at 10 mA cm^{-2} for HER and OER, respectively, in alkaline electrolytes, which originate from the synergistic interaction among the facilitated electron transfer and increasingly exposed active sites due to the modulation of interfaces and construction of heterojunctions.

Keywords: vanadium disulfide; nickel disulfide; layered double hydroxide; interfacial regulation; electrocatalytic overall water splitting; hydrogen/oxygen evolution reaction

1. Introduction

Electrocatalytic overall water splitting, as a promising green hydrogen (ideal zero-carbon energy carrier)-producing strategy, has attracted more attention for developing carbon-neutral alternatives [1–3]. However, electrocatalytic water-splitting activity is severely limited, owing to the sluggish catalytic kinetics for OER and HER [4–7]. Electrocatalysts of excellent efficiency are essential to reducing the overpotential by lowering the activation energy of water splitting [8,9]. At present, although IrO_2, RuO_2 and Pt-based materials have been employed as highly effective electrocatalysts for OER and HER, respectively, their scarcity and high price have hampered their industrialized application on a large scale [10,11]. The preparation of cost-effective bifunctional electrocatalysts with superior activity for OER and HER is crucial to promoting the practical applications and industrialized development of overall electrocatalytic water splitting.

Non-noble transition metal sulfides (TMSs) have presented themselves as appealing catalysts for both HER and OER on the basis of their favorable electrocatalytic activity and electrical conductivity [12,13]. Recently, vanadium disulfide (VS_2) has attracted considerable attention as a candidate electrocatalyst, owing to its intrinsic catalytic properties, which exhibit weak van der Waals interlayer interactions and an interlayer spacing of 5.76 Å [14]. In particular, benefiting from its metallic conductivity and electrocatalytic active sites both on the surface and at the active edge, the 1T phase VS_2 has been widely employed as an electrocatalyst for HER and OER [15]. Nevertheless, previous research reveals that individual TMSs exhibit insufficient active sites due to their strong van der Waals forces [16]. Massive efforts have been directed at constructing multi-component nanomaterial, which has been expected to improve its electrocatalytic activity by providing an abundance of active sites [17,18]. Ru cluster-modified VS_2 can achieve a current density of 50 mA cm^{-2} at an overpotential of 245 mV, and this can be attributed to the optimization of intermediate adsorption/desorption through the interactions between Ru and VS_2 and the sufficient catalytic sites afforded by electro-oxidized of Ru species [19]. Benefiting from induced-layer electron transfer between $NiCo_2S_4$ and ReS_2 interfaces, the $NiCo_2S_4/ReS_2$ nanocomposite showed excellent HER properties in both acidic and alkaline conditions [20]. The CoS_2-ReS_2 heterojunction exhibited improved electrocatalytic activity for HER and OER, which can be accounted for by the rich catalytic sites on the edge and the optimized catalytic kinetics of the interfacial modulation of ReS_2 via modifying CoS_2 [21]. Therefore, constructing binary transition metal sulfide-based electrocatalysts is beneficial to creating abundant catalytic active sites.

Interfacial regulation has been regarded as a powerful method to adjust electronic structure and catalytic kinetics to significantly enhance electrocatalysis [22–26]. NiFe layered double hydroxides (LDHs) have been widely investigated for OER because they can convert into γ-NiOOH by surface reconstruction, which has been proven to be the actual active sites [27,28]. Nevertheless, NiFe LDH exhibits limited electrocatalytic activity for OER due to weak conductivity and insufficient active sites [29]. Constructing heterostructures has been employed to improve its intrinsic catalytic activity by facilitating electron transfer kinetics and promoting the conversion of NiFe LDH into γ-NiOOH active species [30]. Urchin-like Co_9S_8@NiFe layered double-hydroxide heterostructured hollow spheres have been designed to enhance the electrocatalytic OER process by improving electrical conductivity and reducing the reaction energy barrier, requiring a smaller overpotential of 220 mV at 10 mA cm^{-2} [31]. By virtue of the abundant Ni_3S_2-NiFe LDH interfaces, this catalyst is regarded as one of the most productive non-precious metal-based OER catalysts [32]. Furthermore, the in-situ growth of heterogeneous structure electrocatalysts using a binder-free method can provide intimate contact between multi-components and electrodes, which can induce facilitated electron transfer [33,34]. Electrodeposition is a useful technique to rapidly fabricate multi-component nanomaterials through an in-situ growth process [35,36].

Inspired by the above-mentioned facts, a heterogeneous structure-based binary metal sulfide-modified carbon cloth (NiS_2/VS_2) has been constructed via a one-step hydrothermal process. Subsequently, an active OER electrocatalyst, NiFe LDH, was integrated into the surface of NiS_2/VS_2 (NiFe LDH/NiS_2/VS_2) using a facile electrochemical deposition method, which was simultaneously exploited as a bifunctional electrocatalyst for both HER and OER. This designed heterostructure of NiFe LDH/NiS_2/VS_2 nanosheets exhibits interfacial charge transfer from NiS_2 to VS_2 and promotes electron transfer during the electrocatalytic process. Moreover, the formed porous nanoflower-like structure is favorable for exposing more active sites and facilitating bubble releasing, all of which can promote the electrocatalytic performance for overall water splitting. As a bifunctional electrocatalyst, the obtained NiFe LDH/NiS_2/VS_2 can deliver a current density of 10 mA cm^{-2} at an overpotential of 76 and 286 mV severally for HER and OER, together with excellent electrocatalytic stability. Furthermore, the solar-driven electrocatalytic overall splitting system constructed by NiFe LDH/NiS_2/VS_2 demonstrated that it has good potential for practical applications.

2. Results and Discussion

2.1. Characterization of Samples

The synthetic pathway of NiFe LDH/NiS_2/VS_2 is schematically illustrated in Scheme 1, in which VS_2 was in-situ deposited on the surface of carbon cloth (decorated as VS_2/CC) by a facile hydrothermal method. Similarly, the NiS_2/VS_2 decorated CC was constructed by the above-mentioned method with the addition of $NiSO_4{\cdot}6H_2O$. Subsequently, the NiFe LDH was modified onto the surface of NiS_2/VS_2 via a simple electrodeposition method, resulting in NiFe LDH/NiS_2/VS_2.

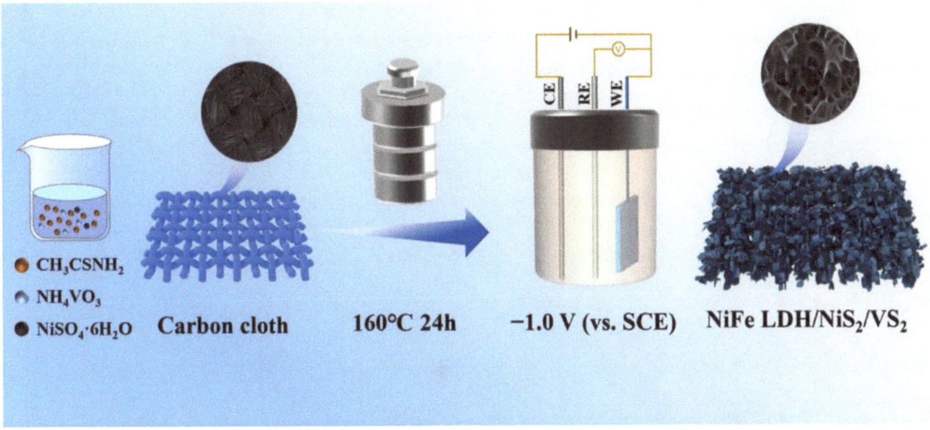

Scheme 1. Schematic illustration of the preparation process of NiFe LDH/NiS_2/VS_2.

The phase and composition of pure VS_2, NiS_2/VS_2 and NiFe LDH/NiS_2/VS_2 were measured by XRD. As shown in Figure S1a, the diffraction peaks located at 2θ values of 15.2, 35.64, 45.07, 47.06 and 57.02° can be ascribed to the (001), (011), (012), (003) and (110) lattice planes of VS_2 (JCPDS No. 89-1640) [37,38]. The additional peaks at 31.58, 38.76, 53.43, 58.50 and 60.91° appeared when NiS_2 was modified on the surface of VS_2, which can be indexed to the (200), (211), (311), (230) and (321) plane of NiS_2 (JCPDS No. 11-0099), respectively [39–41]. Compared with the XRD pattern of NiS_2/VS_2, the characteristic diffraction peaks at 11.26 and 35.28° appeared after the electrodeposition of NiFe LDH, which can be assigned to the (003) and (006) planes of NiFe LDH species (JCPDS No. 51-0463) [42,43]. The above-mentioned results suggest the co-existence of VS_2, NiS_2 and NiFe LDH in NiFe LDH/NiS_2/VS_2.

The morphology of the electrocatalysts was studied using SEM and TEM. As shown in Figure 1a, the pure VS_2 exhibits a nanoflower-like structure, which is distinguished from the NiS_2/VS_2 by a nanosheet morphology (Figure 1b). By contrast, the NiFe LDH/NiS_2/VS_2 has a thin nanosheets structure (Figure 1c), which was beneficial for exposing more active sites. Furthermore, the TEM image (Figure 1d) reveals that NiFe LDH/NiS_2/VS_2 exhibits a nanosheet structure. An interplanar spacing of 2.0, 2.3 and 1.9 Å is shown in the HRTEM image (Figure 1e) and can be identified as the (012) plane of VS_2, (220) plane of NiS_2 and (018) plane of NiFe LDH, respectively [44–46]. As illustrated in Figure 1f, the selected area electron diffraction (SAED) reveals that the (012) plane of NiFe LDH, (012) plane of VS_2 and (321) plane of NiS_2 appeared in the NiFe LDH/NiS_2/VS_2 sample, suggesting the successful construction of the NiFe LDH/NiS_2/VS_2 electrode [44,47,48]. The EDS-mapping measurement (Figure 2) was employed to investigate the element distribution of NiFe LDH/NiS_2/VS_2, revealing an evenly distribution of V, S, Ni, Fe and O elements in the NiFe LDH/NiS_2/VS_2.

To investigate the superficial element composition and the chemical state of the respective elements of the as-prepared electrocatalysts, X-ray photoelectron spectroscopy (XPS) characterization was utilized for the VS_2, NiS_2/VS_2 and NiFe LDH/NiS_2/VS_2. Compared with the XPS survey spectra of the VS_2 (Figure S1b), an obvious signal of the Ni element appears in the survey spectra of the NiS_2/VS_2 and NiFe LDH/NiS_2/VS_2 samples. The XPS survey of NiFe LDH/NiS_2/VS_2 reveals the co-existence of Ni, Fe, V, O and S elements, which corresponds to the result of the EDS-mapping images. As revealed in Figure 3a, the high-resolution XPS spectrum of V for VS_2 contains typical characteristic peaks at 514.12 and 521.62 eV, which can be related to the V^{2+} $2p_{3/2}$ and V^{2+} $2p_{1/2}$, respectively [49,50]. Additionally, there are two dominant peaks at 516.87 and 524.10 eV that can be indexed to the V^{4+} $2p_{3/2}$ and V^{4+} $2p_{1/2}$, respectively [49–51]. By contrast, the V^{4+} $2p_{3/2}$ peak around

517.49 eV and V^{4+} $2p_{1/2}$ around 524.72 eV of NiS_2/VS_2 display a positive shift, demonstrating the electronic interaction between NiS_2 and VS_2. Similarly, the V 2p XPS pattern of NiFe LDH/NiS_2/VS_2 exhibits the two characteristic peaks at 517.61 and 524.86 eV, which can be ascribed to the V^{4+} $2p_{3/2}$ and V^{4+} $2p_{1/2}$, respectively [49–51]. Consequently, this indicates that the valence state of V does not change after the electrodeposition of NiFe LDH, while an obvious positive shift occurs for V^{4+} $2p_{3/2}$ and V^{4+} $2p_{1/2}$ in NiFe LDH/NiS_2/VS_2 compared with NiS_2/VS_2 and VS_2. In the S 2p XPS survey of VS_2, NiS_2/VS_2 and NiFe LDH/NiS_2/VS_2 (Figure 3b), the peaks at 162.2 and 163.4 eV can be fitted to the S^{2-} $2p_{3/2}$ and S^{2-} $2p_{1/2}$, respectively [50,52]. Moreover, the peak at 168.5 eV in the S 2p XPS of NiFe LDH/NiS_2/VS_2 can be attributed to the S-O bond, owing to slight oxidization [38].

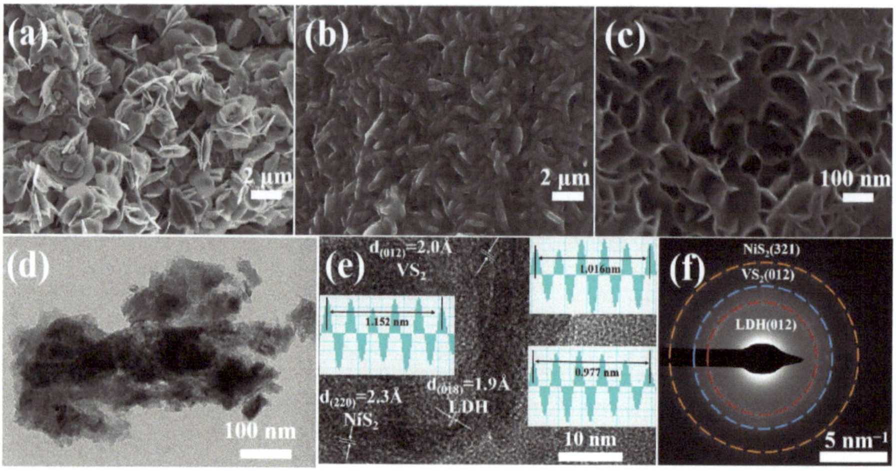

Figure 1. SEM images of (**a**) VS_2, (**b**) 1:5 NiS_2/VS_2, and (**c**) 350 s NiFe LDH/NiS_2/VS_2. (**d**) TEM, (**e**) HRTEM image and (**f**) SAED pattern of 350 s NiFe LDH/NiS_2/VS_2.

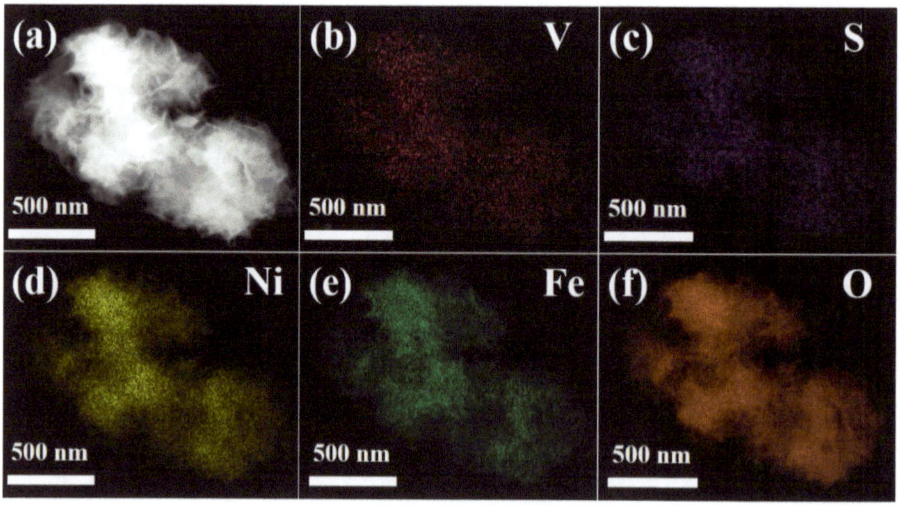

Figure 2. (**a**) High-angle annular dark-field scanning transmission electron microscopy (HAADF-STEM) image of NiFe LDH/NiS_2/VS_2 and the corresponding elemental mapping of (**b**) V, (**c**) S, (**d**) Ni, (**e**) Fe and (**f**) O in NiFe LDH/NiS_2/VS_2.

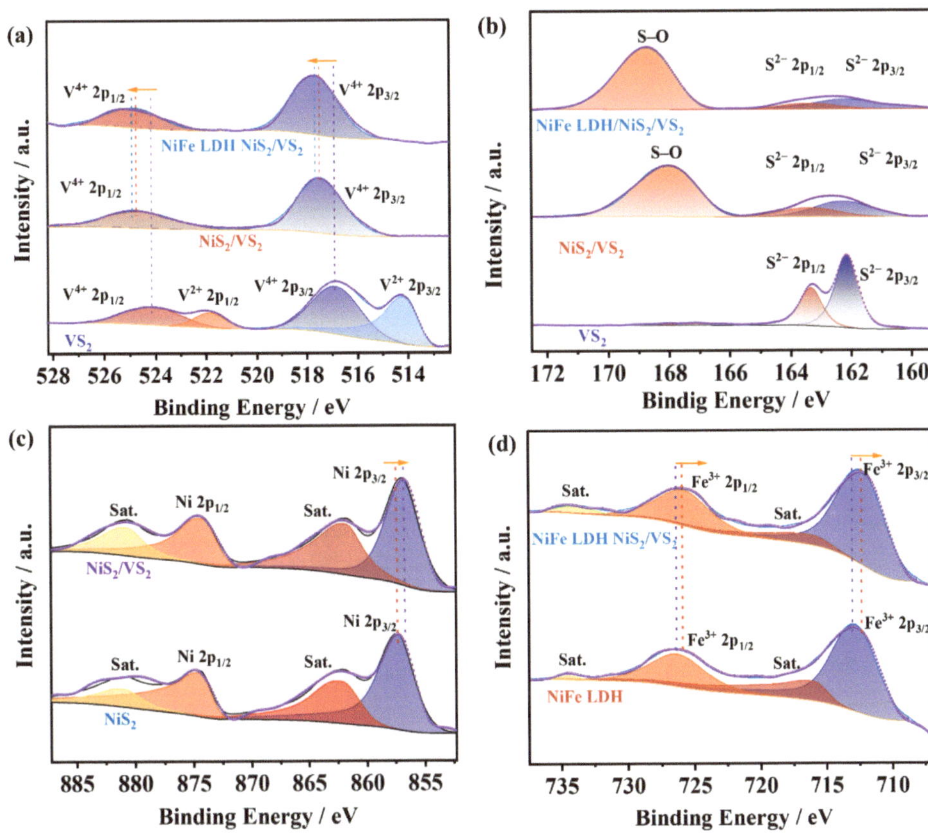

Figure 3. (**a**) High-resolution XPS spectra of (**a**) V 2p, (**b**) S 2p, (**c**) Ni 2p and (**d**) Fe 2p of the prepared VS_2, NiS_2/VS_2 and NiFe LDH/NiS_2/VS_2.

As revealed in Figure 3c, the high-resolution Ni 2p XPS survey of NiS_2/VS_2 reveals a negative shift compared with NiS_2, further suggesting the strong interfacial electronic interaction of NiS_2 and VS_2. In detail, the Ni 2p XPS of NiS_2 displays a peak at 857.23 eV with a satellite peak of 862.45 eV and a peak at 874.44 eV with a satellite peak of 881.11 eV, belonging to the Ni $2p_{3/2}$ and Ni $2p_{1/2}$ of Ni^{2+}, respectively [53–55]. In comparison, the characteristic peaks of Ni $2p_{3/2}$ and Ni $2p_{1/2}$ of NiS_2/VS_2 shift to 856.83 and 874.30 eV, respectively. For further exploring the potential effects of the introduction of NiFe LDH, the high-resolution Fe 2p XPS of NiFe LDH and NiFe LDH/NiS_2/VS_2 was measured. As revealed in Figure 3d, the peaks of Fe^{3+} $2p_{3/2}$ and Fe^{3+} $2p_{1/2}$ in the NiFe LDH/NiS_2/VS_2 are separately located at 712.33 and 725.84 eV, which reveals a negative shift of Fe 2p XPS compared with the peaks of Fe^{3+} $2p_{3/2}$ (713.10 eV) and Fe^{3+} $2p_{1/2}$ (726.5eV) in the NiFe LDH [55–57].

2.2. *Electrocatalytic Oxygen Evolution Reaction Performance*

The electrocatalytic OER performance of the prepared products (VS_2, NiS_2/VS_2, NiFe LDH and NiFe LDH/NiS_2/VS_2) was evaluated in 1.0 M KOH with a typical three-electrode system. As shown in Figure 4a,b, the VS_2 shows poor electrocatalytic performance for OER, which exhibits a high overpotential of 540 mV to reach a current density of 10 mA cm^{-2}, suggesting that the VS_2 is inactive for electrocatalytic OER. Obviously, the NiS_2/VS_2 heterogeneous structure exhibits enhanced electrocatalytic activity for OER, revealing that the formed multi-phase interface is a key factor in boosting the electrocatalytic activity of OER. Figure S2a shows the electrocatalytic OER activity of the VS_2-based electrocatalysts with different additive ratios of $NiSO_4$, in which the 1:5 NiS_2/VS_2 (denoted as NiS_2/VS_2) exhibits an optimal electrocatalytic activity with an overpotential of 386 mV at 10 mA cm^{-2}. Additionally, the electrodeposition time of NiFe LDH was optimized, ranging from 0 to 400 s. Figure S2b reveals that the 350 s NiFe LDH/NiS_2/VS_2 exhibits superior electrocatalytic activity when the electrodeposition time of NiFe LDH is 350 s, which has been denoted as NiFe LDH/NiS_2/VS_2. The NiFe LDH/NiS_2/VS_2 exhibits better electrocatalytic performance for OER compared with VS_2, NiS_2/VS_2 and NiFe LDH, which provides a lower overpotential of 286 mV to deliver a current density of 10 mA cm^{-2}. Additionally, the LSV curves of NiFe LDH/NiS_2/VS_2 and NiFe LDH show small peaks around 1.35 V, which can be interpreted as the redox reactions of the Ni species. These results imply that the formed heterogeneous structure and NiFe LDH decoration are beneficial to improving the electrocatalytic performance of OER.

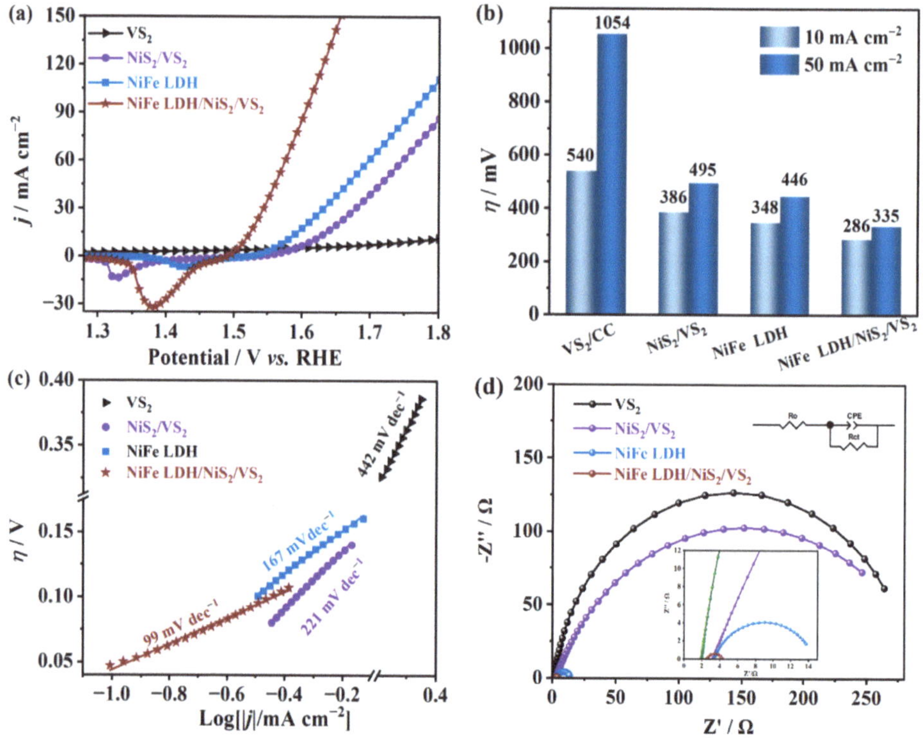

Figure 4. Electrochemical OER performance of the as-prepared VS_2, NiS_2/VS_2, NiFe LDH and NiFe LDH/NiS_2/VS_2 samples: (**a**) polarization curves, (**b**) the corresponding overpotential to deliver a current density of 10 and 50 mA cm^{-2} in 1.0 M KOH, (**c**) the Tafel plots and (**d**) the EIS spectra.

Moreover, the electrocatalytic kinetics of as-prepared catalysts were evaluated by measuring Tafel plots. As shown in Figure 4c, the NiFe LDH/NiS_2/VS_2 exhibits a smaller Tafel slope of 99 mV dec^{-1} than NiFe LDH (167 mV dec^{-1}), NiS_2/VS_2 (221 mV dec^{-1}) and VS_2 (442 mV dec^{-1}), suggesting the facilitated catalytic kinetics of NiFe LDH/NiS_2/VS_2 for OER. The electrochemical impedance spectroscopy (EIS) measurement was utilized to evaluate the charge transfer properties. As Figure 4d depicts, the NiFe LDH/NiS_2/VS_2 exhibits the smallest charge transfer resistance (R_{ct}) among the as-prepared electrocatalysts, indicating that the lower resistance and accelerated electron transfer can be achieved by constructing a heterogeneous structure and decorating NiFe LDH.

The electrochemically active surface area (ECSA) of the as-prepared electrocatalysts was evaluated by calculating the C_{dl} according to the CV curves at various scan rates. As displayed in Figure 5a, the NiFe LDH/NiS_2/VS_2 exhibits a much larger C_{dl} of 3.38 mF cm^{-2} than NiFe LDH (2.55 mF cm^{-2}), NiS_2/VS_2 (1.41 mF cm^{-2}) and VS_2 (0.43 mF cm^{-2}), which can be attributed to the formed heterogeneous structure and the NiFe LDH decoration. Accordingly, the NiFe LDH/NiS_2/VS_2 exhibits an ECSA of 84.80 cm^2, which is higher than that of NiFe LDH (63.75 cm^2), NiS_2/VS_2 (35.25 cm^2) and VS_2 (10.75 cm^2) (Figure S5a). The current density was normalized by the ECSA to evaluate the specific activity. As observed in Figure S5b and Figure 5c, the NiFe LDH/NiS_2/VS_2 exhibits intrinsic activity compared to the others, suggesting that the formed heterogeneous structure and modulated electronic structure are favorable to enhancing electrocatalytic activity. Additionally, the electrocatalytic stability of NiFe LDH/NiS_2/VS_2 was evaluated by i–t curves in 1.0 M KOH. In Figure 5b, we can observe that the NiFe LDH/NiS_2/VS_2 can maintain its superior electrocatalytic activity over 24 h, demonstrating that the prepared NiFe LDH/NiS_2/VS_2 can sustain long-term electrolysis without obvious attenuation. Furthermore, an XPS measurement of the NiFe LDH/NiS_2/VS_2 was employed after the stability test to investigate the structural stability. As observed in Figure S6, the NiFe LDH/NiS_2/VS_2 can maintain its initial chemical composition and chemical state after a long-term stability test. The thin nanosheet structure also can be observed (Figure S7) after long-term stability, demonstrating the structural stability of NiFe LDH/NiS_2/VS_2.

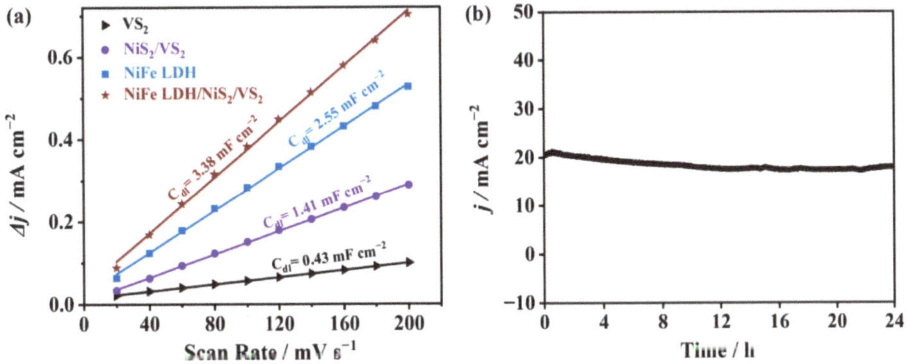

Figure 5. (**a**) The C_{dl} of the as-prepared VS_2, NiS_2/VS_2, NiFe LDH and NiFe LDH/NiS_2/VS_2 samples and (**b**) the long-term stability of NiFe LDH/NiS_2/VS_2 under the potential of 0.5 V (vs. SCE).

2.3. Electrocatalytic Hydrogen Evolution Reaction and Overall Water-Splitting Performance

To investigate its potential for efficient overall water splitting, the electrocatalytic performance of the NiFe LDH/NiS_2/VS_2 for HER was evaluated in 1.0 M KOH. Figure 6a,b shows that the NiFe LDH/NiS_2/VS_2 electrocatalyst exhibits excellent electrocatalytic activity, acquiring a smaller overpotential of 76 mV to deliver a current density of 10 mA cm^{-2}, which is greatly superior to that of the VS_2 (546 mV), NiS_2/VS_2 (270 mV) and NiFe LDH (122 mV). The corresponding Tafel plots are illustrated in Figure 6c. The NiFe LDH/NiS_2/VS_2 has a smaller Tafel slope of 79 mV dec^{-1} than VS_2 (202 mV dec^{-1}), NiS_2/VS_2 (175 mV dec^{-1}) and NiFe LDH (145 mV dec^{-1}). Moreover, the NiFe LDH/NiS_2/VS_2 exhibits remarkable stability during 24 h electrochemical measurement and maintains its electrocatalytic activity for HER (Figure 6d).

Electrocatalytic overall water splitting was measured by a two-electrode system composed of NiFe LDH/NiS_2/VS_2 in 1.0 M KOH. As shown in Figure 7a, the constructed overall water-splitting system is able to deliver a current density of 10 mA cm^{-2} at 1.61 V, confirming its potential to drive the overall water splitting. Furthermore, a commercial solar cell under 2.0 V was employed to explore its solar-drive overall water-splitting performance. As shown in Figure 7b, a number of microbubbles of H_2 and O_2 gas at the cathode and anode, respectively, evolved, which clearly demonstrates the superior electrocatalytic activity of the NiFe LDH/NiS_2/VS_2. Furthermore, the corresponding Video S1 recorded this dynamic state.

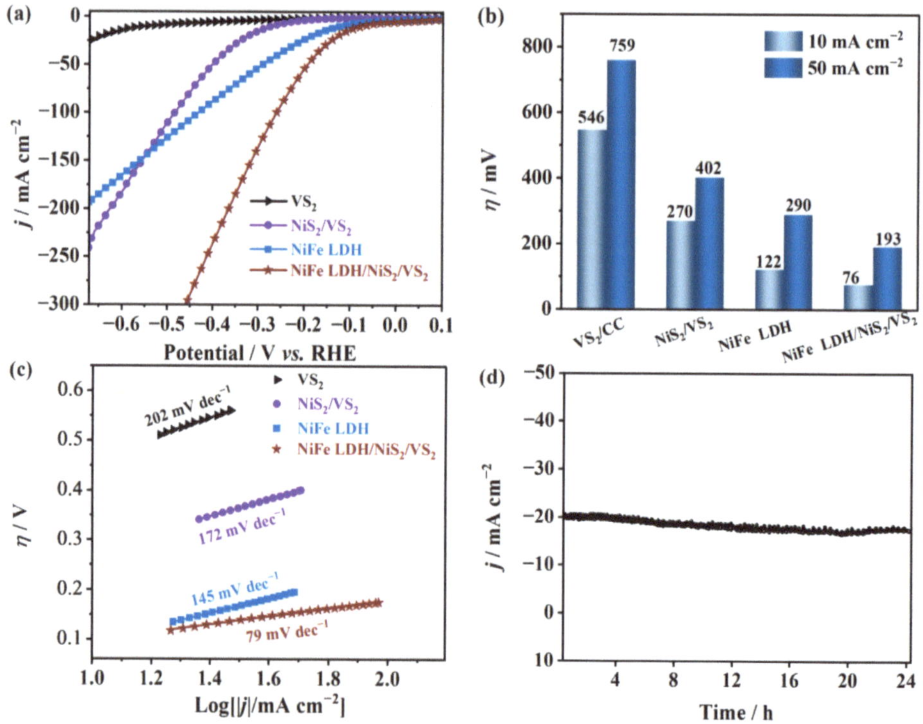

Figure 6. Electrochemical HER performance of the as-prepared VS_2, NiS_2/VS_2, NiFe LDH and NiFe LDH/NiS_2/VS_2 samples: (**a**) polarization curves, (**b**) the corresponding overpotential to deliver a current density of 10 and 50 mA cm^{-2} in 1.0 M KOH, (**c**) the Tafel plots and (**d**) the long-term stability text of NiFe LDH/NiS_2/VS_2 under the potential of −1.2 V (vs. SCE).

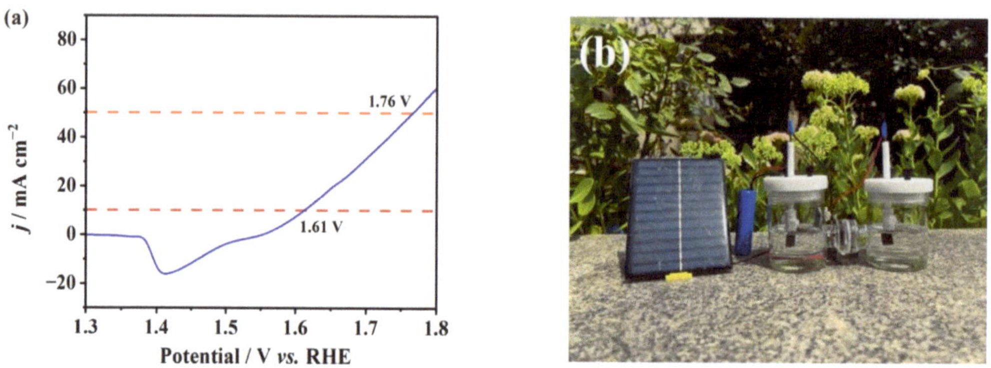

Figure 7. (**a**) The LSV curves of LDH/NiS_2/VS_2 (+, −) device for overall water splitting, (**b**) image of the solar-driven water-splitting system in 1.0 M KOH electrolyte.

3. Experimental Section

3.1. Materials

Ammonium metavanadate (NH_4VO_3, AR, Tianjin Damao Chemical Reagent Factory, Tianjin, China), thioacetamide (CH_3CSNH_2, Aladdin Industrial Corporation, Shanghai, China), nickel sulfate hexahydrate ($NiSO_4{\cdot}6H_2O$, AR, Zhongxin Fine Chemical Co., Ltd., Jinan, China), ($FeSO_4{\cdot}7H_2O$, AR, Sinopharm Chemical Reagent Co., Ltd., Shanghai, China) ammonium hydroxide (NH_4OH, AR, Tianjin Fuyu Fine Chemical Co., Ltd., Tianjin, China), and carbon cloth (CC, SCI Materials Hub, Taiwan, China) were employed as substrates for the in-situ formation of VS_2 and NiS_2/VS_2 on its surface via hydrothermal synthesis. Potassium hydroxide (KOH, Tianjin Kemiou Chemical Reagent Co., Ltd., Tianjin, China), nitric acid (HNO_3, AR, Chengdu Kelong Chemical, Chengdu, China), and ethyl alcohol (C_2H_6O, AR, Tianjin Fuyu Fine Chemical Co., Ltd., Tianjin, China) were obtained to pretreat CC. Deionized water was prepared in our laboratory. All chemicals mentioned were utilized as received without further purification.

3.2. Pretreatment of Carbon Cloth (CC) Substrate

Carbon cloth (CC) with a geometric area of 2×4 cm^2 was sonicated in an ultrasonic bath with deionized water and ethanol (C_2H_5OH) severally for 15 min to clean its surface thoroughly. Then, the hydrophilic surface was obtained via pretreatment in concentrated HNO_3 at 100 °C for 2 h. Finally, the CC surface was fully rinsed with deionized water to remove residual HNO_3, and the pretreated CC was obtained.

3.3. Preparation of 1:5 NiS_2/VS_2

A typical procedure to synthesize 1:5 NiS_2/VS_2 was performed as follows: 1 mmol NH_4VO_3 was added to a weak alkaline solution produced by 25 mL deionized water and 2 mL NH_4OH while stirring constantly. At the same time, 10 mmol CH_3CSNH_2 and 0.2 mmol $NiSO_4{\cdot}6H_2O$ were dispersed in 10 mL deionized water with stirring. The above two solutions were blended and stirred vigorously at room temperature for 1 h. The mixture solution and prepared CC (2×4 cm^2) were sealed and maintained at 160 °C for 24 h in a 50 mL Teflon-lined autoclave. The sample was cooled to an ambient temperature naturally after the reaction. Then, the prepared sample was rinsed several times with absolute ethanol and deionized water, respectively. Finally, it was dried at 60 °C for 6 h, and the 1:5 NiS_2/VS_2 was obtained.

For investigating the effect of NiS_2 on electrocatalytic HER and OER activities, VS_2, 1:10 NiS_2/VS_2, 1:7.5 NiS_2/VS_2, and 1:2 NiS_2/VS_2 were synthesized by a similar method with 1:5 NiS_2/VS_2 in terms of finely varying the additive amount of $NiSO_4{\cdot}6H_2O$. More specifically, 0 mmol, 0.1 mmol, 0.133 mmol, and 0.5 mmol $NiSO_4{\cdot}6H_2O$ were added into the precursor solution for preparing VS_2, 1:10 NiS_2/VS_2, 1:7.5 NiS_2/VS_2, and 1:2 NiS_2/VS_2, individually.

3.4. Preparation of 350 s NiFe LDH/NiS_2/VS_2

The hydrothermal synthetic method for 1:5 NiS_2/VS is based on the above-mentioned method. NiFe LDH was grown through an electrodeposition method on the surface of 1:5 NiS_2/VS_2. Then, 0.025 M $NiSO_4$ and 0.0125 M $FeSO_4$ solutions were homogeneously mixed in a three-electrode system with saturated calomel electrode (SCE), NiS_2/VS_2, and Pt plate as the reference, working, and counter electrode, respectively. The electrodeposition process was operated at a constant voltage of -1.0 V (vs. SCE) at room temperature with different electrodeposition times (250 s, 300 s, 350 s and 400 s), and the corresponding samples were denoted 250 s NiFe LDH/NiS_2/VS_2, 300 s NiFe LDH/NiS_2/VS_2, 350 s NiFe LDH/NiS_2/VS_2 and 400 s NiFe LDH/NiS_2/VS_2, respectively. Finally, these as-obtained catalytic electrodes were gently washed several times using deionized water and then dried naturally.

As a reference, NiFe LDH was grown on pretreated CC using the same electrodeposition synthetic process for 350 s with the absence of 1:5 NiS_2/VS_2.

3.5. Structural Characterization

The phase composition was determined on an XRD-7000 s device (Shimadzu, Kyoto, Japan) with Cu Kα radiation (λ = 1.5418 Å) in a 2 Theta scanning range from 10 to 80° at a current of 40 mA and a voltage of 40 kV. The morphology of the electrocatalysts was measured by means of a scanning electron microscope (SEM, Sigma 300, ZEISS, Oberkochen, Germany), and the energy dispersive spectrum (EDS) was recorded on Oxford Xplore. The transmission electron microscopy (TEM) and high-resolution TEM (HRTEM) images were determined using a TEM instrument (Talos F200×, FEI, Hillsboro, OR, USA) with a field-emission gun operated at 200 kV, and the EDS was recorded. Surface chemical analysis was performed by X-ray photoelectron spectroscopy (XPS, Thermo Scientific, K-Alpha Nexsa, Waltham, MA, USA) using monochromatic Al Kα radiation (1486.6 eV).

3.6. Electrochemical Measurements

The electrocatalytic performances for both HER and OER were carried out on a CHI660E electrochemical workstation using a traditional three-electrode system and an H-type electrochemical cell in an electrolyte of 1.0 M KOH. The as-prepared electrode with a geometric area of 1.0 cm^2, the saturated calomel electrode (SCE), and the Pt plate were individually employed as working, reference and counter electrodes. The working electrode and SCE reference electrode were placed in the same chamber, and the Pt plate was placed in the others. All potentials in the present work were measured compared to reversible hydrogen evolution (RHE) in accordance with the following formula ($E_{RHE} = E_{SCE} + 0.0591pH + E^{\theta}{}_{SCE}$) [58,59]. Before testing, the working electrode was stabilized and the surface was cleaned using a cyclic voltammetry (CV) method. The catalytic activity of each electrocatalyst for HER and OER was estimated through linear sweep voltammetry (LSV) at a scanning rate of 5 mV s^{-1}. According to the formula (ECSA = C_{dl}/C_s), the electrochemical active area (ECSA) was determined by calculating the double-layer capacitance (C_{dl}) values, where C_{dl} can be calculated by CV measurements in the non-Faradaic region with variable scan rates, and C_s represents the specific capacitance [60,61]. The Tafel plots were measured in the same testing system for unraveling the electrocatalytic kinetics of the as-prepared samples. To probe the electron transfer, the electrochemical impedance spectroscopy (EIS) was measured by application of an alternating voltage of 10 mV amplitude with a frequency ranging from 0.1 Hz to 100 kHz. The electrocatalytic stability was evaluated by current–time curves (i–t). Moreover, the potentials related to OER and HER were set as 0.5 V (vs. SCE) and −1.2 V (vs. SCE), respectively. The electrocatalytic overall water-splitting activity of the as-prepared electrocatalyst was measured by a two-electrode configuration, in which the NiFe LDH/NiS_2/VS_2 was employed as an electrocatalyst for the HER cathode and the OER anode in 1.0 M KOH.

4. Conclusions

In conclusion, NiFe LDH/NiS_2/VS_2 heterostructured electrocatalysts have been designed as bifunctional high-efficiency electrocatalysts to split water using a simple method, in which VS_2 with metallic properties and the heterostructured construction endow enhanced electron conductivity and smaller charge transfer resistance to the NiFe LDH/NiS_2/VS_2 electrocatalyst. Furthermore, the formation of a nanoflower structure through decorating NiFe LDH on the surface of NiS_2/VS_2 via a facile electrodeposited method realizes more exposed active sites. The interfacial electron interaction between NiS_2/VS_2 and NiFe LDH was expected to enhance its electrocatalytic activity for overall water splitting. Remarkably, the NiFe LDH/NiS_2/VS_2 exhibits superior electrocatalytic activity compared to VS_2, NiS_2/VS_2 and NiFe LDH; it can achieve a current density of 10 mA cm^{-2} at an overpotential of 76 and 286 mV for HER and OER, respectively. At the same time, the corresponding Tafel plots also decreased significantly. Furthermore, the two-electrode system for overall water splitting was constructed by employing the NiFe LDH/NiS_2/VS_2 as a bifunctional electrocatalyst, which requires an overpotential of 380 mV to achieve a current density of 10 mA cm^{-2}. This study presents a tactic base regarding the interfacial regulation of NiFe LDH, NiS_2 and VS_2 to enhance electrocatalytic OER and HER activity, which is helpful for developing highly efficient electrocatalysts to split water.

Supplementary Materials: The following supporting information can be downloaded at: https://www.mdpi.com/article/10.3390/molecules29050951/s1, Figure S1. (a) XRD patterns and (b) XPS survey spectra of the prepared VS_2, NiS_2/VS_2 and NiFe LDH/NiS_2/VS_2. Figure S2. Electrocatalytic OER activity of (a) the VS_2 and NiS_2/VS_2 with different addition ratios of nickel source and (b) the NiFe LDH/NiS_2/VS_2 with different electrodeposition times. Figure S3. Cyclic voltammetry curves of the (a) VS_2, (b) NiS_2/VS_2, (c) 250 s NiFe LDH/NiS_2/VS_2, (d) 300 s NiFe LDH/NiS_2/VS_2, (e) 350 s NiFe LDH/NiS_2/VS_2, (f) 400 s NiFe LDH/NiS_2/VS_2 and (g) 350 s NiFe LDH in the non-faradaic region with different scan rates. (h) The C_{dl} value of the corresponding samples. Figure S4. Electrocatalytic HER activity of (a) the VS_2 and NiS_2/VS_2 with different addition ratios of nickel source and (b) the NiFe LDH/NiS_2/VS_2 with different electrodeposition times. Figure S5. (a) ECSA and the electrocatalytic activity normalized by ECSA for (b) OER and (c) HER of the as-prepared VS_2, NiS_2/VS_2, NiFe LDH and NiFe LDH/NiS_2/VS2. Figure S6. High-resolution XPS survey of the NiFe LDH/NiS_2/VS_2 before and after stability test; (a) V 2p, (b) S 2p, (c) Ni 2p and (d) Fe 2p. Table S1. Comparisons of HER and OER activity of NiFe LDH/NiS_2/VS_2 with other electrocatalysts in 1.0 M KOH. Figure S7. SEM of NiFe LDH/NiS_2/VS_2 after stability test. Video S1: The dynamic state of the constructed overall water-splitting system. Refs [62–74] are cited in the Supplementary Materials.

Author Contributions: Investigation, T.W. and X.Z.; Methodology, J.L. and K.W.; Supervision, X.Y.; Validation, J.N.; Writing—Original Draft, T.W. and X.Z.; Resources, X.Y. All authors have read and agreed to the published version of the manuscript.

Funding: This work is financially supported by the Doctoral Scientific Research Fund of Xi'an University of Technology, the Natural Science Basic Research Program of Shaanxi (No. 2023-JC-QN-0131, No. 2022JM-082), China Postdoctoral Science Foundation (No. 2020M673615XB).

Institutional Review Board Statement: Not applicable.

Informed Consent Statement: Not applicable.

Data Availability Statement: The data presented in this study are available on request from the corresponding author.

Conflicts of Interest: The authors declare no conflicts of interest.

erences

Xu, H.; Shang, H.; Wang, C.; Du, Y. Surface and interface engineering of noble-metal-free electrocatalysts for efficient overall water splitting. *Coordin. Chem. Rev.* **2020**, *418*, 213374. [CrossRef]

Ahsan, M.; He, T.; Noveron, J.; Reuter, K.; Santiago, A.; Luque, R. Low-dimensional heterostructures for advanced electrocatalysis: An experimental and computational perspective. *Chem. Soc. Rev.* **2022**, *51*, 812–828. [CrossRef] [PubMed]

Tang, J.; Xu, X.; Tang, T.; Zhong, Y.; Shao, Z. Perovskite-Based Electrocatalysts for Cost-Effective Ultrahigh-Current-Density Water Splitting in Anion Exchange Membrane Electrolyzer Cell. *Small Methods* **2022**, *6*, 2201099. [CrossRef]

Abdelghafar, F.; Xu, X.; Jiang, S.P.; Shao, Z. Designing single-atom catalysts toward improved alkaline hydrogen evolution reaction. *Mater. Rep. Energy* **2022**, *2*, 100144. [CrossRef]

Luo, Z.; Zhang, L.; Wu, L.; Wang, L.; Zhang, Q.; Ren, X.; Sun, X. Atomic-scale insights into the role of non-covalent interactions in electrocatalytic hydrogen evolution reaction. *Nano Energy* **2022**, *102*, 107654. [CrossRef]

Yang, Y.; Liu, L.; Chen, S.; Yan, W.; Zhou, H.; Zhang, X.; Fan, X. Tuning Binding Strength of Multiple Intermediates towards Efficient pH-universal Electrocatalytic Hydrogen Evolution by Mo_8O_{26}-NbN_xO_y Heterocatalysts. *Angew. Chem. Int. Ed.* **2023**, *62*, e202306896. [CrossRef] [PubMed]

Hou, Z.; Cui, C.; Li, Y.; Gao, Y.; Zhu, D.; Gu, Y.; Pan, G.; Zhu, Y.; Zhang, T. Lattice-Strain Engineering for Heterogenous Electrocatalytic Oxygen Evolution Reaction. *Adv. Mater.* **2023**, *35*, e2209876. [CrossRef]

Li, C.; Zhao, J.; Xie, L.; Wu, J.; Ren, Q.; Wang, Y.; Li, G. Surface-Adsorbed Carboxylate Ligands on Layered Double Hydroxides/Metal–Organic Frameworks Promote the Electrocatalytic Oxygen Evolution Reaction. *Angew. Chem. Int. Ed.* **2021**, *60*, 18129–18137. [CrossRef]

Jiao, S.; Fu, X.; Wang, S.; Zhao, Y. Perfecting electrocatalysts via imperfections: Towards the large-scale deployment of water electrolysis technology. *Energy Environ. Sci.* **2021**, *14*, 1722–1770. [CrossRef]

Jiao, Y.-Q.; Yan, H.-J.; Tian, C.-G.; Fu, H.-G. Structure Engineering and Electronic Modulation of Transition Metal Interstitial Compounds for Electrocatalytic Water Splitting. *Acc. Mater. Res.* **2022**, *4*, 42–56. [CrossRef]

11. He, X.; Han, X.; Zhou, X.; Chen, J.; Wang, J.; Chen, Y.; Yu, L.; Zhang, N.; Li, J.; Wang, S.; et al. Electronic modulation Pt-incorporated NiFe layered double hydroxide for ultrastable overall water splitting at 1000 mA cm^{-2}. *Appl. Catal. B Env* **2023**, *331*, 122683. [CrossRef]
12. Zahran, Z.N.; Mohamed, E.A.; Tsubonouchi, Y.; Ishizaki, M.; Togashi, T.; Kurihara, M.; Saito, K.; Yui, T.; Yagi, M. Electrocata water splitting with unprecedentedly low overpotentials by nickel sulfide nanowires stuffed into carbon nitride scabbards. *Er Environ. Sci.* **2021**, *14*, 5358–5365. [CrossRef]
13. Guo, Y.; Park, T.; Yi, J.W.; Henzie, J.; Kim, J.; Wang, Z.; Jiang, B.; Bando, Y.; Sugahara, Y.; Tang, J.; et al. Nanoarchitectonic Transition-Metal-Sulfide-Based Electrocatalysts for Water Splitting. *Adv. Mater.* **2019**, *31*, 1807134. [CrossRef] [PubMed]
14. Bhosale, M.; Thangarasu, S.; Murugan, N.; Kim, Y.; Oh, T. Engineering 2D heterostructured VS_2-rGO-Ni nanointerface to tim electrocatalytic water splitting and supercapacitor applications. *J. Energy Storage* **2023**, *73*, 109133. [CrossRef]
15. He, W.; Zheng, X.; Peng, J.; Dong, H.; Wang, J.; Zhao, W. Mo-dopant-strengthened basal-plane activity in VS_2 for accelera hydrogen evolution reaction. *Chem. Eng. J.* **2020**, *396*, 125227. [CrossRef]
16. Wu, Y.; Liu, X.; Han, D.; Song, X.; Shi, L.; Song, Y.; Niu, S.; Xie, Y.; Cai, J.; Wu, S.; et al. Electron density modulation of NiC nanowires by nitrogen incorporation for highly efficient hydrogen evolution catalysis. *Nat. Commun.* **2018**, *9*, 1425. [CrossI
17. Zhang, L.; Rong, J.; Lin, Y.; Yang, Y.; Zhu, H.; Yu, X.; Kang, X.; Chen, C.; Cheng, H.; Liu, G. Mesoporous Single Crystal Microparticles with FeS Clusters Decorated on the Pore Walls for Efficient Electrocatalytic Oxygen Evolution. *Adv. Funct. M* **2023**, *33*, 2307947. [CrossRef]
18. Cheng, Y.; Fu, P.; Yang, X.; Zhang, Y.; Jin, S.; Liu, H.; Shen, Y.; Guo, X.; Chen, L. $MoP_4/Ni_3S_2/MoO_3$ heterogeneous struc nanorod arrays for efficient solar-enhanced overall water splitting. *J. Mater. Chem. A* **2023**, *11*, 24764–24776. [CrossRef]
19. Hou, Z.; Cui, C.; Yang, Y.; Zhang, T. Electrochemical Oxidation Encapsulated Ru Clusters Enable Robust Durability for Effi Oxygen Evolution. *Small* **2023**, *19*, 2207170. [CrossRef]
20. Pei, C.; Kim, M.; Li, Y.; Xia, C.; Kim, J.; So, W.; Yu, X.; Park, H.; Kim, J. Electron Transfer-Induced Metal Spin-Crossov $NiCo_2S_4/ReS_2$ 2D-2D Interfaces for Promoting pH-universal Hydrogen Evolution Reaction. *Adv. Funct. Mater.* **2022**, *33*, 221 [CrossRef]
21. Yu, J.; Qian, Y.; Seo, S.; Liu, Y.; Bui, H.T.; Tran, N.Q.; Lee, J.; Kumar, A.; Wang, H.; Luo, Y.; et al. Exploring catalytic behavic CoS2-ReS2 heterojunction by interfacial engineering. *J. Energy Chem.* **2023**, *85*, 11–18. [CrossRef]
22. Xu, X.; Pan, Y.; Ge, L.; Chen, Y.; Mao, X.; Guan, D.; Li, M.; Zhong, Y.; Hu, Z.; Peterson, V.K.; et al. High-Performance Perov Composite Electrocatalysts Enabled by Controllable Interface Engineering. *Small* **2021**, *17*, 2101573. [CrossRef]
23. Gu, B.S.; Dutta, S.; Hong, Y.; Okello, O.F.N.; Im, H.; Ahn, S.; Choi, S.; Han, J.W.; Ryu, S.; Lee, I.S. Harmonious Heterointer Formed on 2D-Pt Nanodendrites by Facet-Respective Stepwise Metal Deposition for Enhanced Hydrogen Evolution Reac *Angew. Chem. Int. Ed.* **2023**, *62*, e202307816. [CrossRef]
24. Liu, X.; Wang, P.; Liang, X.; Zhang, Q.; Wang, Z.; Liu, Y.; Zheng, Z.; Dai, Y.; Huang, B. Research progress and surface/interf regulation methods for electrophotocatalytic hydrogen production from water splitting. *Mater. Today Energy* **2020**, *18*, 10 [CrossRef]
25. Feng, X.; Jiao, Q.; Chen, W.; Dang, Y.; Dai, Z.; Suib, S.L.; Zhang, J.; Zhao, Y.; Li, H.; Feng, C. Cactus-like $NiCo_2S_4$@NiFe I hollow spheres as an effective oxygen bifunctional electrocatalyst in alkaline solution. *Appl. Catal. B Environ.* **2021**, *286*, 11 [CrossRef]
26. Cao, D.; Shao, J.; Cui, Y.; Zhang, L.; Cheng, D. Interfacial Engineering of Copper-Nickel Selenide Nanodendrites for Enha Overall Water Splitting in Alkali Condition. *Small* **2023**, *19*, 2301613. [CrossRef] [PubMed]
27. Yu, L.; Zhou, H.; Sun, J.; Mishra, I.; Luo, D.; Yu, F.; Yu, Y.; Chen, S.; Ren, Z. Amorphous NiFe layered double hydroxide nanos decorated on 3D nickel phosphide nanoarrays: A hierarchical core-shell electrocatalyst for efficient oxygen evolution. *J. M Chem. A* **2018**, *6*, 13619–13623. [CrossRef]
28. Zhao, Y.; Zhang, X.; Jia, X.; Waterhouse, G.; Shi, R.; Zhang, X.; Zhan, F.; Tao, Y.; Wu, L.; Tung, C.; et al. Sub-3 nm ultra monolayer layered double hydroxide nanosheets for electrochemical water oxidation. *Adv. Energy Mater.* **2018**, *8*, 170 [CrossRef]
29. Zhai, Y.; Ren, X.; Sun, Y.; Li, D.; Wang, B.; Liu, S. Synergistic effect of multiple vacancies to induce lattice oxygen redc NiFe-layered double hydroxide OER catalysts. *Appl. Catal. B Environ.* **2023**, *323*, 122091. [CrossRef]
30. Du, Y.; Liu, D.; Li, T.; Yan, Y.; Liang, Y.; Yan, S.; Zou, Z. A phase transformation-free redox couple mediated electrocata oxygen evolution reaction. *Appl. Catal. B Environ.* **2022**, *306*, 121146. [CrossRef]
31. Feng, X.; Jiao, Q.; Dai, Z.; Dang, Y.; Suib, S.L.; Zhang, J.; Zhao, Y.; Li, H.; Feng, C.; Li, A. Revealing the effect of interfacial ele transfer in heterostructured Co_9S_8@NiFe LDH for enhanced electrocatalytic oxygen evolution. *J. Mater. Chem. A* **2021**, *9*, 1 [CrossRef]
32. Wu, S.-W.; Liu, S.-Q.; Tan, X.-H.; Zhang, W.-Y.; Cadien, K.; Li, Z. Ni_3S_2-embedded NiFe LDH porous nanosheets with abun heterointerfaces for high-current water electrolysis. *Chem. Eng. J.* **2022**, *442*, 136105. [CrossRef]
33. Jia, L.; Du, G.; Han, D.; Hao, Y.; Zhao, W.; Fan, Y.; Su, Q.; Ding, S.; Xu, B. Ni_3S_2/Cu-NiCo LDH heterostructure nanosheet a on Ni foam for electrocatalytic overall water splitting. *J. Mater. Chem. A* **2021**, *9*, 27639–27650. [CrossRef]
34. Li, J.; Liu, X.; Zhao, H.; Zhang, Q.; Du, B.; Lu, L.; Liu, N.; Yang, Y.; Zhao, N.; Pang, X.; et al. Hybrid Nano-Phase Ion/Electron Pathways of Nickel/Cobalt–Boride Cathodes Boosting Intercalation Kinetics for Alkaline Batteries. *ACS Appl. Mater. Inter* **2023**, *15*, 2843–2851. [CrossRef] [PubMed]

Shit, S.; Bolar, S.; Murmu, N.C.; Kuila, T. Tailoring the bifunctional electrocatalytic activity of electrodeposited molybdenum sulfide/iron oxide heterostructure to achieve excellent overall water splitting. *Chem. Eng. J.* **2021**, *417*, 1293333. [CrossRef]

Wang, T.; Liu, X.; Yan, Z.; Teng, Y.; Li, R.; Zhang, J.; Peng, T. Facile Preparation Process of NiCoP-$NiCoSe_2$ Nano-Bilayer Films for Oxygen Evolution Reaction with High Efficiency and Long Duration. *ACS Sustain. Chem. Eng.* **2020**, *8*, 1240–1251. [CrossRef]

Zhu, J.; Cai, L.; Yin, X.; Wang, Z.; Zhang, L.; Ma, H.; Ke, Y.; Du, Y.; Xi, S.; Wee, A.T.S.; et al. Enhanced Electrocatalytic Hydrogen Evolution Activity in Single-Atom Pt-Decorated VS_2 Nanosheets. *ACS Nano* **2020**, *14*, 5600–5608. [CrossRef] [PubMed]

Singh, V.K.; Nakate, U.T.; Bhuyan, P.; Chen, J.; Tran, D.T.; Park, S. Mo/Co doped 1T-VS_2 nanostructures as a superior bifunctional electrocatalyst for overall water splitting in alkaline media. *J. Mater. Chem. A* **2022**, *10*, 9067–9079. [CrossRef]

Xu, D.; Ke, J.; Liu, H.; Bi, Y.; Liu, J. In situ construction of Ni-based N doped porous carbon induced by sulfurization or phosphorization for synergistically enhanced photo/electrocatalytic hydrogen evolution. *J. Mater. Sci. Technol.* **2023**, *179*, 66–78. [CrossRef]

Wang, H.; Liang, M.; Miao, Z. Engineering accordion-like Fe-doped NiS_2 enabling high-performance aqueous supercapacitors and Zn-Ni batteries. *Chem. Eng. J.* **2023**, *470*, 144148. [CrossRef]

Tong, L.; Song, C.; Liu, Y.; Xing, R.; Sekar, K.; Liu, S. Open and porous NiS_2 nanowrinkles grown on non-stoichiometric MoO_x nanorods for high-performance alkaline water electrolysis and supercapacitor. *Int. J. Hydrogen Energy* **2022**, *47*, 14404–14413. [CrossRef]

Yang, Z.; Lin, Y.; Jiao, F.; Li, J.; Wang, J.; Gong, Y. In situ growth of 3D walnut-like nano-architecture Mo-Ni_2P@NiFe LDH/NF arrays for synergistically enhanced overall water splitting. *J. Energy Chem.* **2020**, *49*, 189–197. [CrossRef]

Li, X.; Zheng, L.; Liu, S.; Ouyang, T.; Ye, S.; Liu, Z. Heterostructures of NiFe LDH hierarchically assembled on MoS_2 nanosheets as high-efficiency electrocatalysts for overall water splitting. *Chin. Chem. Lett.* **2022**, *33*, 4761–4765. [CrossRef]

Qu, Y.; Shao, M.; Shao, Y.; Yang, M.; Xu, J.; Kwok, C.T.; Shi, X.; Lu, Z.; Pan, H. Ultra-high electrocatalytic activity of VS_2 nanoflowers for efficient hydrogen evolution reaction. *J. Mater. Chem. A* **2017**, *5*, 15080–15086. [CrossRef]

Chen, X.; Cheng, N.; Ding, Y.-L.; Liu, Z. NiS_2 microsphere/carbon nanotubes hybrids with reinforced concrete structure for potassium ion storage. *J. Electroanal. Chem.* **2021**, *904*, 115852. [CrossRef]

Zheng, Y.; Deng, H.; Feng, H.; Luo, G.; Tu, R.; Zhang, L. Triethanolamine-assisted synthesis of NiFe layered double hydroxide ultrathin nanosheets for efficient oxygen evolution reaction. *J. Colloid Interface Sci.* **2023**, *629*, 610–619. [CrossRef] [PubMed]

Wang, S.; Xiong, D.; Chen, C.; Gu, M.; Yi, F. The controlled fabrication of hierarchical CoS_2@NiS_2 core-shell nanocubes by utilizing prussian blue analogue for enhanced capacitive energy storage performance. *J. Power Sources* **2020**, *450*, 227712. [CrossRef]

Zhang, J.; Zhang, H.; Huang, Y. Electron-rich NiFe layered double hydroxides via interface engineering for boosting electrocatalytic oxygen evolution. *Appl. Catal. B Environ.* **2021**, *297*, 120453. [CrossRef]

Wu, S.; Wang, W.; Shan, J.; Wang, X.; Lu, D.; Zhu, J.; Liu, Z.; Yue, L.; Li, Y. Conductive 1T-VS_2-MXene heterostructured bidirectional electrocatalyst enabling compact Li-S batteries with high volumetric and areal capacity. *Energy Storage Mater.* **2022**, *49*, 153–163. [CrossRef]

Hussain, S.; Vikraman, D.; Sarfraz, M.; Faizan, M.; Patil, S.; Batoo, K.; Nam, K.; Kim, H.; Jung, J. Design of XS_2 (X = W or Mo)-Decorated VS_2 Hybrid Nano-Architectures with Abundant Active Edge Sites for High-Rate Asymmetric Supercapacitors and Hydrogen Evolution Reactions. *Small* **2022**, *19*, 2205881. [CrossRef]

Feng, T.; Ouyang, C.; Zhan, Z.; Lei, T.; Yin, P. Cobalt doping VS_2 on nickel foam as a high efficient electrocatalyst for hydrogen evolution reaction. *Int. J. Hydrogen Energy* **2022**, *47*, 10646–10653. [CrossRef]

Yu, S.H.; Tang, Z.; Shao, Y.; Dai, H.; Wang, H.Y.; Yan, J.; Pan, H.; Chua, D.H.C. In Situ Hybridizing MoS_2 Microflowers on VS_2 Microflakes in a One-Pot CVD Process for Electrolytic Hydrogen Evolution Reaction. *ACS Appl. Energy Mater.* **2019**, *2*, 5799–5808. [CrossRef]

Zhou, W.; Li, X.; Li, X.; Shao, J.; Yang, H.; Chai, X.; Hu, Q.; He, C. Crafting amorphous VO_2–crystalline NiS_2 heterostructures as bifunctional electrocatalysts for efficient water splitting: The different cocatalytic function of VO_2. *Chem. Eng. J.* **2023**, *470*, 144146. [CrossRef]

Zhang, K.; Jia, J.; Tan, L.; Qi, S.; Li, B.; Chen, J.; Li, J.; Lou, Y.; Guo, Y. Morphological and electronic modification of NiS_2 for efficient supercapacitors and hydrogen evolution reaction. *J. Power Sources* **2023**, *577*, 233239. [CrossRef]

Jiang, K.; Li, Q.; Lei, S.; Zhai, M.; Cheng, M.; Xu, L.; Deng, Y.; Xu, H.; Bao, J. Nb-doped NiFe LDH nanosheet with superhydrophilicity and superaerophobicity surface for solar cell-driven electrocatalytic water splitting. *Electrochim. Acta* **2022**, *429*, 140947. [CrossRef]

Li, Y.; Xu, H.; Yang, P.; Li, R.; Wang, D.; Ren, P.; Ji, S.; Lu, X.; Meng, F.; Zhang, J.; et al. Interfacial engineering induced highly efficient CoNiP@NiFe layered double hydroxides bifunctional electrocatalyst for water splitting. *Mater. Today Energy* **2022**, *25*, 100975. [CrossRef]

Zhou, Y.; Guo, Q.; Luo, J.; Wang, X.; Sun, F.; Wang, C.; Wang, S.; Zhang, J. The influence of increased content of Ni(III) in NiFe LDH via Zn doping on electrochemical catalytic oxygen evolution reaction. *Int. J. Hydrogen Energy* **2023**, *48*, 4984–4993. [CrossRef]

Liu, X.; Zhao, X.; Cao, S.; Xu, M.; Wang, Y.; Xue, W.; Li, J. Local hydroxyl enhancement design of NiFe sulfide electrocatalyst toward efficient oxygen evolution reaction. *Appl. Catal. B Environ.* **2023**, *331*, 122715. [CrossRef]

Geng, S.; Sheng, J.; Yang, W.; Tian, F.; Li, M.; Yu, Y.; Liu, Y.; Hou, Y. Activating interfacial S sites of MoS_2 boosts hydrogen evolution electrocatalysis. *Nano Res.* **2021**, *15*, 1809–1816. [CrossRef]

60. Zhou, S.; Wang, J.; Li, J.; Fan, L.; Liu, Z.; Shi, J.; Cai, W. Surface-growing organophosphorus layer on layered double hydro enables boosted and durable electrochemical freshwater/seawater oxidation. *Appl. Catal. B Environ.* **2023**, *332*, 122749. [Cros
61. Jiang, Q.; Wang, S.; Zhang, C.; Sheng, Z.; Zhang, H.; Feng, R.; Ni, Y.; Tang, X.; Gu, Y.; Zhou, X.; et al. Active oxygen sp mediate the iron-promoting electrocatalysis of oxygen evolution reaction on metal oxyhydroxides. *Nat. Commun.* **2023**, *14*, [CrossRef]
62. Bolar, S.; Samanta, P.; Jang, W.; Yang, C.M.; Murmu, N.C.; Kuila, T. Regulating the Metal Concentration for Selective Tuni VS_2/MoS_2 Heterostructures toward Hydrogen Evolution Reaction in Acidic and Alkaline Media. *ACS Appl. Energy Mater.* *5*, 10086–10097. [CrossRef]
63. Patil, S.A.; Rabani, I.; Hussain, S.; Seo, Y.S.; Jung, J.; Shrestha, N.K.; Im, H.; Kim, H. A Facile Design of Solution-Phase Based Multifunctional Electrode for Green Energy Harvesting and Storage. *Nanomaterials* **2022**, *12*, 339. [CrossRef] [PubMed]
64. Patil, S.A.; Shrestha, N.K.; Bui, H.T.; Chavan, V.D.; Kim, D.-K.; Shaikh, S.F.; Ubaidullah, M.; Kim, H.; Im, H. Solvent modu self-assembled VS_2 layered microstructure for electrocatalytic water and urea decomposition. *Int. J. Energy Res.* **202** 8413–8423. [CrossRef]
65. Dhakal, P.P.; Pan, U.N.; Paudel, D.R.; Kandel, M.R.; Kim, N.H.; Lee, J.H. Cobalt-manganese sulfide hybridized Fe-d 1T-Vanadium disulfide 3D-Hierarchical core-shell nanorods for extreme low potential overall water-splitting. *Mater. Today* **2022**, *20*, 100272. [CrossRef]
66. Sekar, K.; Raji, G.; Chen, S.; Liu, S.H.; Xing, R.M. Ultrathin VS_2 nanosheets vertically aligned on $NiCo_2S_4@C_3N_4$ hybri asymmetric supercapacitor and alkaline hydrogen evolution reaction. *Appl. Surf. Sci.* **2020**, *527*, 146856. [CrossRef]
67. Ghorui, U.K.; Mondal, P.; Adhikary, B.; Mondal, A.; Sarkar, A. Newly Designed One-Pot In-Situ Synthesis of VS_2/rGO Nano posite to Explore Its Electrochemical Behavior towards Oxygen Electrode Reactions. *ChemElectroChem* **2022**, *9*, e20220 [CrossRef]
68. Zhang, L.L.; Yang, Y.Q.; Zhu, H.Z.; Cheng, H.M.; Liu, G. Iron-doped NiS_2 microcrystals with exposed {0 0 1} facet electrocatalytic water oxidation. *J. Colloid. Interf. Sci.* **2022**, *608*, 599–604. [CrossRef]
69. Wen, F.S.; Pang, L.; Zhang, T.; Huang, X.L.; Xu, Y.; Li, Y.L. Fe doped NiS_2 derived from metal-organic framework embedded $g\text{-}C_3N_4$ for efficiently oxygen evolution reaction. *Int. J. Hydrogen Energy* **2023**, *48*, 33525–33536. [CrossRef]
70. Zhang, X.Y.; Qiu, Y.L.; Li, Q.; Liu, F.G.; Cui, L.; Li, C.M.; Liu, J.Q. Sandwich structured Fe_3O_4/NiFe LDH/Fe_3O_4 as a bifuncti electrocatalyst with superior stability for highly sustained overall water splitting. *J. Alloys Compd.* **2023**, *932*, 167612. [Cros
71. Wang, J.; Lv, G.; Wang, C. A highly efficient and robust hybrid structure of CoNiN@NiFe LDH for overall water splittin accelerating hydrogen evolution kinetics on NiFe LDH. *Appl. Surf. Sci.* **2021**, *570*, 151182. [CrossRef]
72. Li, Y.T.; Dai, T.Y.; Wu, Q.X.; Lang, X.Y.; Zhao, L.J.; Jiang, Q. Design heterostructure of NiS-NiS_2 on NiFe layered double hydr with Mo doping for efficient overall water splitting. *Mater. Today Energy* **2022**, *23*, 100906. [CrossRef]
73. Zhang, X.; Fan, J.J.; Lu, X.Y.; Han, Z.J.; Cazorla, C.; Hu, L.; Wu, T.; Chu, D. Bridging NiCo layered double hydroxides and N for bifunctional electrocatalysts: The role of vertical graphene. *Chem. Eng. J.* **2021**, *415*, 129048. [CrossRef]
74. Lee, Y.J.; Park, S.K. Metal-Organic Framework-Derived Hollow CoSx Nanoarray Coupled with NiFe Layered Double Hydro as Efficient Bifunctional Electrocatalyst for Overall Water Splitting. *Small* **2022**, *18*, 2200586. [CrossRef]

molecules

MDPI

ew

ılocking the Potential of MBenes in Li/Na-Ion Batteries

n Li [1], Yao Hu [2], Haihui Lan [3] and Huicong Xia [1,4,*]

1 School of Materials Science and Engineering, Zhengzhou University, Zhengzhou 450001, China
2 Materials Science and Engineering Program, Texas Materials Institute, The University of Texas at Austin, Austin, TX 78712, USA
3 Department of Chemistry, Massachusetts Institute of Technology, Cambridge, MA 02139, USA
4 Key Laboratory of Advanced Energy Materials Chemistry (Ministry of Education), Nankai University, Tianjin 300071, China
* Correspondence: hcxia9209@zzu.edu.cn

Abstract

MBenes, an emerging family of two-dimensional transition metal boride materials, are gaining prominence in alkali metal-ion battery research owing to their distinctive stratified architecture, enhanced charge transport properties, and exceptional electrochemical durability. This analysis provides a comprehensive examination of morphological characteristics and fabrication protocols for MBenes, with particular focus on strategies for optimizing energy storage metrics through controlled adjustment of interlayer distance and tailored surface modifications. The discussion highlights these materials' unique capability to host substantial alkali metal ions, translating to exceptional longevity during charge–discharge cycling and remarkable high-current performance in both lithium and sodium battery systems. Current obstacles to materials development are critically evaluated, encompassing precision control in nanoscale synthesis, reproducibility in large-scale production, enhancement of thermodynamic stability, and eco-friendly processing requirements. Prospective research pathways are proposed, including sustainable manufacturing innovations, atomic-level structural tailoring through computational modeling, and expansion into hybrid energy storage-conversion platforms. By integrating fundamental material science principles with practical engineering considerations, this work seeks to establish actionable frameworks for advancing MBene-based technologies toward next-generation electrochemical storage solutions with enhanced energy density and operational reliability.

Keywords: Mbene; metal-ion batteries; two-dimensional materials; energy storage; electrode engineering

1. Introduction

The rapidly growing demand for sustainable and efficient energy storage technologies has become one of the central challenges of the contemporary world [1,2]. As the global community shifts toward renewable energy sources such as solar and wind, which are intermittent by nature, robust energy storage systems are needed to stabilize power supply and maximize energy utilization [3–5]. Among various technologies, rechargeable batteries play a pivotal role due to their high energy density, scalability, and adaptability to diverse applications ranging from portable electronics and electric vehicles to grid-scale energy storage [6]. In this context, lithium-ion batteries (LIBs) have emerged as the dominant technology, owing to their exceptional performance in terms of energy density, cycle life, and efficiency [7,8]. However, concerns regarding Li scarcity, cost, and safety issues have motivated researchers to seek alternative battery chemistries, with sodium-ion batteries (SIBs) attracting substantial attention due to Na's earth abundance and similar electrochemical properties [9,10]. Both LIBs and SIBs rely heavily on electrode materials capable of efficient ion intercalation or storage [11]. Layered materials have proven particularly effective in this regard because their unique structure facilitates ion diffusion and storage while maintaining structural integrity during charge–discharge cycles [12,13]. The past decade has witnessed remarkable advances in two-dimensional (2D) materials, which present high surface areas, tunable interlayer spacing, and plentiful active sites beneficial for electrochemical reactions [14,15].

Since graphene was first isolated, the family of 2D materials has continued to expand, now encompassing a variety of materials with unique physical and chemical properties, such as transition metal dichalcogenides (TMDCs), MXenes, carbon nitrides, hexagonal boron nitride (h-BN), black phosphorus, and phosphorene [5,16]. These materials demonstrate significant application potential in the field of energy storage, with the potential to drive innovations in battery and supercapacitor technologies to meet the growing demand for energy and the requirements of sustainable development (Table 1). Graphene's high conductivity facilitates faster charging and discharging rates in batteries and supercapacitors. Its exceptionally large theoretical specific surface area provides abundant adsorption sites for ions, thereby enhancing the capacity of energy storage devices. TMDCs are a class of two-dimensional materials composed of transition metals (such as Mo, W, Ti, etc.) and chalcogen elements (S, Se, Te), featuring a layered structure with an appropriate interlayer distance that can accommodate the insertion and extraction of various ions [17]. The high specific surface area and abundant surface functional groups of MXenes confer significant advantages in the field of energy storage [18]. Carbon nitrides are a class of two-dimensional materials composed of carbon and nitrogen elements, exhibiting diverse structures and properties. Among these, graphitic carbon nitride (g-C_3N_4) is one of the most extensively studied. g-C_3N_4 possesses an appropriate band structure, making it a promising candidate for applications in photocatalysis and energy storage. h-BN has a layered structure similar to graphene, composed of alternating boron and nitrogen atoms. h-BN exhibits excellent thermal and chemical stability, with high thermal conductivity. Black phosphorus is an allotropic form of phosphorus with a layered structure [19]. When black phosphorus is exfoliated into single or few layers, it is referred to as phosphorene. Phosphorene has a direct bandgap and high carrier mobility, making it a promising candidate for applications in electronic devices and energy storage [20]. Among these, a novel class known as MBenes has recently emerged as a promising candidate for addressing key limitations associated with conventional electrode materials [21,22].

Table 1. Capacity and diffusion energy of several 2D materials.

Materials	Types	Capacity (mA h/g)	Diffusion Energy (eV)	Refs.
Mo_2B_2	LIBs	444	0.270	[23]
TiB	LIBs	480	0.020	[24]
	SIBs	480	0.020	[24]
Ti_2B_2	LIBs	456	0.017	[25]
	SIBs	342	0.008	[25]
Ti_3C_2	LIBs	320	0.280	[26]
	SIBs	370	—	[27]

Table 1. *Cont.*

Materials	Types	Capacity (mA h/g)	Diffusion Energy (eV)	Refs.
Ti_2C	SIBs	359	—	[28]
Graphene	LIBs	372	0.400	[29]
MoS_2	SIBs	669	0.460	[30,31]
Black phosphorus	LIBs	2596	0.080	[32]
Monolayer h-BN	LIBs	762	0.100	[33,34]
	SIBs	571	—	[33]
g-C_3N_4	LIBs	1166	0.570	[35,36]

MBenes are a relatively new family of 2D transition metal borides, analogous to yet distinct from the more widely studied MXenes, which are transition metal carbides or nitrides (Figure 1) [37,38]. While MXenes have gained significant traction in energy storage and conversion due to their metallic conductivity, hydrophilicity, and surface chemistry, MBenes offer unique advantages stemming from their boron-rich composition and structural features [39]. Generally formulated as $M_{n+1}B_n$, where M represents an early transition metal and B stands for boron, MBenes exhibit layered structures that combine high mechanical strength and excellent electrical conductivity with rich surface chemistry [40,41]. These characteristics make them particularly attractive for application as electrode materials in LIBs and SIBs. In LIBs, the MoB MBene synthesized by Xiong et al. maintained a reversible specific capacity of 144.2 mAh/g after 1000 cycles at a current density of 2.0 A/g when used as an anode, outperforming many reported MXene-based anode materials [42]. In SIBs, although the specific capacity of the original MBenes is relatively low, developing MBene-based heterostructures can effectively improve performance. The $Mo_{4/3}B_3T_x$-MoS_2@C composite anode prepared by Liu et al. demonstrated excellent rate capacity (340.6 mAh/g at 1.0 A/g) and durable cycling performance (267.2 mAh/g after 600 cycles at 2.0 A/g) in SIBs [43]. The unique performance exhibited in both battery systems positions MBenes as a promising key material for addressing current energy storage challenges, sparking researchers' enthusiasm for further exploration.

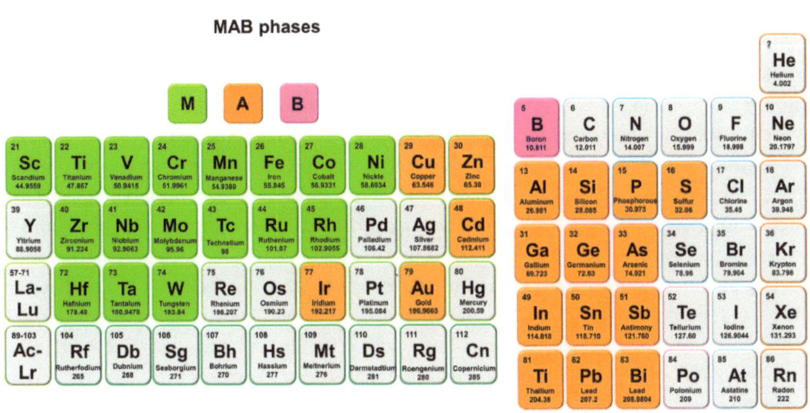

Figure 1. The variety of elements from the periodic table to be used to compose MAB phases.

The conceptual birth of MBenes was closely linked to the exploration of MAX phases and their derivatives (Figure 2) [44]. MAX phases are ternary carbides or nitrides with layered hexagonal structures, featuring alternating layers of early transition metals (M), group A elements (typically from groups 13 or 14), and carbon or nitrogen (X) [45]. The chemical exfoliation of the A-layers from MAX phases results in MXenes, which retain the metal carbide/nitride layers but gain 2D morphology and active surface sites [46]. MBenes differ in that boron layers replace the carbon or nitrogen layers, yielding different bonding environments and electronic properties, thus opening new possibilities for optimizing electrochemical performance. MBenes have attracted intense research interest over the past few years due to their potential to overcome some of the intrinsic challenges faced by other 2D materials in battery applications [47]. These challenges include limited cycling stability, inadequate rate capabilities, and insufficient capacity retention under practical operating conditions. The boron-rich layers in MBenes confer enhanced structural rigidity and chemical stability, which help maintain electrode integrity during repeated ion intercalation and deintercalation. Moreover, the ability to tailor their surface chemistry through functionalization and doping enables optimization of their interaction with electrolyte ions, improving charge transfer kinetics and overall battery efficiency.

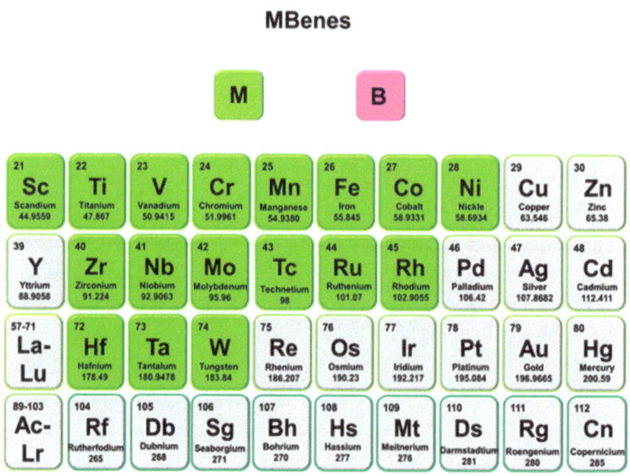

Figure 2. The variety of elements from the periodic table used to compose MBenes.

Early studies have demonstrated promising electrochemical behavior of MBenes when employed as anodes or cathodes in LIBs [48]. Their layered morphology promotes facile Li ion diffusion while providing abundant active sites, resulting in high specific capacities and enhanced rate performance. In SIBs, the relatively larger ionic radius of Na compared to Li introduces additional challenges, such as slower diffusion kinetics and larger volume changes during cycling. MBenes' tunable interlayer spacing can accommodate these larger ions more effectively than some traditional materials, making them strong candidates for advancing SIBs technology. Despite these promising attributes, MBene research is still in its nascent stage, with many fundamental questions regarding their synthesis, structure–property relationships, and long-term cycling behavior yet to be fully explored [49]. The synthesis of high-quality MBene nanosheets with controlled thickness, defect density, and surface terminations remains a key challenge that directly impacts their electrochemical performance [50]. Additionally, understanding the precise mechanisms of ion storage—whether through intercalation, conversion, or pseudocapacitance—is critical for rationally designing MBene-based electrodes with superior durability and capacity [51]. Although MBenes theoretically exhibit excellent electrical conductivity and mechanical properties, these properties often fall short of expectations in practical experiments. This may be attributed to defects or impurities present during the preparation process [21]. A large number of defects can cause a sharp increase in the resistivity of the material, thereby weakening its conductive properties. This is inconsistent with the theoretically predicted high electrical conductivity, limiting its wide application in electronics and other fields [52].

This review provides a comprehensive overview of MBenes, with a particular focus on their roles in LIBs and SIBs (Figure 3). First, we discuss the unique structural features and synthesis routes of MBenes, emphasizing recent advances in scalable and environmentally friendly production methods. Next, we analyze the physicochemical properties that contribute to their suitability as battery electrodes, including electrical conductivity, ion transport characteristics, and surface chemistry. Subsequently, we delve into the state-of-the-art applications of MBenes in Li and Na storage, covering material performance metrics, mechanistic insights, and the benefits of MBene-based composites. Finally, we present the current challenges faced by the field and outline prospective research directions that could unlock the full potential of MBenes for next-generation energy storage systems. By systematically reviewing the progress and prospects of MBenes in these two critically battery technologies, this work aims to underscore their transformative potential and promote further understanding and innovation in sustainable energy storage solutions.

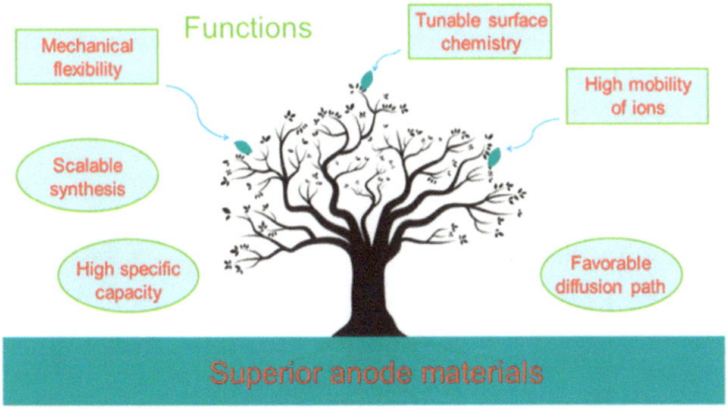

Figure 3. Schematic diagram illustrating the relationship among synthesis, structure, properties, and functions described in this review.

2. Structure, Synthesis, and Physicochemical Properties of MBenes

With the rapid evolution of rechargeable battery technologies, the exploration of innovative electrode materials has become a central focus of materials science. MBenes—a relatively new group in the family of 2D compounds—have garnered widespread attention for their unique structures, versatile synthesis paths, and attractive physicochemical properties [53,54]. These factors combine to make MBenes strong candidates for advancing LIBs and SIBs technologies. This section outlines the fundamental structural characteristics of MBenes, reviews their principal synthesis approaches, and analyzes their core physicochemical attributes in the context of energy storage applications.

2.1. Structural Characteristics

MBenes are defined by the general chemical formula $M_{n+1}B_n$, where M is a transition metal (commonly Mo, Ti, V, Cr, Nb, etc.), B is boron, and 'n' is an integer (usually 1 or 2) (Figure 4) [55,56]. This results in a 2D layered structure; each layer comprises boron atoms sandwiched between layers of transition metal atoms. The strong in-plane metal–boron bonding and relatively weak out-of-plane van der Waals interactions enable these materials to exist as nanoscale sheets, reminiscent of the structural motifs found in other 2D materials such as graphene and MXenes.

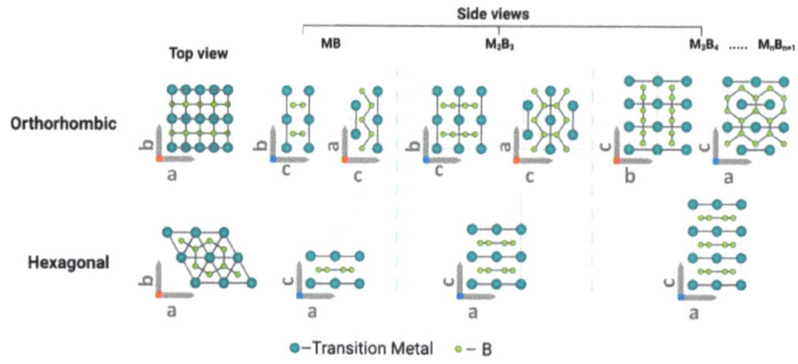

Figure 4. Schematic atomic arrangements predicted for typical MBenes [56]. Copyright 2022, Wiley-VCH.

Structurally, MBenes are closely related to MXenes, which have a general formula of $M_{n+1}X_n$, where X is carbon or nitrogen [57]. The primary distinction lies in the replacement of the carbon or nitrogen atoms with boron; this seemingly small substitution brings significant changes in both electronic structure and chemical reactivity. For instance, while MXenes inherit a combination of metallic conductivity, hydrophilicity, and certain catalytic activities from their parent MAX phases ($M_{n+1}AX_n$, where A is an 'A-site' element like Al or Si), MBenes generally exhibit higher theoretical electrical conductivities, greater mechanical robustness, and a broader range of electron-rich sites due to the boron's unique electron configuration [58].

A key structural benefit of MBenes is their flexible and tunable interlayer spacing [59,60]. This interlayer distance can be precisely controlled by selecting transition metals, surface modifications, or post-synthetic treatments, making it possible to accommodate the varying sizes of guest ions—most notably, Na^+ ions that are relatively large compared with Li^+. This flexibility is an advantage over traditional layered materials such as graphite in Na ion systems, where the insufficient interlayer gap often prevents efficient Na^+ intercalation.

In addition to their layered geometry and variable interlayer spacing, the surfaces of MBenes can be terminated by a variety of functional groups (e.g., –OH, –O, –F), introduced either during synthesis or via post-synthetic treatments [61,62]. These groups play a key role in tuning surface wettability, interfacial reactivity, and the chemical affinity toward ions or electrolytes. Furthermore, like all 2D materials, MBenes are susceptible to lattice defects, edge sites, and vacancies [63]. While excessive defects can negatively impact their long-term stability or electrical conductivity, controlled introduction of vacancies and edge dislocations can create new active sites for ion storage and enhance pseudocapacitive behavior.

2.2. Synthesis Methods

The synthesis of MBenes is foundational to unlocking their full technological promise. Reliable, scalable, and environmentally conscious production methods are required for both academic research and industrial-scale deployment. Several methods have been developed and refined in recent years, each offering unique advantages and posing specific challenges.

Traditional selective etching (de-A Method) (Figure 5a) [64–67] borrows from MXene synthesis. The classical route to MBenes involves selectively removing "A-site" elements from their MAX phase precursors (if such exist), typically using strong acids or molten salts. However, boron's strong bonding with transition metals often demands more aggressive reaction conditions and careful control to avoid structural collapse or contamination.

Via chemical etching (Figure 5b) [68,69], many MBenes can be synthesized directly from binary transition metal borides by using halogen elements (e.g., Cl2 or Br2) or halide salts under high temperatures. This method is less reliant on MAX-like precursors and can yield high-purity products with adjustable thickness and lateral size. However, it often involves hazardous reagents, elevated temperatures, and precise post-processing to remove byproducts or terminate surfaces appropriately.

Topochemical modification and surface functionalization techniques (Figure 5c) [70–73], such as electrochemical treatment, can be used to modify surface chemistry, expand interlayer spacing, and introduce desired terminal groups. This approach is particularly valuable for optimizing ion transport properties or hydrophilicity for battery electrode applications.

Mechanical exfoliation (Figure 5d) [74] is similar to graphene production; MBenes can be mechanically exfoliated from layered bulk precursors by sonication or shear force. While this green and straightforward method avoids toxic reagents, its scalability is limited, and controlling the sheet thickness or size distribution remains challenging.

Regarding green and scalable synthesis (Figure 5e) [75–77], recent attention has focused on eco-friendly and scalable approaches such as low-temperature molten salt synthesis, hydrothermal processes, or even mechanochemical ball milling in the presence of benign solvents. These routes minimize hazardous waste and can be more easily upscaled while maintaining structural and functional diversity. As the field matures, integration of continuous flow reactors, modular process control, and recycling strategies are paving the way toward large-scale industrial production.

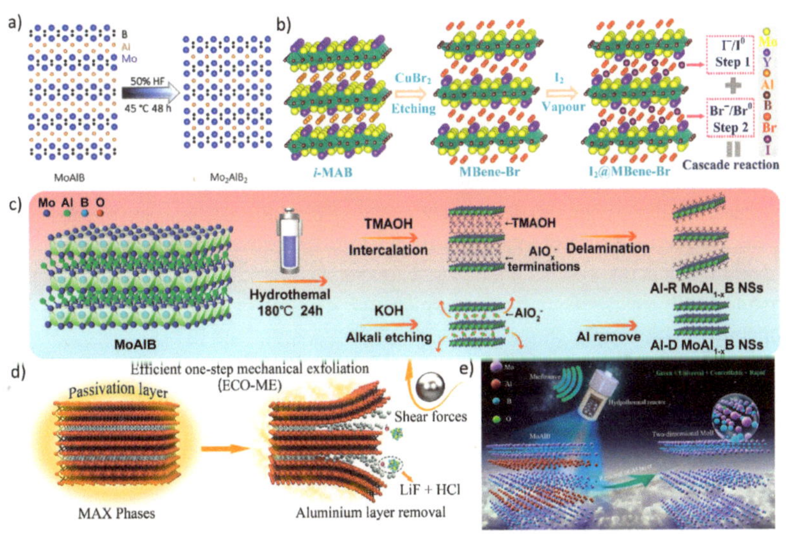

Figure 5. (**a**) Schematic illustration of the structural evolution from MoAlB to Mo_2AlB_2 [78]. Copyright 2023, Wiley-VCH. (**b**) Diagram showing the synthesis of I_2@MBene-Br and the concept of a cascade reaction [79]. Copyright 2024, Wiley-VCH. (**c**) Illustration of the exfoliation method for Al-R and Al-D $MoAl_{1-x}B$ nanosheets in organic and strong alkaline solutions, respectively [80]. Copyright 2025, American Chemical Society. (**d**) Schematic depiction of the efficient one-step mechanical exfoliation (ECO-ME) synthesis pathway for MXenes [74]. Copyright 2023, Elsevier. (**e**) Illustration of the synthesis of 2D MoB MBenes using a microwave-assisted hydrothermal etching process in an alkaline solution [81]. Copyright 2025, Elsevier.

The chosen synthesis method deeply influences the MBene's morphology, defect concentration, and surface terminations, and ultimately, its electrochemical performance in batteries. The future of MBene-based devices hinges on reproducible, safe, and cost-effective manufacturing.

2.3. Physicochemical Properties

The intrinsic physicochemical properties of MBenes provide the foundation for their rapid rise as next-generation battery materials. These features are pivotal to their superior performance in LIBs and SIBs.

Physical properties include electrical conductivity and ion transport [82–84]. Thanks to metallic bonding within the transition metal–boron sheets, MBenes boast excellent in-plane electrical conductivity—often surpassing that of MXenes and rivaling the best-known 2D materials. High electronic mobility ensures reduced internal resistance, enabling rapid charge/discharge capabilities and supporting fast ion transport at electrode/electrolyte interfaces.

With regard to specific surface area and layer spacing [85], the enlarged surface area and customizable interlayer distance provide abundant sites for ion adsorption and intercalation. This enhances reversible capacity and rate performance. Especially in the context of SIBs, where Na ions require larger galleries for migration, the MBene's tunable structure provides a unique platform for efficient and rapid ion diffusion.

Surface chemistry and functional groups are relevant in this context [86]. The presence of polar functional groups (e.g., –OH, –O) further improves hydrophilicity, facilitating intimate contact with aqueous or organic electrolytes. These surface terminations can also participate in faradaic reactions, contributing to pseudocapacitive behavior and elevating the total storage capacity. At the same time, careful control of such chemical moieties is essential to optimize solid electrolyte interphase (SEI) formation, suppress side reactions, and extend cycle life—challenges especially pronounced during long-term cycling in lithium- and sodium-ion batteries.

With regard to their thermal and chemical stability [87], MBenes exhibit robust thermal and chemical stability under a broad range of operating conditions. The strong metal–boron lattice and well-defined surface terminations confer resistance to oxidation, hydrolysis, and other degradation pathways that often plague less stable layered materials. This durability is particularly advantageous for battery applications demanding high safety margins and long operational lifetimes.

The effects of defects should be considered [88–90]. While excessive defects (e.g., vacancies, grain boundaries) can compromise conductivity or structural integrity, a controlled amount of defects can create new redox-active sites, boosting capacity and sometimes even enhancing ion transport. Advanced MBene preparation and post-treatment methods aim to carefully balance these effects for optimal electrochemical outcomes.

In sum, the exceptional physicochemical profile of MBenes is centered on their electrical conductivity, adjustable layer structure, rich surface chemistry, and stability, and sets the stage for their powerful role in the next generation of LIBs and SIBs.

3. MBenes' Applications in Lithium-Ion Batteries

The ongoing advancement of LIBs is essential to meeting the growing energy demands of modern technologies, from electric vehicles to renewable energy storage [91–93]. A primary focus of battery materials research is the development of novel electrode materials that offer higher capacity, enhanced cycling stability, and improved rate capability [94,95]. In recent years, MBenes, a family of 2D transition metal borides, have emerged as an innovative option for advancing LIB performance [96,97]. Their outstanding properties, including high electrical conductivity, robust mechanical strength, adjustable interlayer spacing, and rich surface chemistry, have positioned MBenes as promising candidates to overcome the limitations of conventional materials such as graphite or metal oxides. This section provides a comprehensive overview of the application potential of MBenes in LIBs, discussing their role as electrode materials, the impact of composites, and the complex Li storage mechanisms involved.

3.1. MBene as Electrode Materials: Capacity, Stability and Performance

Among the most appealing features of MBenes are their superior theoretical capacities. While graphite, the archetypal anode in commercial LIBs, is limited to a theoretical capacity of 372 mAh/g, various MBene compositions have demonstrated much higher specific capacities [98,99]. This arises primarily from their unique layered structures. MBenes typically adopt the $M_{n+1}B_n$ formula, where transition metal and boron layers alternate, yielding numerous accessible sites for Li ion intercalation [56]. Some experimental reports have indicated that MBene-based anodes can attain reversible capacities exceeding 500 mAh/g, with studies on Mo_2B_2, Ti_2B_2, and related MBenes revealing both high capacity and excellent reversibility over hundreds of cycles [100].

Theoretical studies predict reversible Li storage capacities for Ti_2B_2 up to 456 mAh/g [25], approaching or surpassing the theoretical maximum of graphite. Wang et al. employed a simple and effective ball milling method to introduce functional indium (In) vacancies into the *h*-MAB phase Ti_2InB_2 (Figure 6) [101]. The In-deficient Ti_2InB_2 *h*-MAB phase (V_{In}-Ti_2InB_2), used as an anode material for LIBs, achieved a high capacity (600 mAh/g) and excellent long-term cycling stability. Compared to recently reported MAX/MAB phase materials, the V_{In}-Ti_2InB_2 electrode in this study stands out in terms of reversible capacity and cycling stability across different current densities. Furthermore, through a combination of theoretical calculations and experimental tests, the researchers proposed a dual redox reaction Li storage mechanism for V_{In}-Ti_2InB_2, involving the Li-In alloying reaction facilitated by indium vacancies on the (010) and (001) surfaces of V_{In}-Ti_2InB_2, and the redox reaction of Li-TiB occurring at the (010) edge and (001) surface of V_{In}-Ti_2InB_2. Research has found that not only can Ti_2InB_2-like configurations stably exist in the Zr-A-B and Hf-A-B systems, but two new configurations, "211" and "314", have also been discovered (Figure 7) [102]. Notably, the identification of Hf_2PbB and Hf_2BiB structures provides a potential pathway for extending traditional MAX phases toward borides. High-precision theoretical calculations indicate that ZrB monolayers are excellent candidate materials for Li-ion battery electrodes. In addition, Wang and colleagues combined materials genome prediction with experimental validation to discover a series of novel hexagonal ternary MAB phases and their two-dimensional derivatives, hexagonal MBenes (Figure 8) [52]. Theoretical calculations revealed that the hexagonal crystal system ternary MAB phases (*h*-MAB) served as superior precursors for MBenes compared with the conventional orthorhombic crystal system MAB phases (ort-MAB). Based on theoretical predictions, the authors successfully synthesized three representative structural prototype materials of the h-MAB phase (Hf_2InB_2, V_3PB_4, and Hf_2PbB). Taking *h*-MAB Hf_2InB_2as an example, they selectively etched away the In layer to successfully prepare the first hexagonal MBene (*h*-MBene), namely HfBO. Further application studies found that the 2D HfBO material showed great potential as an anode for LIBs.

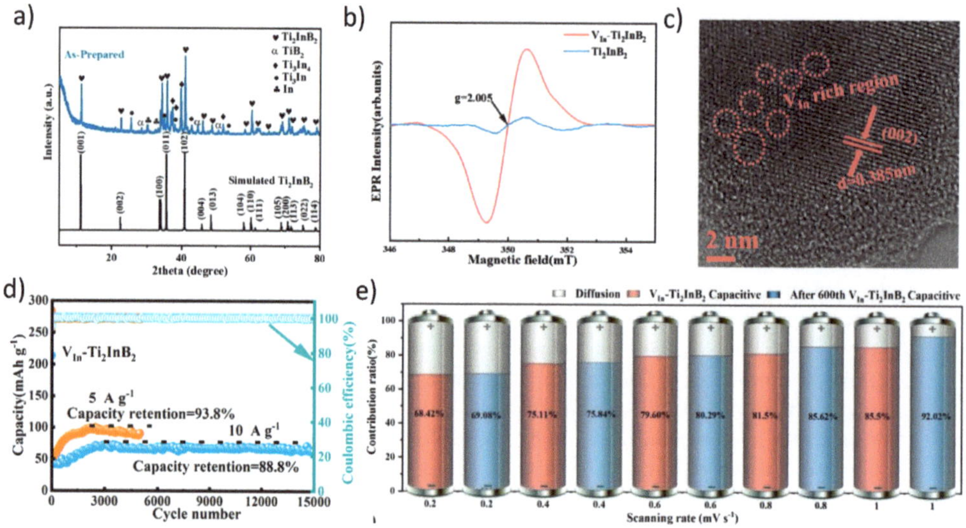

Figure 6. (**a**) XRD patterns and (**b**) EPR spectra of as-prepared Ti_2InB_2. (**c**) HRTEM image of VIn-Ti_2InB_2. (**d**) Long-term cycling performance of V_{In}-Ti_2InB_2 electrodes tested at current densities of 5 and 10 A/g. (**e**) Capacitive contribution ratios of the V_{In}-Ti_2InB_2 electrode at various rates, measured before and after 600 cycles [101]. Copyright 2024, Wiley-VCH.

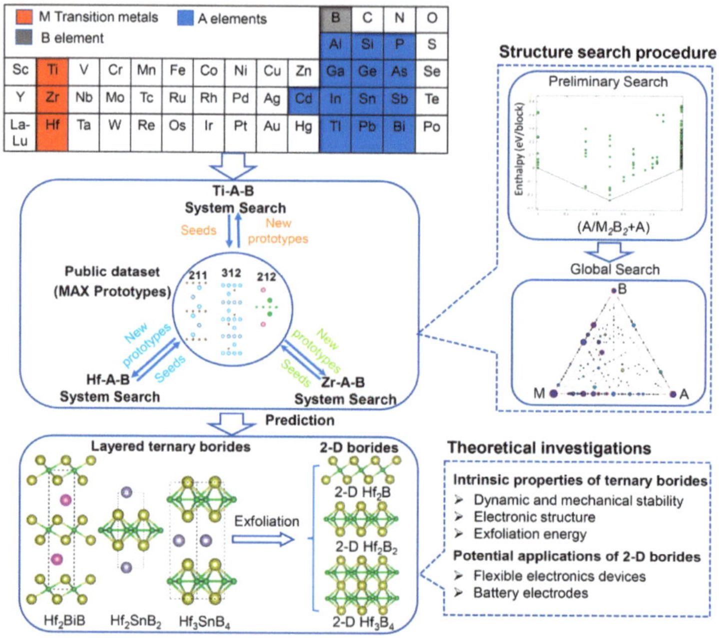

Figure 7. Computational strategy for discovering ternary borides. Each search for ternary compound structures involves two stages—a preliminary pseudo-binary structure search followed by a global ternary structure search. Theoretical analyses were conducted on each predicted layered ternary compound and its corresponding 2D structures [102]. Copyright 2020, American Chemical Society.

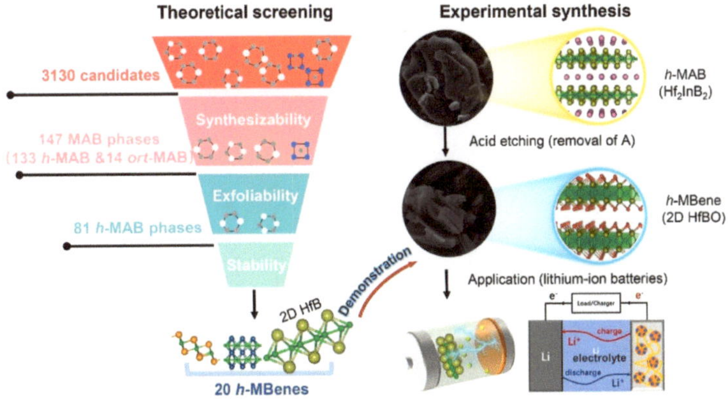

Figure 8. Overview of the calculation-driven approach used to discover *h*-MAB phases and h-MBenes, combining extensive high-throughput computational screening with experimental validation [52]. Copyright 2023, Wiley-VCH.

The metallic nature of MBenes further supports rapid electron transfer during charging and discharging, contributing to their pronounced rate capability. Notably, many MBenes exhibit significant capacity retention even at high current densities, a property essential for fast-charging applications [79,103,104]. This combination of electrical conductivity and structural resilience sets MBenes apart from many emerging materials that suffer from slow kinetics or dramatic fading of capacity under demanding conditions. Cycling stability, a persistent challenge for many high-capacity electrodes, is another area where MBenes excel [89,105]. Owing to the robust covalent bonding within the metal–boron lattice, MBenes can endure the strain associated with Li insertion and extraction without undergoing the mechanical degradation that plagues materials like silicon or some metal oxides. Their planar and flexible nature further allows for minor volume changes during cycling, reducing the risk of cracking or pulverization. Consequently, MBene-based LIBs often demonstrate high Coulombic efficiencies and capacity retention above 90% after several hundred cycles.

Identifying novel anode materials with high energy density, excellent cycle stability, and outstanding rate performance has become a key research direction in LIBs development. Chen et al. [106] successfully prepared Mo_xB_y (MBene) derivatives with an accordion-like structure by completely etching the Al layer in MoAlB using the molten salt method. The b values for Peak 1 and Peak 2 were calculated to be 0.95 and 0.91, respectively, indicating that the energy storage kinetics of the MBene anode were dominated by a pseudocapacitive mechanism (Figure 9a,b). When used as an advanced anode material for LIBs, this MBene exhibited outstanding electrochemical performance, achieving a reversible specific capacity of 638.2 mAh/g after 100 cycles at a current density of 0.1 A/g (Figure 9c–e). Wang et al. [107] successfully synthesized a novel Co_9S_8-MoB MBene heterostructure material. The Co_9S_8-MoB MBene electrode exhibited outstanding performance, characterized by significant pseudocapacitive behavior. It maintained a specific capacity of 756.34 mAh/g after 200 cycles at a current density of 100 mA/g, with a cycle retention rate as high as 91.27%. The excellent cycle stability and rate performance were attributed to the two-dimensional layered structure of the conductive MoB MBene (Figure 9f–i).

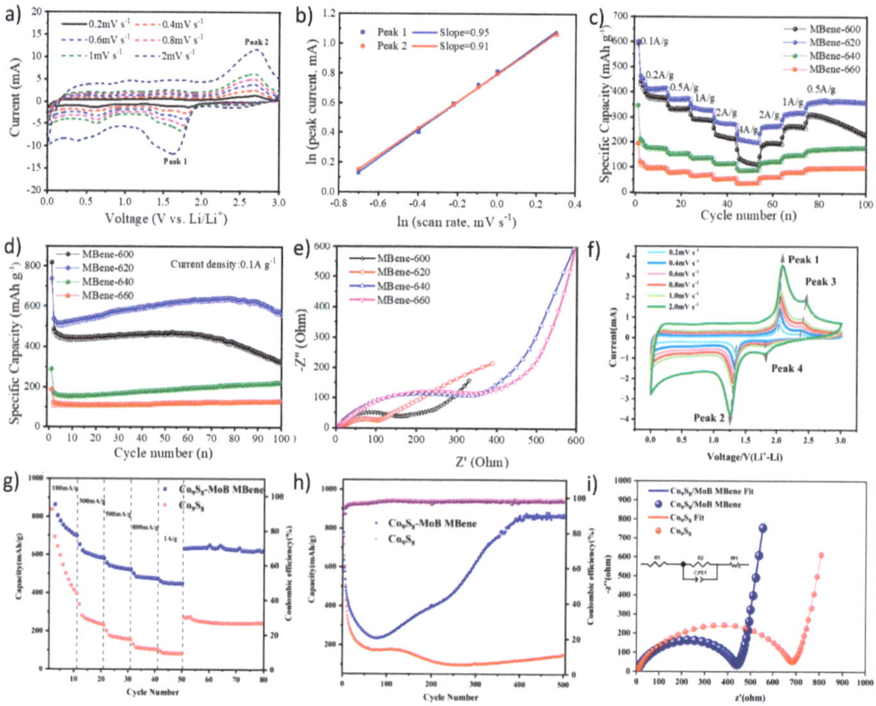

Figure 9. (**a**) CV curves of MBene-620 with the gradually increasing scanning rate. (**b**) Fitting diagram of relationship between ln (peak current) and ln (scan rate). (**c**) Rate capability of MBenes prepared at different temperatures, (**d**) cycle performance, and (**e**) EIS spectra of MBene anodes in LIBs [106]. Copyright 2024. The Royal Society of Chemistry. (**f**) CV curves of Co_9S_8-MoB MBene electrode at different scan rates (0.2–2.0 mV/s). (**g**) Rate performance, (**h**) cyclic specific capacity profiles at a current density of 300 mA/g, and (**i**) EIS of Co_9S_8-MoB MBene and Co_9S_8 electrodes in LIBs [107]. Copyright 2025. Elsevier Inc.

3.2. MBene-Based Composite Systems: Synergy and Engineering

Despite their outstanding intrinsic properties, pure MBene electrodes are not immune to practical challenges, such as tendencies toward aggregation, restacking of nanosheets, and limited first-cycle efficiency due to SEI formation [108,109]. To mitigate these issues and enhance performance further, researchers have developed a range of composite electrode architectures that harness the synergistic effects of MBene with other functional materials.

Integrating MBenes with conductive carbon materials like graphene, carbon nanotubes, or amorphous carbon, carbon composites [110–112] benefit from enhanced electronic connectivity, flexibility, and dispersion. The carbon matrix acts to prevent restacking, thus preserving the accessible surface area for Li storage and facilitating fast charge–discharge kinetics. Hybrid structures such as MBene-graphene [113] or MBene-CNT [114] electrodes have been shown to achieve not only higher capacities but also better rate capability and cycling stability compared with pure MBene electrodes.

MBenes can also be combined with electroactive polymers such as polyaniline or PEDOT, as well as with redox-active metal oxides [115,116]. Polymers provide additional mechanical resilience and can buffer electrode expansion during cycling, while metal oxides introduce extra Li storage sites and support hybrid storage mechanisms. These composite designs effectively improve electrode robustness, enhance active material utilization, and suppress unwanted side reactions at the interface, collectively resulting in prolonged cycle life and stable operation under demanding charge–discharge regimes.

Structural design is very important [117–122]. The success of composite approaches relies heavily on precise control over composition, architecture, and nanoscale interface. Recent advances include layer-by-layer assembly, solution-phase co-precipitation, and in-situ hybridization methods that produce electrodes with optimized porosity, hierarchical order, and integrated conductive pathways. Smart design of MBene-based hybrids thus holds the key to balancing high capacity, mechanical integrity, and electrochemical stability in high-performance LIBs.

3.3. Lithium Storage Mechanisms in MBene Electrodes

Understanding the Li storage mechanism in MBene electrodes is crucial for unlocking their full potential and guiding rational material and device design. These mechanisms encompass both traditional intercalation and more complex surface and conversion reactions, the balance of which can be tuned by composition and surface engineering. In LIBs, the energy storage mechanism of MBenes primarily involves efficient Li^+ intercalation/deintercalation, combined with some surface capacitive behavior. Its high conductivity and open layered structure allow for rapid Li^+ diffusion, while changes in transition metal oxidation states (e.g., Mo^{2+}/Mo^{4+}) contribute to redox capacity. Compared with graphite, MBenes have larger interlayer spacing, reducing Li^+ insertion resistance, and surface defects or functional groups may provide additional active sites, improving rate performance and cycle stability.

With regard to intercalation [123,124], the principal Li storage pathway in most MBene electrodes is the reversible insertion of Li ions between their 2D layers. This process minimally disrupts the MBene framework due to its robust metal–boron bonding and adjustable interlayer spacing. In situ X-ray diffraction and electron microscopy studies provide direct evidence that Li intercalation in many MBenes causes uniform, reversible lattice expansion that is distinctly less destructive than the phase transformations observed in alloying materials. Because this intercalation is highly reversible, MBene electrodes show good long-term capacity retention and cycling stability.

The surface chemistry of MBenes, including surface adsorption and pseudocapacitance, significantly impacts their storage behavior [125,126]. The presence of terminal groups like –OH, –O, or –F on MBene surfaces offers abundant sites for Li adsorption, contributing to a pseudocapacitive storage mechanism. Pseudocapacitance enhances high-rate capability as it involves rapid, surface-driven Faradaic reactions as opposed to slow bulk diffusion. Through careful tuning of surface functionalization, researchers can tailor the relative contributions of intercalation and pseudocapacitive processes, optimizing both energy and power densities.

Conversion reactions are also relevant [127–129]. For certain MBene compositions, particularly those containing multi-valent transition metals, Li may participate in partial conversion reactions involving temporary phase changes or redox-driven alloying. While these mechanisms offer opportunities for even higher capacities, they require careful management to avoid irreversible structural changes or capacity fade.

SEI formation and interface stability should also be considered [130,131]. Formation and control of the SEI is a vital consideration for all high-performance LIB electrodes, including MBene-based ones. The nature of MBene surfaces and their interactions with the electrolyte strongly influence SEI composition, uniformity, and stability, which in turn affect initial Coulombic efficiency, cycle life, and safety. Recent research suggests that surface-functionalized MBenes can promote formation of stable, compliant SEI layers that resist ongoing decomposition while maintaining efficient Li transport.

In summary, MBenes herald a new era of electrode material for LIBs, owing to their remarkable intrinsic properties, adaptable structures, and chemical versatility. Their high theoretical capacity, excellent rate performance, and robust cycling stability position them as promising candidates for advancing LIB technology beyond the limits of conventional electrodes. The development of MBene-based composites and hybrid architectures further expands design possibilities, allowing for the customization of electrode systems to meet diverse performance requirements. Ongoing research to elucidate the dynamic, multi-mechanistic storage behavior of MBenes, with a focus on optimizing capacity, stability, and safety. As synthetic methods advance and understanding deepens, MBenes are set to play a pivotal role in the next generation of high-performance, durable, and safe LIBs.

4. MBenes Applications in Sodium-Ion Batteries

SIBs have emerged as strong contenders in the race for next-generation energy storage systems, primarily due to Na's abundance, low cost, and environmental compatibility [132–134]. However, the successful development of SIBs has been hampered by several technical challenges, chief among them being the larger ionic radius of Na compared with Li and the resulting difficulty in identifying electrode materials with compatible structures [135]. In this context, MBenes—a new class of 2D boride materials—offer remarkable promise, thanks to their tunable interlayer spacing, exceptional conductivity, and robust mechanical framework [136]. This section explores the structural advantages of MBenes for SIBs, provides a critical comparison with other anode materials, delves into Na storage and migration mechanisms, and highlights the considerable performance enhancements achieved through strategic use of composite materials.

4.1. Structural Advantages of MBene for Accommodating Larger Sodium Ions

One of the defining hurdles in SIB technology is the accommodation of Na ions, which possess a substantially larger ionic radius (1.02 Å) compared with that of Li^+ (0.76 Å) [137]. This size difference leads to sluggish Na ion diffusion and significant structural deformation or even collapse in many conventional layered electrode materials [138]. Most traditional anode materials used in LIBs—such as graphite—are poorly suited to storing Na ions in their host structures, as the narrow interlayer spacing and insufficient flexibility restrict the insertion of Na ions and result in poor reversibility [139].

MBene materials, characterized by their $M_{n+1}B_n$ composition and 2D morphology, inherently offer more adaptable and tunable interlayer spacing than many oxide or carbonaceous materials [140–142]. The strong yet flexible metal–boron framework of MBenes not only stabilizes their structure during repeated ion insertion and extraction but also allows controlled expansion of the interlayer space, which is critical for accommodating the bulkier Na ions. Surface chemistry modifications, including the introduction of polar functional groups or the intercalation of organic/inorganic spacers, are often employed to further increase gallery height and improve ion accessibility [143]. Gao et al. presented a series of orthogonal 2D MBenes based on first-principles density functional theory, which were obtained by mechanically exfoliating an MBene layer from a bulk MAB phase (Figure 10) [144]. These included V_2B_2, Cr_2B_2, Mn_2B_2, Ti_2B_2, Zr_2B_2, and Nb_2B_2. The thermodynamic and kinetic stability of monolayer MBenes at room temperature was confirmed through AIMD simulations and phonon spectrum analysis. In additionally, Wang et al. reported two novel Mo_2B_2 monolayers and investigated their stability, electronic structures, lattice dynamics, and potential as anode materials for energy storage, using crystal structure prediction techniques and first-principles calculations (Figure 11) [145]. The calculated phonon spectra and electronic structures revealed that the predicted tetragonal and trigonal Mo_2B_2 monolayers possessed excellent electronic conductivity and strong stability.

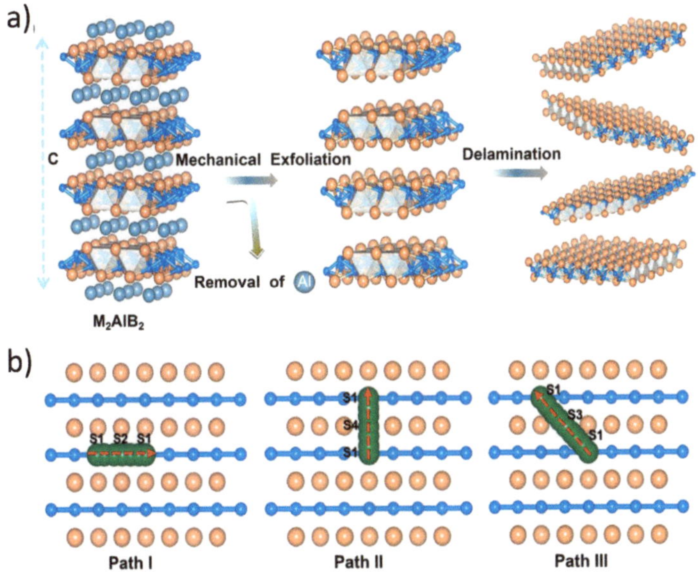

Figure 10. (**a**) Schematic illustration of Al removal to create MBenes through mechanical exfoliation of the MAB phase. (**b**) Diagram showing the metal cation diffusion pathways on monolayer MBenes, including S1 → S2 → S1, S1 → S4 → S1, and S1 → S3 → S1 routes [144]. Copyright 2019, Royal Society of Chemistry.

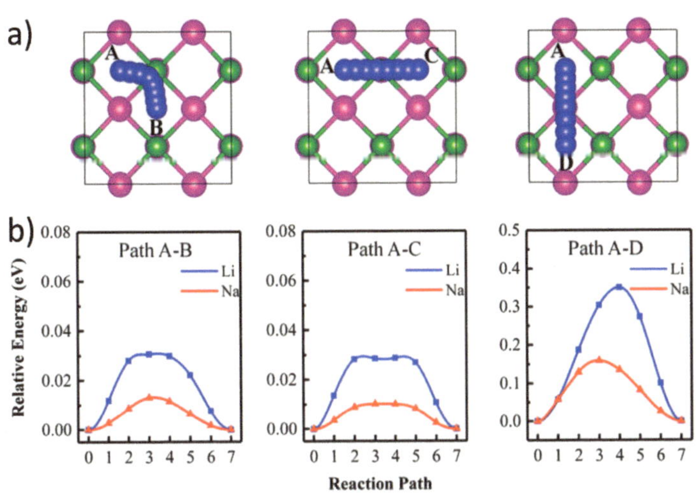

Figure 11. (**a**) Top-down views of three diffusion pathways for Li/Na on the tetr-Mo_2B_2 monolayer. Mo, B, and Li/Na atoms are shown in violet, green, and blue, respectively. (**b**) Diffusion energy barriers for Li and Na on tetr-Mo_2B_2 [145]. Copyright 2019, Royal Society of Chemistry.

Recent studies have shown that MBenes' unique atomic configurations can effectively mitigate common degradation mechanisms such as exfoliation, restacking, and lattice collapse, which are factors that otherwise compromise SIBs' long-term cycling performance [21]. The combination of high mechanical rigidity and structural tunability makes MBenes especially attractive not only for Na ion storage but also for sustaining stable operation over hundreds or thousands of cycles. Xiong et al. [81] proposed an efficient microwave-assisted hydrothermal alkaline solution etching strategy to peel off the MoAlB MAB phase into two-dimensional MoB MBenes with a good wrinkled structure. This material exhibited excellent electrochemical performance in sodium-ion batteries (SIBs), achieving a reversible specific capacity of 196.5 mAh/g at a current density of 50 mA/g (Figure 12a–c). Liu et al. [146] proposed a sodium storage strategy combining interface engineering with Na_2S adsorption, vertically growing SnS@C nanosheets on MBenes. The MBenes-SnS@C anode exhibited a high capacity of 411 mAh/g at a current density of 1 A/g and maintained a capacity of 420 mAh/g after 100 cycles at a current density of 0.5 A/g (Figure 12d,e). The CV curves of the MBenes-SnS@C-2 electrode at a scan rate of 0.1–1 mV/s show good redox characteristics and minimal peak displacement, indicating low polarization and fast electrochemical kinetics (Figure 12f–h).

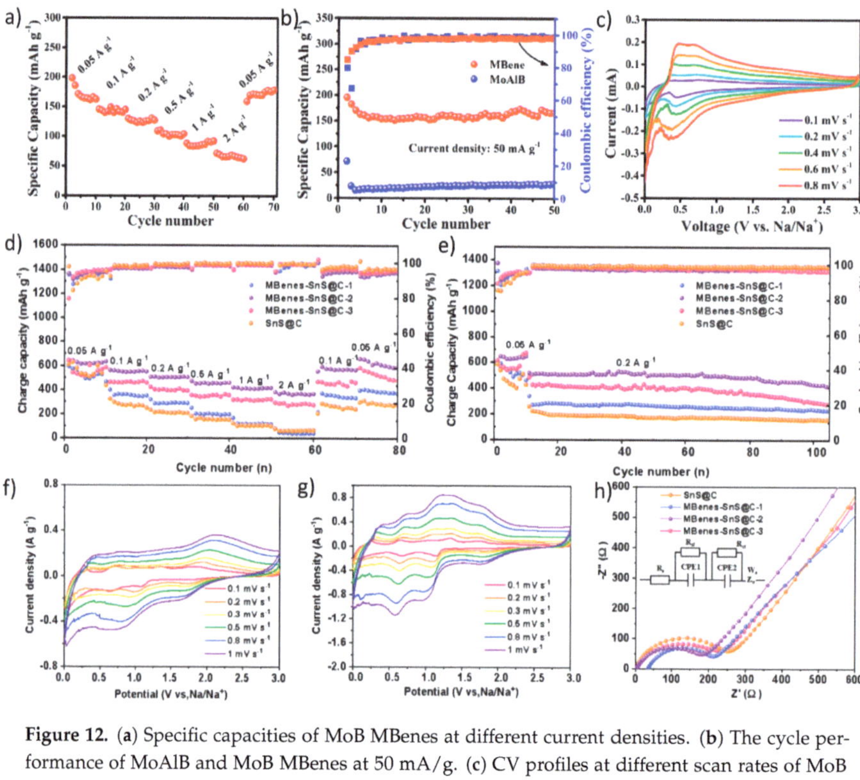

Figure 12. (**a**) Specific capacities of MoB MBenes at different current densities. (**b**) The cycle performance of MoAlB and MoB MBenes at 50 mA/g. (**c**) CV profiles at different scan rates of MoB MBenes [81]. Copyright 2025. Elsevier Ltd. (**d**) Rate performance and (**e**) cyclic performance of SnS@C and MBenes-SnS@C in SIBs. CV profiles of (**f**) SnS@C and (**g**) MBenes-SnS@C-2 at scanning rates of 0.1–1 mV/s. (**h**) EIS of SnS@C and MBenes-SnS@C [146]. Copyright 2025. Elsevier Inc.

4.2. *Performance Comparison: MBene vs. Other Anode Materials*

The practical deployment of any electrode material depends not only on its ability to host Na ions but also on its electrochemical performance, including reversible capacity, rate capability, cyclability, and safety [136,147]. Compared with more conventional materials such as hard carbon, transition metal oxides, or MXenes, MBenes have begun to distinguish themselves with several noteworthy advantages [148–150].

For example, hard carbon is commonly used as a reference anode material in SIBs, offering reversible capacities of ~300 mAh/g and reasonable cycling stability [151–153]. However, its poor rate performance and limited electrical conductivity often restrict its application in fast-charging scenarios. MBene anodes have demonstrated higher intrinsic electrical conductivity, which enables rapid electron transfer and supports higher charge–discharge rates [154]. Laboratory reports reveal that certain MBene compositions (e.g., Mo_2B_2 or Ti_2B_2-based MBenes) can deliver initial reversible capacities between 350–500 mAh/g with excellent rate capability and retention of 80–90% of initial capacity after several hundred cycles under practical current densities [40,81].

Compared with MXenes, MBenes sometimes offer superior cycling stability given their stronger metal–boron bonds (as opposed to metal–carbon/nitrogen) that can better withstand repeated sodiation and desodiation [155–157]. While MXene anodes may suffer from rapid capacity fade due to structural fatigue, MBenes' robust lattice provides resilience and supports more stable SEI formation [158]. Metal oxides, though they may promise high capacity, often experience significant irreversible capacity loss and poor rate capability due to volume expansion and sluggish Na ion diffusion, which are drawbacks that MBene architecture can help ameliorate [159].

Despite these advantages, MBene anodes are not without drawbacks. Their initial Coulombic efficiency can be somewhat lower than that of traditional hard carbons due to the formation of SEI or irreversible trapping of Na at defect or functional group sites [160]. Still, ongoing research on surface engineering and electrolyte optimization is addressing these limitations, gradually aligning MBenes' practical and theoretical performance.

4.3. Sodium Storage and Migration Mechanisms in MBenes

A nuanced understanding of how Na ions are stored, migrate, and interact within host materials is essential for optimizing performance and predicting long-term durability. Na storage in MBene electrodes is widely recognized to follow a hybrid mechanism that encompasses intercalation between layers, pseudocapacitive surface adsorption, and, in certain cases, conversion reactions involving active transition metals. In SIBs, the charge storage mechanism of MBenes primarily relies on interlayer insertion reactions and surface adsorption/redox reactions. Na^+ can be inserted into the interlayer gaps of MBenes, while the abundant transition metal sites on the surface (e.g., Mo, Ti, etc.) provide additional pseudocapacitive capacity through redox reactions. The layered structure and tunable interlayer spacing of MBenes help mitigate volume expansion during Na^+ intercalation, while surface functional groups (e.g., –O, –F) may participate in ion adsorption, enhancing interfacial charge transfer.

Intercalation [161,162] is the primary Na storage mechanism in most MBenes, wherein Na ions insert into galleries between MBene layers during discharge. The process is largely reversible, especially when the interlayer spacing is optimized. In situ X-ray diffraction and atomic force microscopy have shown that sodiation leads to moderate, reversible expansion of the MBene lattice, in a process far less destructive than observed in many oxide or alloying-type anodes. The strong metal–boron bonds and the planar geometry of MBene flakes help contain the structural changes associated with sodiation and desodiation, preventing pulverization or exfoliation.

Regarding surface adsorption and pseudocapacitance [163,164], MBene sheets are typically terminated with surface functionalities such as –OH, –O, or –F due to synthesis protocols. These groups provide abundant high-energy adsorption sites for Na ions, leading to substantial pseudocapacitive contributions. Such surface-driven storage mechanisms enhance the high-rate performance and capacitance of MBene anodes, facilitating the fast-charging and discharging crucial for modern SIB applications.

In some cases, Na storage may involve partial conversion reactions [165,166], forming transient nanophases or alloys within the material during cycling, especially for MBenes containing redox-active transition metals. While these reactions can provide extra capacity, they must be carefully managed to prevent irreversible phase changes and capacity loss over long cycles.

Na ion migration through MBene electrodes is facilitated by both the intrinsic conductivity of the material and the accessible, low-energy diffusion pathways within and on the surfaces of the layers. Computational investigations using density functional theory have highlighted low Na ion diffusion barriers on MBene surfaces, often lower than for analogous carbon or oxide electrodes, thus confirming the theoretical advantages of MBene frameworks for SIB operation.

4.4. MBene-Based Composite Materials: Enhancing Sodium-Ion Battery Performance

To overcome remaining limitations such as layer stacking, agglomeration, and sub-optimal Coulombic efficiency, numerous studies have explored the integration of MBenes with other functional materials to yield high-performance composites. These composite designs have often been inspired by analogous efforts in MXene and graphene research, leveraging synergies between the constituents for higher overall battery performance.

With regard to MBene/carbon composites [167,168], the incorporation of conductive carbon frameworks (e.g., graphene, carbon nanotubes, amorphous carbon) with MBene nanosheets is highly effective for preventing restacking, improving electrode flexibility, and promoting electron/ion transport. Such hybrid electrodes exhibit increased surface area, suppressed aggregation, and enhanced rate capability, enabling practical capacities and long-cycle durability under demanding conditions.

Polymer and metal oxide composites have been used [169,170]. Polymers, especially those with elastic or self-healing properties, can buffer volume changes during repeated sodiation/desodiation and improve structural robustness. Metal oxides, when homogeneously integrated with MBenes, may introduce additional Na storage sites while acting as spacers to keep the MBene layers apart, further enhancing cycling stability and rate performance.

For optimization of surface chemistry [171,172], surface engineering remains a key method for tuning Na affinity, controlling SEI formation, and improving first-cycle efficiency. Rational choice of composite type, composition ratio, and synthesis method—whether in situ assembly, physical mixing, or solution-phase hybridization—offers a powerful toolkit for tailoring MBene-based electrodes to the specific needs of SIBs technologies.

MBene materials have rapidly established themselves as among the most exciting 2D platforms for next-generation SIBs. Their unique structural adaptability, robust bonding, and outstanding electrical properties address several fundamental challenges facing this battery technology. However, to reach full industrial relevance, further research is needed to optimize synthesis, engineer surfaces and interfaces, and master the design of multifunctional composite electrodes. With continued innovation, MBene-based materials may soon enable production of high-performance, cost-effective, and long-lived SIBs with far-reaching impact for grid storage and beyond.

5. Challenges and Development Trends

The emergence of MBenes as novel 2D boride materials has created exciting opportunities to revolutionizing LIB and SIB technologies. Despite their unique structures and outstanding electrochemical properties, significant scientific and technological challenges remain before MBenes can be realistically implemented in commercial electrochemical storage systems. Addressing these issues is essential for translating the impressive laboratory results of MBenes into meaningful real-world battery advancements. In this section, we discuss the main challenges facing the application of MBenes in these batteries and highlight research trends aimed at overcoming them (Figure 13).

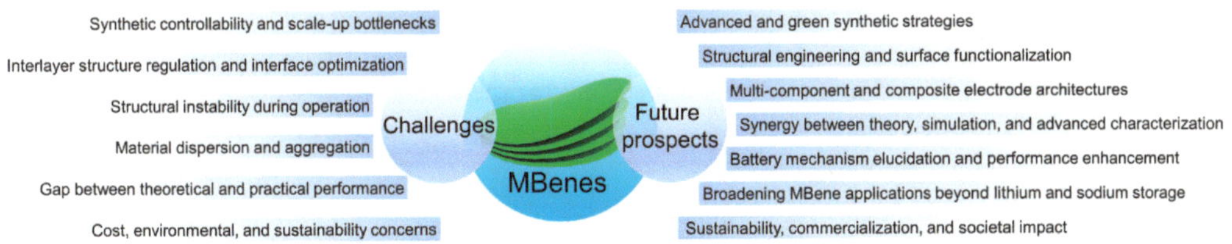

Figure 13. Schematic representation of current challenges and future prospects for MBenes.

(I) Synthetic controllability and scale-up bottlenecks

A major obstacle hindering the widespread adoption of MBene-based electrodes is the challenge of achieving highly controllable, reproducible, and scalable synthetic methods. Most conventional approaches for producing MBenes, such as selective chemical etching and top-down exfoliation of bulk precursors, typically require harsh chemicals (e.g., strong acids or fluorides), stringent environmental controls, and labor-intensive procedures. Research has found that a 2 M LiF/6 M HCl solution causes corrosion of the Mo_2B_2 layer, while even orthogonal Mo_2B_2 cannot be completely etched out of MoAlB using a 10% NaOH solution, because low-concentration alkaline solutions can etch only the surface layer of the MAX phase [71,173]. These constraints lead to problems such as low yield, poor batch-to-batch consistency, and limited control over flake thickness, lateral size, defect density, and surface terminations. Scaling up for industrial application presents an even greater challenge, as processes must deliver high throughput, impurity control, and cost efficiency. Current research trends are addressing these bottlenecks by exploring greener synthetic protocols, such as molten salt methods, electrochemical exfoliation, and mechanochemical approaches that significantly reduce environmental hazards and operational complexity. Developing closed-loop, solvent-free, or recyclable processes aligns with global trends toward sustainable manufacturing. Additionally, optimization precursor selection, process automation, and standardizing process parameters are key directions for enabling large-scale production of MBenes with consistent quality, which is critical for battery manufacturing.

(II) Interlayer structure regulation and interface optimization

Another significant challenge involves fine-tuning the interlayer structure and optimizing interfaces within MBene materials, both of which are essential for efficient ion transport and stable electrochemical performance. While MBenes offer unique advantages including their layered architecture and high specific surface area, these features can also be problematic; uncontrolled stacking or aggregation diminishes ion accessibility and hinders charge transfer. To overcome the self-weight stacking problem and improve electrochemical performance, Li et al. [110] reported the use of CNTs in combination with a single-layer

MBene (forming mono-MBene/CNT). This composite material effectively prevented the nanoplates from stacking due to their weight, thereby significantly improving the rate performance and cycling stability of the electrode material.

Additionally, the interlayer spacing in pristine MBenes may not be ideally suited to the ionic radii of Li or Na, particularly in SIBs, which require broader galleries to facilitate Na diffusion. Recent research has focused on modifying interlayer chemistry through approaches such as ion intercalation (e.g., inserting organic cations or small molecules), surface functionalization (e.g., with oxygen, hydroxyl, or other moieties), and the intentional introduction of defects and dopants. These modifications can expand interlayer spacing, improve wettability, and facilitate faster, more reversible ion migration. At the electrolyte interface, engineering the chemical environment (by modifying terminal groups or using protective coatings) can mitigate the formation of unstable SEI and suppress side reactions. The advent of in situ and operando characterization tools has also accelerated insights into how MBene structures evolve during battery operation, providing a sound foundation for interface and structure optimization.

(III) Structural instability during operation

Stable cycling at high charge–discharge rates and over many cycles is essential for any practical battery material. For MBenes, preserving mechanical and electrochemical integrity during extended use is challenging. Repeated ion insertion and extraction cause volumetric changes that, if too great, can result in structural distortion, particle fragmentation, nanosheet delamination, and even electrical disconnection within the electrode. Current strategies to address these issues include enhancing interlayer interactions via cross-linking, integrating flexible or elastic matrices, and designing MBene-based composite electrodes. Incorporation of carbon nanostructures, conductive polymers, or hybrid binders can alleviate mechanical stress, prevent sheet restacking, and help maintain continuous conductive networks. Additionally, optimization of the electrode architecture, including controlling porosity, flake orientation, and thickness, can better accommodate expansion and contraction, thereby maximizing cycle life without compromising capacity or rate performance.

(IV) Material dispersion and aggregation

Achieving molecular-scale dispersion of MBene nanosheets is essential for maximizing the active surface area and improving overall battery performance. However, MBenes naturally tend to agglomerate or restack due to strong van der Waals forces and high surface energy, resulting in reduced accessible surface, fewer active sites, and hindered ion transport. To address this, surface engineering and composite design strategies are increasingly being utilized. Surface functionalization with charged groups, incorporation of spacers (such as organic molecules or nanoparticles), and integration with dispersive matrices like graphene or conductive polymers have all demonstrated effectiveness in preventing aggregation. Furthermore, in situ exfoliation methods, where MBenes are generated and processed in a single step or within a composite, can improve processability and ensure uniform dispersion of electroactive components throughout the electrode.

(V) Gap between theoretical and practical performance

Although computational studies and small-scale laboratory tests highlight MBenes' remarkable theoretical properties, such as high capacity and exceptional conductivity, the translation of these findings to commercial-level performance has often lagged behind. Several factors contribute to this gap; real-world electrodes are typically thicker, polydisperse, and incorporate binders or conductive additives, all of which can hinder rapid ion transport, electron conduction, and mechanical stability. Additionally, parasitic side reactions and imperfect interfaces further diminish effective utilization of materials. Bridging this divide requires close feedback between theory, modeling, and practical engineering. Increasingly sophisticated multi-scale simulations (DFT, molecular dynamics, continuum models) are now being intergrated with high-fidelity experimental validations to refine both material design and cell assembly protocols. Furthermore, the adoption of advanced operando diagnostics provides richer datasets to elucidate failure modes, capacity fade, and real-world performance limitations. Interdisciplinary collaboration among theoretical chemists, material scientists, and process engineers is essential to align predictions and outcomes for next-generation battery design.

(VI) Cost, environmental, and sustainability concerns

The main technical bottlenecks hindering the transformation of MBenes from laboratory-scale demonstration to commercial battery technology lie in the reproducibility and cost issues during large-scale production. Currently, the synthesis process of MBenes is complex and energy-consuming, leading to high production costs and making it difficult to achieve economic feasibility in mass production. As the main precursor, the cost of MAB varies depending on the type of transition metal contained, its purity, and market supply-demand relationships. If rare or expensive transition metals are involved, the cost may be relatively high. In addition, the long-term stability and cycle life of MBenes still need further optimization, especially in terms of their durability performance under actual usage environments. For industrial applications, balancing high performance with low costs, improving production efficiency, and ensuring material consistency and reliability are the key technical challenges faced by this field. Although certain progress has been made in laboratory synthesis methods, the further development of efficient, reproducible, and precisely controlled nanoscale synthesis technologies is required to achieve industrial-scale production. Meanwhile, the establishment of recycling pathways is also crucial. Waste materials, defective products, and other by-products generated during the production process need effective recycling treatment to improve resource utilization, reduce costs, and minimize environmental pollution.

In summary, the application of MBenes in LIBs and SIBs holds significant promise, but realizing this potential will require focused research to overcome synthetic, structural, interfacial, and practical engineering challenges. The ongoing integration of advanced synthetic chemistry, surface engineering, theory-driven design, and sustainable manufacturing is expected to accelerate the transition from laboratory to industry, ensuring that MBenes make a meaningful contribution to the future of clean energy storage. Emerging solutions and research directions not only address current limitations but also pave the way for technological breakthroughs in high-performance, durable, and sustainable battery systems.

6. Future Prospects

As the global pursuit of sustainable, efficient, and high-capacity energy storage solutions intensifies, the development and optimization of advanced materials such as MBenes are expected to play a pivotal role in shaping the next generation of batteries. While significant progress has been made in understanding and applying these novel 2D boride materials in LIBs and SIBs, numerous opportunities and challenges persist for both fundamental research and technological implementation. Looking forward, several strategic directions are set to unlock the full potential of MBenes, ranging from innovative synthesis methods and theoretical investigations to hybrid material design, mechanistic insights, and expanded application prospects (Figure 13).

(I) Advanced and green synthetic strategies

Efficient and scalable synthesis remains fundamental to the widespread adoption of any new material. For MBenes, particular attention must be given to advancing synthetic methods that can reliably produce high-quality, defect-controlled nanosheets while meeting environmental and industrial standards. Conventional synthesis techniques, such as chemical etching or top-down exfoliation, though effective, often involve hazardous chemicals, yield low quantities, or offer limited control over layer thickness and surface composition. The transition toward green synthesis and environmentally friendly protocols is both a scientific necessity and an industrial priority. Alternative approaches may include mechanochemical synthesis, electrochemical exfoliation, and bio-inspired methods that reduce energy consumption and environmental impact. Additionally, fine-tuning synthetic parameters—such as temperature, precursor concentration, and reaction time—will be essential for producing MBenes with uniform size, desirable surface functionality, and minimal defects. The development of closed-loop, recyclable, and energy-efficient synthesis platforms will not only minimize waste but also lower production costs, thereby accelerating commercial deployment.

(II) Structural engineering and surface functionalization

Fully harnessing the exceptional properties of MBenes demands ongoing advancements in material engineering at the atomic and molecular scales. Rationally designing the interlayer spacing, through ion exchange, organic molecule intercalation, or targeted doping, offers a pathway to optimize ion mobility and accommodate the significant volume changes induced by charge–discharge cycles. These approaches are particularly crucial for SIBs, where the larger ionic radius presents greater challenges than in Li systems. Surface functionalization represents another essential direction. The intentional introduction, modification, or removal of terminal groups—such as hydroxyl, oxygen, or fluorine—can tailor the chemical reactivity, electronic structure, and interfacial behavior of MBenes. By tuning these surface characteristics, it becomes possible to optimize electrode-electrolyte compatibility, enhance solid electrolyte interphase stability, and suppress parasitic side reactions that compromise long-term cycling. Notably, controlled surface modifications may also render MBenes suitable for multifunctional hybrid systems, enabling the exploitation of synergistic properties across different material classes.

(III) Multi-component and composite electrode architectures

Looking beyond the use of pristine MBene nanosheets, future research will increasingly emphasize the development of multi-component composite materials that integrate MBenes with other advanced functional phases. For instance, MBene–carbon, MBene–polymer, or MBene–metal oxide composites have demonstrated significant potential for achieving high conductivity, enhanced mechanical strength, and diverse charge–storage mechanisms simultaneously. Through the rational design of heterostructures or hybrid electrodes, scientists can further improve rate capabilities, mitigate structural degradation, and preserve electrolyte accessibility. Moreover, incorporating flexible or self-healing polymers could enable applications in flexible and wearable electronics, while strategic combinations with metal oxides may facilitate dual intercalation/conversion-type storage for exceptional capacity gains. Leveraging synergistic effects through composite engineering will therefore be central for unlocking the practical benefits of MBenes in next-generation battery technologies.

(IV) Synergy between theory, simulation, and advanced characterization

The convergence of theoretical modeling, computational screening, and advanced in situ characterization techniques is poised to transform the understanding and optimization of MBene-based systems. Atomistic simulations, such as density functional theory and molecular dynamics, will play a crucial role in predicting optimal compositions, surface configurations, and reaction pathways, thereby guiding targeted synthesis and property modulation. These theoretical frameworks, when combined with high-throughput screening methods, can dramatically accelerate material discovery and customization. Experimentally, the application of in situ and operando techniques—including synchrotron magnetic, spectroscopic, and X-ray synchrotron radiation—will illuminate the dynamic evolution of MBenes' structure, interfaces, and chemical environments during real-time battery operation. Achieving a deeper mechanistic understanding of Li and Na storage, degradation, and interfacial reactions will enable more precise material and device engineering. Therefore, the integration of theory and experiment remains fundamentals to advancing next-generation MBene research.

(V) Elucidating battery mechanisms and performance enhancement

One of the most promising areas for future research is the ongoing elucidation of ion storage mechanisms in MBenes. Whether through classical intercalation, pseudocapacitive surface storage, or conversion reactions, unraveling these processes at the atomic level will guide the rational design of materials with tailored performance characteristics. Special attention must be given to balancing high energy and power densities with exceptional cycling stability, a necessity for electric vehicles and large-scale stationary storage. Additionally, active control of the SEI remains a critical issue, as SEI stability governs Coulombic efficiency and cell longevity. Investigating MBene–electrolyte interactions, identifying additives that stabilize the interphase, or engineering artificial SEIs could provide pathways to reliable, high-performance batteries.

(VI) Broadening MBene applications beyond lithium and sodium storage

While LIBs and SIBs represent the immediate frontiers of MBene research, the unique properties of these materials suggest far broader application possibilities. For instance, MBenes could serve key roles in aqueous batteries, magnesium/zinc-ion batteries, supercapacitors, electrocatalysis, and sensors, owing to their excellent electrical conductivity, chemical tunability, and mechanical resilience. Additionally, the high surface area and chemical reactivity of MBenes make them promising candidates for environmental remediation, water splitting, and hydrogen storage. Harnessing these functionalities while drawing insights from battery-focused research will foster cross-disciplinary progress and drive the development of multifunctional MBene-based devices.

(VII) Sustainability, commercialization, and societal impact

For real-world application, MBene research must increasingly focus on environmental sustainability and economic feasibility. This involves not only green synthesis but also the recyclability of MBene-containing electrodes and their integration into existing manufacturing ecosystems. Evaluating resource availability, lifecycle impacts, and end-of-life scenarios should inform industrial scaling strategies. Finally, fostering collaboration among academia, industry, and policymakers will be crucial to translating MBene science into tangible societal benefits, from affordable, high-performance batteries that support renewable energy grids to safer, longer-lasting batteries for consumer devices and transportation.

MBenes, as an emerging family of two-dimensional transition metal borides, boast core advantages such as tunable layered structures, high electrical conductivity, and excellent electrochemical stability. In LIBs and SIBs, MBenes demonstrate exceptional capacity, long-term cycling durability, and high-rate performance. Defect engineering and composite strategies further enhance their practicality. However, key challenges remain, including scalable and precise synthesis methods, improvements in thermodynamic stability, and environmentally friendly processing techniques. Looking ahead, integrating sustainable manufacturing, achieving atomic-level structural design through computational modeling, and expanding into hybrid energy platforms represent promising research directions. By combining fundamental materials science with engineering considerations, MBenes holds the potential to drive the development of next-generation electrochemical energy storage systems, achieving higher energy density and operational reliability, and ultimately contributing to the global transition toward sustainable energy solutions.

Author Contributions: H.X.: writing—original draft preparation, review, and editing; Z.L.: writing—review and editing; Y.H.: writing—review and editing; H.L.: writing—review and editing. All authors have read and agreed to the published version of the manuscript.

Funding: This work was partially supported by the Key Research Projects of University in Henan Province (24A150041), the Science and Technology R&D Program Joint Fund of Henan Province (242301420042), the Henan Province Postdoctoral Research Funding Project (HN2024016), the Key R&D and Promotion Special (Scientific Problem Tackling) Project of Henan Province (242102240106), the Postdoctoral Fellowship Program of China Postdoctoral Science Foundation (GZC20232382), and the China Postdoctoral Science Foundation (2024M752924).

Institutional Review Board Statement: Not applicable.

Informed Consent Statement: Not applicable.

Data Availability Statement: Source data are available from the corresponding authors upon reasonable request.

Conflicts of Interest: The authors declare no conflicts of interest.

References

1. Nieto, K.; Windsor, D.S.; Vishnugopi, B.S.; Mukherjee, P.P.; Prieto, A.L. Performance metrics and mechanistic consideration the development of 3D batteries. *Nat. Rev. Chem.* **2025**, *9*, 118–133. [CrossRef] [PubMed]
2. Ma, M.; Huang, Z.; Li, L.; Zhang, W.; Guo, R.; Zhang, R.; Fa, W.; Han, C.; Cao, Y.; Yu, S.; et al. Modulating Photogener Electron Density of Pr Single-Atom Sites by Coordination Environment Engineering for Boosting Photoreduction of CC CH_3OH. *Appl. Catal. B* **2023**, *330*, 122626. [CrossRef]
3. Su, T.; Zhao, J.; Gomez-Exposito, A.; Chen, Y.; Terzija, V.; Gentle, J.P. Grid-enhancing technologies for clean energy systems. *Rev. Clean Technol.* **2025**, *1*, 16–31. [CrossRef]
4. Kong, Z.J.; Wu, J.C.; Liu, Z.J.; Yan, D.F.; Wu, Z.P.; Zhong, C.J. Advanced electrocatalysts for fuel cells: Evolution of active sites synergistic properties of catalysts and carrier materials. *Exploration* **2025**, *5*, 20230052. [CrossRef]
5. Li, T.; Jing, T.; Rao, D.; Mourdikoudis, S.; Zuo, Y.; Wang, M. Two-dimensional materials for electrocatalysis and energy sto applications. *Inorg. Chem. Front.* **2022**, *9*, 6008–6046. [CrossRef]
6. Liu, X.; Dong, X.; Adenusi, H.; Wu, Y.; Passerini, S. Co-solvent strategy for rechargeable post-lithium metal batteries. *Nat. Chem.* **2025**, *9*, 415–426. [CrossRef]
7. Yang, S.; Wang, Y.; Pan, H.; He, P.; Zhou, H. Lithium extraction from low-quality brines. *Nature* **2024**, *636*, 309–321. [CrossR
8. Armand, M.; Tarascon, J.M. Building Better Batteries. *Nature* **2008**, *451*, 652–657. [CrossRef]
9. Tarascon, J.M.; Armand, M. Issues and challenges facing rechargeable lithium batteries. *Nature* **2001**, *414*, 359–367. [CrossR
10. Service, R.F. Sodium batteries power up. *Science* **2025**, *387*, 813–814. [CrossRef]
11. Tang, Z.; Zhou, S.; Huang, Y.; Wang, H.; Zhang, R.; Wang, Q.; Sun, D.; Tang, Y.; Wang, H. Improving the Initial Coulo Efficiency of Carbonaceous Materials for Li/Na-Ion Batteries: Origins, Solutions, and Perspectives. *Electrochem. Energy Rev.* *6*, 8. [CrossRef]
12. Zhang, T.; Li, Y.; Song, Z.; Huang, Y.; Li, F.; Cheng, S.; Li, F. Insights into chemical-mechanical degradation and modifica strategies of layered oxide cathode materials of sodium ion batteries. *J. Energy Chem.* **2025**, *103*, 294–315. [CrossRef]

Zheng, Y.; Meng, Y.; Hu, X.; Peng, H.; Feng, L.; Wang, Y.; Li, B. Synthesis-Structure-Property of High-Entropy Layered Oxide Cathode for Li/Na/K-Ion Batteries. *Adv. Mater.* **2024**, *37*, 2413202. [CrossRef]
Javed, M.S.; Zhang, X.; Ahmad, T.; Usman, M.; Shah, S.S.A.; Ahmad, A.; Hussain, I.; Majeed, S.; Khawar, M.R.; Choi, D.; et al. to MBenes: Latest development and opportunities for energy storage devices. *Mater. Today* **2024**, *74*, 121–148. [CrossRef]
Liu, Z.H.; Liu, C.; Chen, Z.T.; Huang, H.L.; Liu, Y.F.; Xue, L.; Sun, J.W.; Wang, X.; Xiong, P.; Zhu, J.W. Recent advances in two-dimensional materials for hydrovoltaic energy technology. *Exploration* **2023**, *3*, 20220061. [CrossRef] [PubMed]
Nisar, S.; Dastgeer, G.; Shazad, Z.M.; Zulfiqar, M.W.; Rasheed, A.; Iqbal, M.Z.; Hussain, K.; Rabani, I.; Kim, D.-k.; Irfan, A.; et al. 2D materials in advanced electronic biosensors for point-of-care devices. *Adv. Sci.* **2024**, *11*, 2401386. [CrossRef]
Tan, C.; Lai, Z.; Zhang, H. Ultrathin Two-Dimensional Multinary Layered Metal Chalcogenide Nanomaterials. *Adv. Mater.* **2017**, *29*, 1701392. [CrossRef]
Zhu, J.; Ha, E.; Zhao, G.; Zhou, Y.; Huang, D.; Yue, G.; Hu, L.; Sun, N.; Wang, Y.; Lee, L.Y.S.; et al. advance in MXenes: A promising 2D material for catalysis, sensor and chemical adsorption. *Coord. Chem. Rev.* **2017**, *352*, 306–327. [CrossRef]
Rabiei Baboukani, A.; Khakpour, I.; Drozd, V.; Wang, C. Liquid-Based Exfoliation of Black Phosphorus into Phosphorene and Its Application for Energy Storage Devices. *Small Struct.* **2021**, *2*, 2000148. [CrossRef]
Carvalho, A.; Wang, M.; Zhu, X.; Rodin, A.S.; Su, H.; Castro Neto, A.H. Phosphorene: From theory to applications. *Nat. Rev. Mater.* **2016**, *1*, 16061. [CrossRef]
Xu, T.; Wang, Y.; Xiong, Z.; Wang, Y.; Zhou, Y.; Li, X. A Rising 2D Star: Novel MBenes with Excellent Performance in Energy Conversion and Storage. *Nano-Micro Lett.* **2022**, *15*, 6. [CrossRef]
Yang, Y.; Wang, H.; Wang, C.; Liu, J.; Wu, H.; Liu, N.; Wang, Q.; Shang, Y.; Zheng, J. Novel 2D Material of MBenes: Structures, Synthesis, Properties, and Applications in Energy Conversion and Storage. *Small* **2024**, *20*, 2405870. [CrossRef] [PubMed]
Guo, Z.; Zhou, J.; Sun, Z. New two-dimensional transition metal borides for Li ion batteries and electrocatalysis. *J. Mater. Chem. A* **2017**, *5*, 23530–23535. [CrossRef]
Wang, J.; Ye, T.-N.; Gong, Y.; Wu, J.; Miao, N.; Tada, T.; Hosono, H. Discovery of hexagonal ternary phase Ti2InB2 and its evolution to layered boride TiB. *Nat. Commun.* **2019**, *10*, 2284. [CrossRef] [PubMed]
Bo, T.; Liu, P.-F.; Xu, J.; Zhang, J.; Chen, Y.; Eriksson, O.; Wang, F.; Wang, B.-T. Hexagonal Ti_2B_2 monolayer: A promising anode material offering high rate capability for Li-ion and Na-ion batteries. *Phys. Chem. Chem. Phys.* **2018**, *20*, 22168–22178. [CrossRef] [PubMed]
Tang, Q.; Zhou, Z.; Shen, P. Are MXenes Promising Anode Materials for Li Ion Batteries? Computational Studies on Electronic Properties and Li Storage Capability of Ti_3C_2 and $Ti_3C_{2\times2}$ (X = F, OH) Monolayer. *J. Am. Chem. Soc.* **2012**, *134*, 16909–16916. [CrossRef]
Xie, Y.; Dall'Agnese, Y.; Naguib, M.; Gogotsi, Y.; Barsoum, M.W.; Zhuang, H.L.; Kent, P.R.C. Prediction and Characterization of MXene Nanosheet Anodes for Non-Lithium-Ion Batteries. *ACS Nano* **2014**, *8*, 9606–9615. [CrossRef]
Wang, X.; Kajiyama, S.; Iinuma, H.; Hosono, E.; Oro, S.; Moriguchi, I.; Okubo, M.; Yamada, A. Pseudocapacitance of MXene nanosheets for high-power sodium-ion hybrid capacitors. *Nat. Commun.* **2015**, *6*, 6544. [CrossRef]
Persson, K.; Sethuraman, V.A.; Hardwick, L.J.; Hinuma, Y.; Meng, Y.S.; van der Ven, A.; Srinivasan, V.; Kostecki, R.; Ceder, G. Lithium Diffusion in Graphitic Carbon. *J. Phys. Chem. Lett.* **2010**, *1*, 1176–1180. [CrossRef]
Hu, X.; Zhang, W.; Liu, X.; Mei, Y.; Huang, Y. Nanostructured Mo-based electrode materials for electrochemical energy storage. *Chem. Soc. Rev.* **2015**, *44*, 2376–2404. [CrossRef]
Peng, H.; Miao, W.; Zeng, J.; Wang, Z.; Yan, C.; Ma, G.; Lei, Z. Electronic Modulation and Built-in Electric Field Strategies in Heterostructures Together Induce 1T-Rich MoS_2 Conversion for Advanced Sodium Storage. *Adv. Sci.* **2025**, *12*, 2417288. [CrossRef] [PubMed]
Jin, H.; Xin, S.; Chuang, C.; Li, W.; Wang, H.; Zhu, J.; Xie, H.; Zhang, T.; Wan, Y.; Qi, Z.; et al. Black phosphorus composites with engineered interfaces for high-rate high-capacity lithium storage. *Science* **2020**, *370*, 192–197. [CrossRef]
Kansara, S.; Gupta, S.K.; Sonvane, Y.; Pajtler, M.V.; Ahuja, R. Inquisitive Geometric Sites in h-BN Monolayer for Alkali Earth Metal Ion Batteries. *J. Phys. Chem. C* **2019**, *123*, 19340–19346. [CrossRef]
Shi, L.; Xu, A.; Zhao, T. First-Principles Investigations of the Working Mechanism of 2D h-BN as an Interfacial Layer for the Anode of Lithium Metal Batteries. *ACS Appl. Mater. Interfaces* **2017**, *9*, 1987–1994. [CrossRef]
Yang, X.; Peng, J.; Zhao, L.; Zhang, H.; Li, J.; Yu, P.; Fan, Y.; Wang, J.; Liu, H.; Dou, S. Insights on advanced g-C_3N_4 in energy storage: Applications, challenges, and future. *Carbon Energy* **2024**, *6*, e490. [CrossRef]
Pan, H. Graphitic Carbon Nitride Nanotubes As Li-Ion Battery Materials: A First-Principles Study. *J. Phys. Chem. C* **2014**, *118*, 9318–9323. [CrossRef]
Liu, P.; Xu, H.; Wang, X.; Tian, G.; Yu, X.; Wang, C.; Zeng, C.; Wang, S.; Fan, F.; Liu, S.; et al. 2D MXene/MBene superlattice with narrow bandgap as superior electrocatalyst for high-performance lithium-oxygen battery. *Small* **2024**, *20*, 2404483. [CrossRef]

38. Si, J.; Lan, H.; An, J.; Ge, S.; Liu, W.; Zhang, Y.; Mao, X.; He, W. Emerging reactive oxygen and nitrogen species regulator biomedical therapy: 2D MXenzymes. *Mater. Today* **2025**, *85*, 112–140. [CrossRef]
39. Li, Y.; Zhao, T.; Li, L.; Huang, R.; Wen, Y. Computational evaluation of ScB and TiB MBenes as promising anode material high-performance metal-ion batteries. *Phys. Rev. Mater.* **2022**, *6*, 045801. [CrossRef]
40. Lv, X.; Han, T.; Liu, R.; Li, F.; Gong, J.; Chen, Z. High-throughput theoretical exploration of multifunctional planar MB Magnetism, topology, superconductivity, and anode applications. *Adv. Powder Mater.* **2025**, *4*, 100297. [CrossRef]
41. Si, J.; Yu, J.; Lan, H.; Niu, L.; Luo, J.; Yu, Y.; Li, L.; Ding, Y.; Zeng, M.; Fu, L. Chemical Potential-Modulated Ultrahigh-Phase-P Growth of Ultrathin Transition-Metal Boride Single Crystals. *J. Am. Chem. Soc.* **2023**, *145*, 3994–4002. [CrossRef]
42. Xiong, W.; Feng, X.; Xiao, Y.; Huang, T.; Li, X.; Huang, Z.; Ye, S.; Li, Y.; Ren, X.; Wang, X.; et al. Fluorine-free prep two-dimensional molybdenum boride (MBene) as a promising anode for lithium-ion batteries with superior electrochen performance. *Chem. Eng. J.* **2022**, *446*, 137466. [CrossRef]
43. Liu, G.; Yuan, W.; Zhao, Z.; Li, J.; Wu, N.; Guo, D.; Liu, X.; Liu, Y.; Cao, A.; Liu, X. $Mo_{4/3}B_2T_x$ induced hierarchical structure rapid reaction dynamics in MoS_2 anode for superior sodium storage. *Chem. Eng. J.* **2024**, *493*, 152576. [CrossRef]
44. Ozkan, M.; Quiros, K.A.M.; Watkins, J.M.; Nelson, T.M.; Singh, N.D.; Chowdhury, M.; Namboodiri, T.; Talluri, K.R.; Yua Curbing pollutant CO_2 by using two-dimensional MXenes and MBenes. *Chem* **2024**, *10*, 443–483. [CrossRef]
45. Jiang, M.; Wang, D.; Kim, Y.H.; Duan, C.; Talapin, D.V.; Zhou, C. Evolution of Surface Chemistry in Two-Dimensional MX From Mixed to Tunable Uniform Terminations. *Angew. Chem. Int. Ed.* **2024**, *63*, e202409480. [CrossRef] [PubMed]
46. Fei, L.; Lei, L.; Xu, H.; Guo, X.; Chen, B.; Han, X.; Chen, X.; Huang, Q.; Wang, D. Ion transport behaviors in MXene electrochemical energy storage and conversion. *Carbon Energy* **2025**, *7*, e678. [CrossRef]
47. Ahmad, S.; Xu, H.; Chen, L.; Din, H.U.; Zhou, Z. Functionalized MBenes as promising anode materials for high-perform alkali-ion batteries: A first-principles study. *Nanotechnology* **2024**, *35*, 285401. [CrossRef]
48. Hayat, A.; Bashir, T.; Ahmed, A.M.; Ajmal, Z.; Alghamdi, M.M.; El-Zahhar, A.A.; Sohail, M.; Amin, M.A.; Al-Hadeethi, Y.; Gh E.; et al. Novel 2D MBenes-synthesis, structure, properties with excellent performance in energy conversion and storag review. *Mater. Sci. Eng. R* **2024**, *159*, 100796. [CrossRef]
49. Akgul, E.T.; Altıncı, O.C.; Umay, A.; Aghamohammadi, P.; Farghaly, A.A.; Ma, P.; Chen, Y.; Demir, M. Nanoengineering c MBenes for energy storage applications: A review. *J. Energy Storage* **2024**, *84*, 110882. [CrossRef]
50. Zhu, Z.; Jiang, T.; Ali, M.; Meng, Y.; Jin, Y.; Cui, Y.; Chen, W. Rechargeable Batteries for Grid Scale Energy Storage. *Chem.* **2022**, *122*, 16610–16751. [CrossRef]
51. Li, X.; Huang, Z.; Shuck, C.E.; Liang, G.; Gogotsi, Y.; Zhi, C. MXene chemistry, electrochemistry and energy storage applicat *Nat. Rev. Chem.* **2022**, *6*, 389–404. [CrossRef] [PubMed]
52. Miao, N.; Gong, Y.; Zhang, H.; Shen, Q.; Yang, R.; Zhou, J.; Hosono, H.; Wang, J. Discovery of Two-dimensional Hexag MBene HfBO and Exploration on its Potential for Lithium-Ion Storage. *Angew. Chem. Int. Ed.* **2023**, *62*, e202308436. [CrossR
53. Zhang, B.; Zhou, J.; Sun, Z. MBenes: Progress, challenges and future. *J. Mater. Chem. A* **2022**, *10*, 15865–15880. [CrossRef]
54. Jing, T.; Zhang, N.; Zhang, C.; Mourdikoudis, S.; Sofer, Z.; Li, W.; Li, P.; Li, T.; Zuo, Y.; Rao, D. Improving C–N–FeOx Oxy Evolution Electrocatalysts through Hydroxyl-Modulated Local Coordination Environment. *ACS Catal.* **2022**, *12*, 7443– [CrossRef]
55. Khan, A.J.; Shah, S.S.; Khan, S.; Mateen, A.; Iqbal, B.; Naseem, M.; He, L.; Zhang, Y.; Che, Y.; Tang, Y.; et al. 2D metal bor (MBenes): Synthesis methods for energy storage applications. *Chem. Eng. J.* **2024**, *497*, 154429. [CrossRef]
56. Nair, V.G.; Birowska, M.; Bury, D.; Jakubczak, M.; Rosenkranz, A.; Jastrzębska, A.M. 2D MBenes: A novel member in the flat *Adv. Mater.* **2022**, *34*, 2108840. [CrossRef]
57. Palei, S.; Murali, G.; Kim, C.-H.; In, I.; Lee, S.-Y.; Park, S.-J. A Review on Interface Engineering of MXenes for Perovskite S Cells. *Nano-Micro Lett.* **2023**, *15*, 123. [CrossRef]
58. Bi, W.; Gao, G.; Li, C.; Wu, G.; Cao, G. Synthesis, properties, and applications of MXenes and their composites for electrical en storage. *Prog. Mater. Sci.* **2024**, *142*, 101227. [CrossRef]
59. Yuan, H.; Hua, J.; Wei, W.; Zhang, M.; Hao, Y.; Chang, J. Progress and prospect of flexible MXene-based energy storage. *Ca Energy* **2024**, *7*, e639. [CrossRef]
60. Qin, R.; Nong, J.; Wang, K.; Liu, Y.; Zhou, S.; Hu, M.; Zhao, H.; Shan, G. Recent Advances in Flexible Pressure Sensors Base MXene Materials. *Adv. Mater.* **2024**, *36*, 2312761. [CrossRef]
61. VahidMohammadi, A.; Rosen, J.; Gogotsi, Y. The world of two-dimensional carbides and nitrides (MXenes). *Science* **2021**, eabf1581. [CrossRef] [PubMed]
62. Kamysbayev, V.; Filatov, A.S.; Hu, H.; Rui, X.; Lagunas, F.; Wang, D.; Klie, R.F.; Talapin, D.V. Covalent surface modifications superconductivity of two-dimensional metal carbide MXenes. *Science* **2020**, *369*, 979–983. [CrossRef]

Lin, Y.-C.; Torsi, R.; Younas, R.; Hinkle, C.L.; Rigosi, A.F.; Hill, H.M.; Zhang, K.; Huang, S.; Shuck, C.E.; Chen, C.; et al. Recent Advances in 2D Material Theory, Synthesis, Properties, and Applications. *ACS Nano* **2023**, *17*, 9694–9747. [CrossRef] [PubMed]

Xia, H.; Zan, L.; Dong, H.; Wei, Y.; Yu, Y.; Shu, J.; Zhang, J.-N.; Shan, C.-X. Work-function effect of Ti_3C_2/Fe-N-C inducing solid electrolyte interphase evolution for ultra-stable sodium storage. *Nano Res.* **2024**, *17*, 7163. [CrossRef]

Jakubczak, M.; Szuplewska, A.; Rozmysłowska-Wojciechowska, A.; Rosenkranz, A.; Jastrzębska, A.M. Novel 2D MBenes—Synthesis, Structure, and Biotechnological Potential. *Adv. Funct. Mater.* **2021**, *31*, 2103048. [CrossRef]

Zhang, G.; Li, X.; Chen, K.; Guo, Y.; Ma, D.; Chu, K. Tandem Electrocatalytic Nitrate Reduction to Ammonia on MBenes. *Angew. Chem. Int. Ed.* **2023**, *62*, e202300054. [CrossRef]

Park, S.J.; Nguyen, T.H.; Tran, D.T.; Dinh, V.A.; Lee, J.H.; Kim, N.H. Delaminated MBene sheets beyond usual 2D transition metal materials for securing Pt single atoms to boost hydrogen evolution. *Energ. Environ. Sci.* **2023**, *16*, 4093–4104. [CrossRef]

De, S.K.; Aparna, T.K.; Krishna, K.R.G.; Bhattacharyya, S. Enhancing the electrochemical performance of $TiVCT_x$ MXene by tuning termination groups through different synthesis routes. *Chem. Eng. J.* **2025**, *504*, 158882. [CrossRef]

Huang, P.; Han, W.-Q. Recent Advances and Perspectives of Lewis Acidic Etching Route: An Emerging Preparation Strategy for MXenes. *Nano-Micro Lett.* **2023**, *15*, 68. [CrossRef]

Hou, J.; Ji, S.; Ma, X.; Gong, B.; Wang, T.; Xu, Q.; Cao, H. Functionalized MXene composites for protection on metals in electric power. *Adv. Colloid Interface Sci.* **2025**, *341*, 103505. [CrossRef]

Alameda, L.T.; Moradifar, P.; Metzger, Z.P.; Alem, N.; Schaak, R.E. Topochemical Deintercalation of Al from MoAlB: Stepwise Etching Pathway, Layered Intergrowth Structures, and Two-Dimensional MBene. *J. Am. Chem. Soc.* **2018**, *140*, 8833–8840. [CrossRef] [PubMed]

Bhaskar, G.; Gvozdetskyi, V.; Batuk, M.; Wiaderek, K.M.; Sun, Y.; Wang, R.; Zhang, C.; Carnahan, S.L.; Wu, X.; Ribeiro, R.A.; et al. Topochemical Deintercalation of Li from Layered LiNiB: Toward 2D MBene. *J. Am. Chem. Soc.* **2021**, *143*, 4213–4223. [CrossRef]

Alameda, L.T.; Lord, R.W.; Barr, J.A.; Moradifar, P.; Metzger, Z.P.; Steimle, B.C.; Holder, C.F.; Alem, N.; Sinnott, S.B.; Schaak, R.E. Multi-Step Topochemical Pathway to Metastable Mo_2AlB_2 and Related Two-Dimensional Nanosheet Heterostructures. *J. Am. Chem. Soc.* **2019**, *141*, 10852–10861. [CrossRef] [PubMed]

Liu, Y.; Tang, Q.; Xu, M.; Ren, J.; Guo, C.; Chen, C.; Geng, W.; Lei, W.; Zhao, X.; Liu, D. Efficient mechanical exfoliation of MXene nanosheets. *Chem. Eng. J.* **2023**, *468*, 143439. [CrossRef]

Thakur, A.; Highland, W.J.; Wyatt, B.C.; Xu, J.; Chandran, B.S.N.; Zhang, B.; Hood, Z.D.; Adhikari, S.P.; Oveisi, E.; Pacakova, B.; et al. Synthesis of a 2D tungsten MXene for electrocatalysis. *Nat. Synth.* **2025**. [CrossRef]

Liu, Y.; Liu, S.; Zhao, G.; Shen, M.; Gao, X.; Zhao, Y.; Liu, X.; Hou, L.; Yuan, C. General and Fast Gas–Solid Synthesis of Functional MXenes and Derivatives on the Scale of Tens of Grams. *Angew. Chem. Int. Ed.* **2024**, *64*, e202420287. [CrossRef]

Zhu, X.; Zhou, X.; Jing, Y.; Li, Y. Electrochemical synthesis of urea on MBenes. *Nat. Commun.* **2021**, *12*, 4080. [CrossRef]

Majed, A.; Torkamanzadeh, M.; Nwaokorie, C.F.; Eisawi, K.; Dun, C.; Buck, A.; Urban, J.J.; Montemore, M.M.; Presser, V.; Naguib, M. Toward MBenes Battery Electrode Materials: Layered Molybdenum Borides for Li-Ion Batteries. *Small Methods* **2023**, *7*, 2300193. [CrossRef]

Zhang, Z.; Li, Y.; Mo, F.; Wang, J.; Ling, W.; Yu, M.; Huang, Y. MBene with Redox-Active Terminal Groups for an Energy-Dense Cascade Aqueous Battery. *Adv. Mater.* **2024**, *36*, 2311914. [CrossRef]

Chen, Z.; Wang, H.; Zhang, C.; Gou, Y.; Gong, Z.; Jiang, Y.; Zeng, H.; Wang, J.; Meng, F.; Cui, Y. MBene Brønsted Acid Catalyst for Hydrogen Evolution Reaction in Alkaline Electrolyte. *ACS Catal.* **2025**, *15*, 2885–2895. [CrossRef]

Xiong, W.; Feng, X.; Huang, T.; Huang, Z.; He, X.; Liu, J.; Xiao, Y.; Wang, X.; Zhang, Q. Rapid synthesis of two-dimensional MoB MBene anodes for high-performance sodium-ion batteries. *J. Mater. Sci. Technol.* **2025**, *212*, 67–76. [CrossRef]

Jiang, Y.; Lao, J.; Dai, G.; Ye, Z. Advanced Insights on MXenes: Categories, Properties, Synthesis, and Applications in Alkali Metal Ion Batteries. *ACS Nano* **2024**, *18*, 14050–14084. [CrossRef] [PubMed]

Zarepour, A.; Ahmadi, S.; Rabiee, N.; Zarrabi, A.; Iravani, S. Self-Healing MXene- and Graphene-Based Composites: Properties and Applications. *Nano-Micro Lett.* **2023**, *15*, 100. [CrossRef] [PubMed]

Yin, Y.; Wang, Y.; Yang, N.; Bi, L. Unveiling the importance of the interface in nanocomposite cathodes for proton-conducting solid oxide fuel cells. *Exploration* **2024**, *4*, 20230082. [CrossRef] [PubMed]

Lakmal, A.; Thombre, P.B.; Shuck, C.E. Solid-Solution MXenes: Synthesis, Properties, and Applications. *Acc. Chem. Res.* **2024**, *57*, 3007–3019. [CrossRef]

Yong, B.; Wang, Y.; Zhao, H.; Wang, T.; Zhu, J.; Tai, J.; Ma, D.; Sun, S.; Mi, H.; He, T.; et al. A Trifunctional Hydroxylated Borophene-Mediated MXene Enabled Super-Stable and Fast-Kinetics Interface Storage. *Adv. Funct. Mater.* **2024**, *34*, 2316127. [CrossRef]

Chen, J.; Zhang, W.; Chen, R.; Dai, Y.; Zhang, J.; Yang, H.; Zong, W.; Jiang, Z.; Zhong, Y.; Wang, J.; et al. From Synthesis to Energy Storage, The Microchemistry of MXene and MBene. *Adv. Energy Mater.* **2024**, *15*, 2403757. [CrossRef]

88. Liu, Z.; Tian, Y.; Li, S.; Wang, L.; Han, B.; Cui, X.; Xu, Q. Revealing High-Rate and High Volumetric Pseudo-Intercalation Ch Storage from Boron-Vacancy Doped MXenes. *Adv. Funct. Mater.* **2023**, *33*, 2301994. [CrossRef]
89. Pu, J.; Fan, S.; Shen, Z.; Yin, J.; Tan, Y.; Zhang, K.; Wu, B.; Hong, G.; Yao, Y. F-Free Fabrication Novel 2D Mo-Based MBene Cat for Advanced Lithium–Sulfur Batteries. *Adv. Funct. Mater.* **2025**, *35*, 2424215. [CrossRef]
90. Zhang, X.; Bai, X.; Wei, C.; Xi, B.; Xiong, S.; Feng, J. Unraveling Electronic Microenviroment of Iron Single Atom Modulate Nitrogen-Bridged Ligands in MBene Toward Bilateral Sulfur Redox Chemistry. *Adv. Funct. Mater.* **2024**, *34*, 2410583. [Cross
91. Peng, C.; Liang, S.; Yu, Y.; Cao, L.; Yang, C.; Liu, X.; Guo, K.; Müller-Buschbaum, P.; Cheng, Y.J.; Wang, C. A chronicle of tita niobium oxide materials for high-performance lithium-ion batteries: From laboratory to industry. *Carbon Neutraliz.* **20** 1036–1091. [CrossRef]
92. Xia, H.C.; Qu, G.; Yin, H.B.; Zhang, J.A. Atomically Dispersed Metal Active Centers as a Chemically Tunable Platform for En Storage Devices. *J. Mater. Chem. A* **2020**, *8*, 15358–15372. [CrossRef]
93. Rana, M.; Alghamdi, N.; Peng, X.; Huang, Y.; Wang, B.; Wang, L.; Gentle, I.R.; Hickey, S.; Luo, B. Scientific issues of zinc-bro flow batteries and mitigation strategies. *Exploration* **2023**, *3*, 20220073. [CrossRef] [PubMed]
94. Kim, J.-H.; Kim, N.-Y.; Ju, Z.; Hong, Y.-K.; Kang, K.-D.; Pang, J.-H.; Lee, S.-J.; Chae, S.-S.; Park, M.-S.; Kim, J.-Y.; et al. Upsc high-areal-capacity battery electrodes. *Nat. Energy* **2025**, *10*, 295–307. [CrossRef]
95. Park, C.Y.; Kim, J.; Lim, W.G.; Lee, J. Toward maximum energy density enabled by anode-free lithium metal batteries: Re progress and perspective. *Exploration* **2023**, *4*, 20210255. [CrossRef]
96. Wang, S.; Gao, Y.; Yang, Z.; Zhou, H.; Ni, D.; Li, C.; Li, Q.; Zhang, X. Heterostructured CoO/MoB MBene composites for performance lithium-ion batteries anode. *iScience* **2025**, *28*, 112133. [CrossRef]
97. Tao, H.; Xiong, Z.; Yabin, A.; Shasha, Z.; Chen, L.; Xianzhong, S.; Kai, W.; Yanwei, M. Nanoarchitectonics and applicatio two-dimensional materials as anodes for lithium-ion capacitors. *Energy Mater.* **2025**, *5*, 500003.
98. He, R.; Zhong, W.; Wu, Y.; Liu, W.; Cai, C.; Cheng, S.; Huang, L.; Xie, J. Two-Layer Graphite Anode for Energy and Po Densified $LiFePO_4$ Battery. *Adv. Mater.* **2025**, 2501185. [CrossRef]
99. Xia, H.C.; Li, K.X.; Guo, Y.Y.; Guo, J.H.; Xu, Q.; Zhang, J.N. CoS_2 nanodots trapped within graphitic structured N-doped ca spheres with efficient performances for lithium storage. *J. Mater. Chem. A* **2018**, *6*, 7148–7154. [CrossRef]
100. Rout, C.S.; Shinde, P.V.; Patra, A.; Jeong, S.M. Recent Developments and Future Perspectives of Molybdenum Borides and MB *Adv. Sci.* **2024**, *11*, 2308178. [CrossRef]
101. Shen, Q.; Shi, Y.; He, Y.; Wang, J. Defect Engineering of Hexagonal MAB Phase Ti_2InB_2 as Anode of Lithium-Ion Battery Excellent Cycling Stability. *Adv. Sci.* **2024**, *11*, 2308589. [CrossRef] [PubMed]
102. Miao, N.; Wang, J.; Gong, Y.; Wu, J.; Niu, H.; Wang, S.; Li, K.; Oganov, A.R.; Tada, T.; Hosono, H. Computational Predicti Boron-Based MAX Phases and MXene Derivatives. *Chem. Mater.* **2020**, *32*, 6947–6957. [CrossRef]
103. Hou, Y.; Chen, Z.; Li, X.; Wang, Y.; Li, P.; Cui, H.; Zhang, R.; Yang, S.; Zhang, S.; Zhi, C. MBene promoted Zn peroxide chem in rechargeable near-neutral Zn–air batteries. *Energ. Environ. Sci.* **2023**, *16*, 3407–3415. [CrossRef]
104. Wang, Y.; Sun, Y.; Wu, F.; Zou, G.; Gaumet, J.-J.; Li, J.; Fernandez, C.; Wang, Y.; Peng, Q. Nitrogen-Anchored Boridene Ena $Mg–CO_2$ Batteries with High Reversibility. *J. Am. Chem. Soc.* **2024**, *146*, 9967–9974. [CrossRef]
105. Zhang, Z.; Ling, W.; Ma, N.; Wang, J.; Chen, X.; Fan, J.; Yu, M.; Huang, Y. Ultralong Cycle Life and High Rate of Zn||I2 Ba Enabled by MBene-Hosted I2 Cathode. *Adv. Funct. Mater.* **2024**, *34*, 2310294. [CrossRef]
106. Chen, Y.-Z.; Mao, T.-T.; Liao, S.-Y.; Yao, S.X.; Min, Y.-G. Layered Mo_xB_y (MBenes) derived by a molten-salt method and application in advanced LIB anodes. *J. Mater. Chem. A* **2024**, *12*, 12163–12172. [CrossRef]
107. Wang, S.; Zhou, H.; Wei, H.; Wang, H.; Xie, H.; Xu, Z.; Ni, D.; Li, C.; Bollella, P.; Li, J.; et al. Superior performance lithiur battery anode based on Co_9S_8 nanoparticles layered in-situ growth with capacitive synergy. *J. Colloid Interface Sci.* **2025**, 505–516. [CrossRef]
108. Zhang, W.; He, L.; Chen, Y.; Wu, Z.; Yu, P.; Chen, K.; Ge, F.; Li, M.; Yu, L.; Lin, N.; et al. Unraveling the potential of MXen multifunctional cathodes: Innovations and challenges for next-generation energy storage systems. *Mater. Sci. Eng. R* **2025**, 100975. [CrossRef]
109. He, Y.; Wang, L.; Wang, A.; Zhang, B.; Pham, H.; Park, J.; He, X. Insight into uniform filming of LiF-rich interphase via syner adsorption for high-performance lithium metal anode. *Exploration* **2023**, *4*, 20230114. [CrossRef]
110. Li, Z.; Zeng, Q.; Yu, Y.; Liu, Y.; Chen, A.; Guan, J.; Wang, H.; Liu, W.; Liu, X.; Liu, X.; et al. Application of transition metal b nanosheet as sulfur host in high loading Li-S batteries. *Chem. Eng. J.* **2023**, *452*, 139366. [CrossRef]
111. Liu, Z.; Wu, D.; Wei, S.; Xing, K.; Li, M.; Jiang, Y.; Yuan, R.; Chen, G.; Hu, Z.; Huang, Y.; et al. MXene Hollow Microsphere-Boo Nanocomposite Electrodes for Thermocells with Enhanced Thermal Energy Harvesting Capability. *ACS Nano* **2025**, *19*, 3392– [CrossRef] [PubMed]

Ding, M.C.; Zhao, D.M.; Wei, R.; Duan, Z.Y.; Zhao, Y.X.; Li, Z.Y.; Lin, T.H.; Li, C.W. Multifunctional elastomeric composites based on 3D graphene porous materials. *Exploration* **2024**, *4*, 20230057. [CrossRef] [PubMed]
Luo, T.; Peng, Q.; Yang, M.; Hu, H.; Ding, J.; Chen, Y.; Gong, X.; Yang, J.; Qu, Y.; Zhou, Z.; et al. Boosted Li_2CO_3 reversible conversion utilizing Cu-doped TiB MBene/graphene for Li–CO_2 batteries. *J. Mater. Chem. A* **2024**, *12*, 25887–25895. [CrossRef]
Wang, Q.; Yuan, H.; Zhang, M.; Yang, N.; Cong, S.; Zhao, H.; Wang, X.; Xiong, S.; Li, K.; Zhou, A. A Highly Conductive and Supercapacitive MXene/N-CNT Electrode Material Derived from a MXene-Co-Melamine Precursor. *ACS Appl. Electron. Mater.* **2023**, *5*, 2506–2517. [CrossRef]
Fan, S.; Ai, D.; Zhang, W.; Yang, R.; Shen, G.; Yang, X.; Cheng, Y.; Yu, X. In-plane aligned doping pattern in electrospun PEI/MBene nanocomposites for high-temperature capacitive energy storage. *Mater. Horiz.* **2025**, *12*, 2267–2278. [CrossRef] [PubMed]
Wang, Y.; Shi, Y.; Qiu, J.; Cheng, J.; Xu, Y.; Wang, Y. Insights into molecular interactions at organic-MBene heterointerfaces for efficient Zn-ion storage. *J. Colloid Interface Sci.* **2025**, *678*, 95–104. [CrossRef]
Jing, C.; Huang, L.; Tao, S.; Chen, Y.; Zhang, S.; Dong, W.; Ling, F.; Tang, X.; Li, Y.; Feng, L.; et al. Construction of MoB@LDH heterojunction and its derivates through phase and interface engineering for advanced supercapacitor applications. *J. Colloid Interface Sci.* **2024**, *660*, 10–20. [CrossRef]
Nam, S.W.; Nguyen, T.H.; Tran, D.T.; Dinh, V.A.; Ta, T.T.N.; Dong, C.-L.; Kim, N.H.; Lee, J.H. Molecularly engineered potential of d-orbital modulated iron-bridged delaminated MBene for rechargeable Zn–air batteries. *Energ. Environ. Sci.* **2024**, *17*, 6559–6570. [CrossRef]
Shi, Y.; Song, G.; Yang, B.; Tang, Y.; Liu, Z.; Zhang, Z.; Shakouri, M.; Cheng, J.; Pang, H. Prussian Blue Analogues "Dressed" in MXene Nanosheets Tightly for High Performance Lithium-Ion Batteries. *Adv. Mater.* **2025**, *37*, 2416665. [CrossRef]
Hou, Y.; Chen, Z.; Zhang, R.; Cui, H.; Yang, Q.; Zhi, C. Recent advances and interfacial challenges in solid-state electrolytes for rechargeable Li-air batteries. *Exploration* **2023**, *3*, 20220051. [CrossRef]
Li, T.; Zhu, C.; Yang, X.; Gao, Y.; He, W.; Yue, H.; Zhao, H. Co_3O_4 nanoneedle@electroactive nickel boride membrane core/shell arrays: A novel hybrid for enhanced capacity. *Electrochim. Acta* **2017**, *246*, 226–233. [CrossRef]
Li, Y.; Wang, W.; Huang, B.; Mao, Z.; Wang, R.; He, B.; Gong, Y.; Wang, H. Abundant heterointerfaces in MOF-derived hollow CoS_2–MoS_2 nanosheet array electrocatalysts for overall water splitting. *J. Energy Chem.* **2021**, *57*, 99–108. [CrossRef]
Luo, Y.; Handy, J.V.; Das, T.; Ponis, J.D.; Albers, R.; Chiang, Y.-H.; Pharr, M.; Schultz, B.J.; Gobbato, L.; Brown, D.C.; et al. Effect of pre-intercalation on Li-ion diffusion mapped by topochemical single-crystal transformation and operando investigation. *Nat. Mater.* **2024**, *23*, 960–968. [CrossRef]
Zhang, H.; Yang, Y.; Ren, D.; Wang, L.; He, X. Graphite as anode materials: Fundamental mechanism, recent progress and advances. *Energy Storage Mater.* **2021**, *36*, 147–170. [CrossRef]
Fleischmann, S.; Mitchell, J.B.; Wang, R.; Zhan, C.; Jiang, D.-E.; Presser, V.; Augustyn, V. Pseudocapacitance: From Fundamental Understanding to High Power Energy Storage Materials. *Chem. Rev.* **2020**, *120*, 6738–6782. [CrossRef]
Augustyn, V.; Simon, P.; Dunn, B. Pseudocapacitive oxide materials for high-rate electrochemical energy storage. *Energ. Environ. Sci.* **2014**, *7*, 1597. [CrossRef]
Wu, F.; Yushin, G. Conversion cathodes for rechargeable lithium and lithium-ion batteries. *Energ. Environ. Sci.* **2017**, *10*, 435–459. [CrossRef]
Yun, Q.; Li, L.; Hu, Z.; Lu, Q.; Chen, B.; Zhang, H. Layered Transition Metal Dichalcogenide-Based Nanomaterials for Electrochemical Energy Storage. *Adv. Mater.* **2020**, *32*, e1903826. [CrossRef] [PubMed]
Chen, J.; Ren, L.; Chen, X.; Wang, Q.; Chen, C.; Fan, J.; Wang, S.; Binas, V.; Shen, S. Well-defined nanostructures of high entropy alloys for electrocatalysis. *Exploration* **2024**, *5*, 20230036. [CrossRef]
Dopilka, A.; Larson, J.M.; Cha, H.; Kostecki, R. Synchrotron Near-Field Infrared Nanospectroscopy and Nanoimaging of Lithium Fluoride in Solid Electrolyte Interphases in Li-Ion Battery Anodes. *ACS Nano* **2024**, *18*, 15270–15283. [CrossRef]
Hobold, G.M.; Wang, C.; Steinberg, K.; Li, Y.; Gallant, B.M. High lithium oxide prevalence in the lithium solid–electrolyte interphase for high Coulombic efficiency. *Nat. Energy* **2024**, *9*, 580–591. [CrossRef]
Dou, Y.; Zhao, L.; Liu, Y.; Zhang, Z.; Zhang, Y.; Li, R.; Liu, X.; Zhou, Y.; Wang, J.; Wang, J. High-abundance and low-cost anodes for sodium-ion batteries. *Carbon Neutraliz.* **2024**, *3*, 954–995. [CrossRef]
Zhu, Y.-F.; Xiao, Y.; Dou, S.-X.; Kang, Y.-M.; Chou, S.-L. Spinel/Post-spinel engineering on layered oxide cathodes for sodium-ion batteries. *eScience* **2021**, *1*, 13–27. [CrossRef]
Wang, Y.; Huang, R.; Zhang, Q.; Ma, Y.; Gao, B.; Li, Z. Theoretical investigation of the V_2BX_2 (X = S, Se, and Te) monolayers as anode materials for Na-ion batteries. *Mater. Today Commun.* **2023**, *35*, 105923. [CrossRef]
Mao, H.; Yang, S.; Yang, Y.; Yang, J.; Yuan, G.; Zheng, M.; Hu, H.; Liang, Y.; Yu, X. Hybrid catalyst-assisted synthesis of multifunctional carbon derived from Camellia shell for high-performance sodium-ion batteries and sodium-ion hybrid capacitors. *Carbon Neutraliz.* **2024**, *3*, 673–688. [CrossRef]

136. Liang, B.; Ma, N.; Wang, Y.; Wang, T.; Fan, J. N-functionalized Ti2B MBene as high-performance anode materials for sodium batteries: A DFT study. *Appl. Surf. Sci.* **2022**, *599*, 153927. [CrossRef]
137. Xia, H.; Yuan, P.; Zan, L.; Qu, G.; Tu, Y.; Zhu, K.; Wei, Y.; Wei, Z.; Zheng, F.; Zhang, M.; et al. Probing the active sites o nanosheets with Fe-N-C carbon shell encapsulated FexC/Fe species for boosting sodium-ion storage performances. *Nanc* **2021**, *15*, 7154–7162. [CrossRef]
138. Xia, H.C.; Zan, L.X.; Wei, Y.F.; Guo, K.; Yan, W.F.; Deng, D.H.; Zhang, J.N. Catalytic Effect of Carbon-Based Electrode Materia Energy Storage Devices. *Sci. China Mater.* **2022**, *65*, 3229–3242. [CrossRef]
139. Wei, Y.; Zan, L.; Xia, H.; Yan, W.; Zhang, J.-N. Mechanism of interfacial effects in sodium-ion storage devices. *Nano Res.* **202** 1313–1326. [CrossRef]
140. Huang, W.G.; Bin Zulkifli, M.Y.; Chai, M.; Lin, R.J.; Wang, J.J.; Chen, Y.L.; Chen, V.; Hou, J.W. Recent advances in enzyn biofuel cells enabled by innovative materials and techniques. *Exploration* **2023**, *3*, 20220145. [CrossRef]
141. Bi, W.; Li, S.; Wang, W.; Liu, Y.; Shen, J.; Gao, G.; Zhang, Z.; Wu, G.; Cao, G. MXenes and their composites as electrodes for soc ion batteries. *Energy Storage Mater.* **2024**, *71*, 103568. [CrossRef]
142. Cheng, W.D.; Zhao, M.Y.; Lai, Y.C.; Wang, X.; Liu, H.Y.; Xiao, P.; Mo, G.; Liu, B.; Liu, Y.P. Recent advances in battery characteriz using in situ XAFS, SAXS, XRD, and their combining techniques: From single scale to multiscale structure detection. *Explor* **2024**, *4*, 20230056. [CrossRef]
143. Guan, Y.; Jiang, S.; Ding, Y.; Xiao, B.; Pi, Y.; Wang, Z.; Cong, Y. Insight into the mechanism of nitrogen doping in MXenes controllable surface chemistry. *Mater. Today Energy* **2024**, *44*, 101642. [CrossRef]
144. Jia, J.; Li, B.; Duan, S.; Cui, Z.; Gao, H. Monolayer MBenes: Prediction of anode materials for high-performance lithium/soc ion batteries. *Nanoscale* **2019**, *11*, 20307–20314. [CrossRef]
145. Bo, T.; Liu, P.-F.; Zhang, J.; Wang, F.; Wang, B.-T. Tetragonal and trigonal Mo_2B_2 monolayers: Two new low-dimensional mate for Li-ion and Na-ion batteries. *Phys. Chem. Chem. Phys.* **2019**, *21*, 5178–5188. [CrossRef]
146. Liu, G.; Xu, M.; Cao, R.; Zhao, Z.; Yuan, W.; Liu, Y.; Cao, A.; Wang, L.; Liu, X. $Mo_{4/3}B_2T_x$ boosting the electrochemical kin and Na_2S adsorption of SnS anode in sodium ion batteries. *J. Colloid Interface Sci.* **2025**, *695*, 137801. [CrossRef] [PubMed]
147. Kashif Masood, M.; Wang, J.; Song, J.; Liu, Y. A novel two-dimensional whorled CrB_4 and MoB_4 as high-performance a material for metal ion batteries. *Appl. Surf. Sci.* **2024**, *652*, 159301. [CrossRef]
148. Xue, Y.; Chen, Y.; Liang, Y.; Shi, L.; Ma, R.; Qiu, X.; Li, Y.; Guo, N.; Zhuang, Q.; Xi, B.; et al. Substitution Index-Prediction Rule Low-Potential Plateau of Hard Carbon Anodes in Sodium-Ion Batteries. *Adv. Mater.* **2025**, 2417886. [CrossRef]
149. Cai, C.; Li, X.; Li, J.; Yu, R.; Hu, P.; Zhu, T.; Li, T.; Lee, S.; Xu, N.; Fan, H.; et al. Transition metal vacancy and position enginee enables reversible anionic redox reaction for sodium storage. *Nat. Commun.* **2025**, *16*, 100. [CrossRef]
150. Tang, S.; Li, J.; Yuan, Q.; Wang, T.; Xiang, W.; Yu, J.S. Robust Sodium Storage Enabled by Heterogeneous Engineering Electrolyte Modification. *Adv. Energy Mater.* **2025**, *15*, 2404418. [CrossRef]
151. Wang, F.; Chen, L.; Wei, J.; Diao, C.; Li, F.; Du, C.; Bai, Z.; Zhang, Y.; Malyi, O.I.; Chen, X.; et al. Pushing slope- to plateau-behavior in hard carbon for sodium-ion batteries via local structure rearrangement. *Energ. Environ. Sci.* **2025**, *18*, 4312– [CrossRef]
152. Zhang, Y.; Zhang, S.-W.; Chu, Y.; Zhang, J.; Xue, H.; Jia, Y.; Cao, T.; Qiu, D.; Zou, X.; Wang, D.-W.; et al. Redefining closed por carbons by solvation structures for enhanced sodium storage. *Nat. Commun.* **2025**, *16*, 3634. [CrossRef] [PubMed]
153. Hou, W.; Yi, Z.; Yu, H.; Jia, W.; Dai, L.; Yang, J.; Chen, J.; Xie, L.; Su, F.; Chen, C.-M. Fractal dimension revealed from SAXS descriptor of structural disorder in hard carbon anodes of sodium ion battery. *Chin. Chem. Lett.* **2025**, 111124. [CrossRef]
154. Dong, S.; Chen, Z.; Liu, H.; Zhang, Y.; Yan, X.; Cui, Z.; Jiang, Z.; Li, J.; Xu, H.; Peng, W.; et al. Accordion-like structure MBene ordered metal vacancies as catalytic host material to enhance Se reaction kinetics in aqueous Cu-Se batteries. *Adv. Funct. M* **2025**, 2425475. [CrossRef]
155. Xia, Y.; Que, L.; Yu, F.; Deng, L.; Liang, Z.; Jiang, Y.; Sun, M.; Zhao, L.; Wang, Z. Tailoring Nitrogen Terminals on MXene Ena Fast Charging and Stable Cycling Na-Ion Batteries at Low Temperature. *Nano-Micro Lett.* **2022**, *14*, 143. [CrossRef] [PubMe
156. Naguib, M.; Mochalin, V.N.; Barsoum, M.W.; Gogotsi, Y. 25th anniversary article: MXenes: A new family of two-dimensi materials. *Adv. Mater.* **2014**, *26*, 992–1005. [CrossRef] [PubMed]
157. Zhao, Y.; Zhang, J.; Guo, X.; Cao, X.; Wang, S.; Liu, H.; Wang, G. Engineering strategies and active site identificatio MXene-based catalysts for electrochemical conversion reactions. *Chem. Soc. Rev.* **2023**, *52*, 3215–3264. [CrossRef]
158. Ma, P.; Zhang, Z.; Wang, J.; Li, H.; Yang, H.Y.; Shi, Y. Self-Assembled 2D $VS_2/Ti_3C_2T_x$ MXene Nanostructures with Ultr Kinetics for Superior Electrochemical Sodium-Ion Storage. *Adv. Sci.* **2023**, *10*, 2304465. [CrossRef]
159. Kuang, J.; Liu, Z.; Fu, L.; Shi, Y.; Zhang, M.; Wang, Y.; Ding, N.; Sun, D.; Tang, Y.; Wang, H. Charge Tuning and Anchor E Achieving Stable High-Voltage Layered Metal Oxides for Sodium-Ion Battery. *Angew. Chem. Int. Ed.* **2025**, *137*, e202500 [CrossRef]

Jin, J.; Hu, L.; Hu, T.; He, X.; Xia, Y.; Zhang, J.; Gan, Y.; Fang, R.; Xia, X.; Zhang, W.; et al. MXene-Induced 2D Hard Carbon with In Situ Embedding of TiC Scaffolds Enabling Fast Na+ Diffusion and Interfacial Stabilization. *Small* **2025**, *21*, 2412196. [CrossRef]

Pan, F.; Li, Z.; Yao, S.; Liu, J.; Wei, Z.; Chen, X.; Xie, Y.; Du, F. Combined intercalation and space-charge mechanism enabled high-capacity, ultrafast and long-lifespan sodium-ion storage for chalcogenide anodes. *Energ. Environ. Sci.* **2025**, *18*, 1856–1866. [CrossRef]

Liu, Z.; Peng, H.; Wang, X.; Xie, X.; Li, Y.; Ma, G.; Lei, Z. Micropore filling and sodium cluster formation in optimized hard carbon for robust sodium storage. *J. Energy Chem.* **2025**, *108*, 118–128. [CrossRef]

Xia, H.; Zan, L.; Qu, G.; Tu, Y.; Dong, H.; Wei, Y.; Zhu, K.; Yu, Y.; Hu, Y.; Deng, D.; et al. Evolution of a solid electrolyte interphase enabled by FeNx/C catalysts for sodium-ion storage. *Energ. Environ. Sci.* **2022**, *15*, 771–779. [CrossRef]

Wen, X.; Tang, D.; Li, J.; Li, R.; Li, S.; Zhang, J.; Fu, T.; Fan, S.; Lu, Y.; Wei, Q.; et al. Ultrathin Mesoporous Sandwiched Junctions with Monolayered Mesopores for Ultrahigh-Rate Sodium-Ion Storage. *Nano Lett.* **2025**, *25*, 8003–8011. [CrossRef]

Xia, H.; Zan, L.; Yuan, P.; Qu, G.; Dong, H.; Wei, Y.; Yu, Y.; Wei, Z.; Yan, W.; Hu, J.S.; et al. Evolution of Stabilized 1T-MoS_2 by Atomic-Interface Engineering of 2H-MoS_2/Fe−Nx towards Enhanced Sodium Ion Storage. *Angew. Chem. Int. Ed.* **2023**, *62*, e202218282. [CrossRef] [PubMed]

Wu, X.; Yu, X.; Tian, Z.; Yang, X.; Xu, J. Carbonaceous catalyst boosting conversion kinetics of Na_2S in Na-ion batteries. *Energy Storage Mater.* **2025**, *74*, 103899. [CrossRef]

Liu, Y.; Qiu, M.; Hu, X.; Yuan, J.; Liao, W.; Sheng, L.; Chen, Y.; Wu, Y.; Zhan, H.; Wen, Z. Anion Defects Engineering of Ternary Nb-Based Chalcogenide Anodes Toward High-Performance Sodium-Based Dual-Ion Batteries. *Nano-Micro Lett.* **2023**, *15*, 104. [CrossRef]

Yin, H.; Xiao, B.; Yu, Z.-P.; Cunha, J.; Çaha, İ.; Zhang, T.-Q.; Hou, Z.-H.; Li, G.-Y. Embedding CoS_2 nanoparticles within hierarchically porous carbon matrix for enhanced sodium-ion storage and cyclic stability. *Rare Met.* **2025**. [CrossRef]

Wu, S.; Tang, F.; Zhang, K.; Zhang, L.; Huang, F. Synergistic Long- and Short-Range Sodium-Ion Transport Pathways for Enhanced Low-Temperature Performance in Ceramic-DEE-Polymer Electrolytes. *Adv. Funct. Mater.* **2025**, 2501107. [CrossRef]

Nie, Z.; Liu, C.; Lai, Q.-S.; Li, W.; Li, Q.; Yang, R.; Gao, X.-W.; Gu, Q.; Luo, W.-B. Constructing multiphase junction towards layer-structured cathode material for enhanced sodium ion batteries. *Energy Storage Mater.* **2025**, *74*, 103971. [CrossRef]

Zhang, G.; Fu, C.; Gao, S.; Zhao, H.; Ma, C.; Liu, Z.; Li, S.; Ju, Z.; Huo, H.; Zuo, P.; et al. Regulating Interphase Chemistry by Targeted Functionalization of Hard Carbon Anode in Ester-Based Electrolytes for High-Performance Sodium-Ion Batteries. *Angew. Chem. Int. Ed.* **2025**, *64*, e202424028. [CrossRef] [PubMed]

Feng, S.; Lu, Y.; Lu, X.; Chen, H.; Wu, X.; Wu, M.; Xu, F.; Wen, Z. Surface Engineering through In Situ Construction of Co_xB-Spinel Dual Coating Layers for High-Voltage Stable Sodium-Ion Batteries. *Adv. Energy Mater.* **2024**, *14*, 2303773. [CrossRef]

Wei, Y.; Zhang, P.; Soomro, R.A.; Zhu, Q.; Xu, B. Advances in the Synthesis of 2D MXenes. *Adv. Mater.* **2021**, *33*, 2103148. [CrossRef] [PubMed]

molecules

Review

Correlation of Structure and Electrocatalytic Performance of Bulk Oxides for Water Electrolysis

Chuanhui Zhu [1,†], Changming Zhao [1,†], Hao Tian [1,*] and Shuk-Yin Tong [1,2,*]

[1] School of Science and Engineering, The Chinese University of Hong Kong, Shenzhen 518172, China
[2] Institute of Materials Science and Devices, Suzhou University of Science and Technology, Suzhou 215009, China
* Correspondence: tianhao@cuhk.edu.cn (H.T.); tongsy@cuhk.edu.cn (S.-Y.T.)
† These authors contributed equally to this work.

Abstract: Hydrogen-centered electrochemical technologies play a vital role in sustainable energy conversion and storage. One of the challenges in achieving cheap hydrogen is to bridge the gap between advanced electrocatalysts and highly effective electrodes. The key lies in designing electrocatalysts with high intrinsic activity and understanding the structure–activity relationship in water electrolysis. Being proposed as promising electrocatalysts, bulk oxides, with their compositional and crystal structure flexibility, provide a good platform for studying the correlation between intrinsic activity and electronic structure and also for screening superior catalysts for water electrolysis. In this review, we discuss the recent developments of oxide electrocatalysts in understanding the structure–activity relationship. Firstly, we present a thorough overview of recent advances from both theoretical and experimental aspects. Subsequently, we highlight the design principles to provide guidance for promoting performance. Finally, the remaining challenges and perspectives about this field are presented. This review aims to provide guidance for the design of highly advanced oxide electrocatalysts for water electrolysis and large-scale green energy supply.

Keywords: bulk oxides; water electrolysis; intrinsic activity; crystal structure; electronic structure

1. Introduction

With the growing demand for decarbonizing the industrial sector, increasing the share of clean and sustainable energy resources has become one of the top priorities for the science community in the coming decades. Therein, hydrogen economy coupling with water electrolysis has emerged as a key path to develop sustainable energy [1–3]. However, the high cost resulting from sluggish kinetics largely limits water electrolysis market penetration. The introduction of an electrocatalyst can effectively reduce the activation barrier, thus reducing the applied voltage and energy consumption of the reaction. To achieve low overpotential of the water electrolysis, it is of great urgency to develop electrocatalysts with excellent performance. To date, the widely used electrocatalysts for water electrocatalysis are dominated by noble metal electrocatalysts due to the significantly reduced overpotentials. However, the large-scale application of noble metal electrocatalysts is impeded by the rare reserves on earth and concomitantly high cost [4–6]. Therefore, the design of low-cost catalysts has been the research focus. Great attention, for example, was dedicated to developing earth-abundant transition metal-based nanomaterials, aiming to achieve high catalytic performance at a low cost. While the widely studied nanocatalysts surpass the commercial catalysts in the three-electrode configuration at laboratories, their performance can hardly be transferred to commercial electrolyzers operated in industrial conditions. Additionally, the nanocatalysts typically suffer from unsatisfactory repeatability, especially from the lack of a scaled-up synthesis strategy for industrial manufacture, and the instability for electrode fabrication of water electrolyzer techniques. For example, fabricating the electrode in alkaline water electrolyzers (AWEs) is generally prepared by platting the powder catalyst on the nickel net at an elevated temperature (plasma spraying), during which process the catalytic activity stemming from the quantum size effect, facet-dependent effect, and other nanostructured morphology benefits are largely removed. Therefore, research efforts on catalyst design are still urgently demanded.

Bulk oxides synthesized using a solid-state reaction demonstrate approachable scaled-up synthesis and are stable under industrial plasma spraying [7]. Especially, the advances in solid-state chemistry have facilitated the synthesis of numerous new oxides with enriched chemical space. The diversity of bulk oxides affords numerous degrees of freedom to tailor the physicochemical properties for enhancing the catalytic activity [8–10]. Here, the term 'bulk oxides' refers to materials that can be distinguished from nanoscale materials by their ability to be synthesized in large quantities via simple and cost-effective solid-state reactions, resulting in micrometer-sized particles as opposed to the more complex and expensive methods required for nanoscale oxides. In particular, bulk oxides address some critical aspects of water electrolysis (especially scaled-up synthesis using solid-state reactions), although they still have inherent drawbacks, such as low electronic conductivity and severely limited performance by virtue of the finite active sites in the bulk materials [11–13]. Therefore, bulk oxide electrocatalysts have emerged as promising electrocatalyst candidates to possess a highly efficient water electrolysis performance. More importantly, advancements in solid-state chemistry have facilitated the discovery of numerous oxides, and over 90% of the elements in the periodic table can be incorporated into the compositions and structures, such as perovskite oxides (ABO_3), spinel oxides (AB_2O_4), and pyrochlore oxides ($A_2B_2O_7$). Thus, bulk oxides with a flexible composition and crystal structure provide a good platform for studying the correlation between the intrinsic activity and electronic structure and screening for superior catalysts for water electrolysis [14–16]. In-depth understanding and significant progress have been achieved for bulk oxides in water electrolysis, such as the e_g-filling descriptor, emphasizing the great potential of bulk oxides for achieving hydrogen economy. Focusing on the merits and disadvantages of bulk oxides in catalyzing water splitting, the need to review this exciting research field is motivated by the urgent need to unveil highly active electrocatalysts and intensify the systematic understanding of the structure–activity correlations.

In this review, we present a focused review of the correlation of the structure and electrocatalytic performance of bulk oxides for water electrolysis and summarize the rational guidelines for advanced electrocatalyst design. Firstly, we briefly introduce the recent developments in the design of highly efficient oxide electrocatalysts based on comprehensive theoretical and experimental advancements. Then, we introduce several rational strategies for catalyst design, including regulating the electronic structure and/or increasing active sites on the oxide electrocatalysts. Finally, we present the perspectives and challenges for the development of highly advanced oxide electrocatalysts for water electrolysis.

2. The Introduction of Water Electrolysis

2.1. Fundamental Concepts of Water Electrolysis

As a key technology for hydrogen production, water electrolysis typically operates within an electrolyzer, which consists of an electrolyte, cathode, and anode [17]. Under an applied voltage, water molecules would decompose into hydrogen and oxygen at the surface of electrodes, as shown in the following reaction:

$$H_2O \rightarrow H_2 + 1/2O_2$$

Based on electron transfer, the total reaction can be divided into two half-reactions: a hydrogen evolution reaction (HER) at the cathode and oxygen evolution reaction (OER) at the anode, where hydrogen and oxygen are generated at the respective electrodes. In addition, the half-reactions process varies depending on the electrolyte environment:

In acidic electrolytes:

$$\text{Cathode: } 2H^+ + 2e^- \rightarrow H_2$$

$$\text{Anode: } H_2O \rightarrow 2H^+ + 2e^- + 1/2O_2$$

In alkaline and neutral electrolytes:

$$\text{Cathode: } 2H_2O + 2e^- \rightarrow H_2 + 2OH^-$$

$$\text{Anode: } 2OH^- \rightarrow 1/2O_2 + H_2O + 2e^-$$

Water electrolysis is an endothermic reaction, where the theoretical reversible (E_r) and thermoneutral (E_t) voltage are 1.229 and 1.481 V (at 25 °C, 1 atm), respectively. As shown in Figure 1, the E_r and E_t depend on the temperature, where the sudden change at 100 °C can be ascribed to the evaporation of water [18]. In addition, in practical applications, a much higher voltage than the E_r is required to bypass the activation barriers at the cathode (η_c) and anode (η_a), as well as various overpotentials (η_v) caused by other factors, such as mass transport and other resistance. Thus, the overall working voltage (E_w) is expressed as follows:

$$E_w = E_r + \eta_c + \eta_a + \eta_v$$

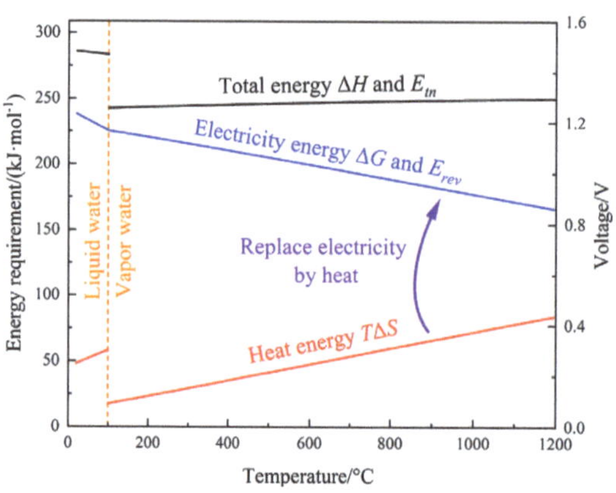

Figure 1. Thermodynamics of water electrolysis at atmospheric pressure [18].

Therefore, the actual working voltage for water electrolysis is usually over 1.6 V. Here, the overpotentials η_c and η_a arising from the slow reaction kinetics are largely dependent on the intrinsic activity of electrocatalysts as well as the structure of electrodes. To achieve efficient production of green hydrogen, developing highly efficient electrocatalysts for practical devices, such as alkaline water electrolyzers (AWE) and proton exchange membrane water electrolyzers (PEMWE), is critical for achieving hydrogen-based economies with zero-carbon emissions.

2.2. Oxygen Evolution Reaction and Hydrogen Evolution Reaction

As a critical half-reaction in water electrolysis, the OER occurs at the anode and involves a four-electron transfer process, resulting in more sluggish kinetics compared with the HER [18]. With respect to the OER mechanism, it is widely accepted that the OER can proceed through two typical mechanisms: the adsorbate evolution mechanism (AEM) and lattice oxygen-mediated mechanism (LOM) [19–21]. Currently, the AEM is the most commonly used OER mechanism, and the reaction steps in the different electrolyte environments are shown below:

In alkaline electrolytes:

$$OH^- + * \rightarrow HO^* + e^-$$

$$HO^* + OH^- \rightarrow H_2O + O^* + e^-$$

$$O^* + OH^- \rightarrow HOO^* + e^-$$

$$HOO^* + OH^- \rightarrow O_2 + * + H_2O + e^-$$

In acidic or neutral electrolytes:

$$H_2O + * \rightarrow HO^* + e^- + H^+$$

$$HO^* \rightarrow O^* + e^- + H^+$$

$$O^* + H_2O \rightarrow HOO^* + e^- + H^+$$

$$HOO^* \rightarrow O_2 + * + e^- + H^+$$

Here, * represents the active site (generally metal site for AEM) at the surface. With the increase in covalency of electrocatalysts, the binding energy of metal and oxygen intermediates would become weak; consequently, the LOM process for the OER may become favorable. In general, the lattice oxygens on the electrocatalyst surface would participate in the LOM-mediated OER process, which involves different lattice oxygen-associated pathways and nonconcerted proton–electron transfer. Therefore, the LOM-dominated electrocatalysts can overcome the adsorption scaling relation (HO* and HOO*) and the limitations of the minimum theoretical overpotential for the AEM process (~0.37 V) [22]. Nevertheless, the participation of lattice oxygen inevitably leads to the decline in stability, which needs to be resolved for the catalyst design [23]. In addition, the intrinsic origin of the LOM has not been fully understood, such as the bindings between lattice oxygen and oxygen intermediates are still in dispute [24]. Recently, more specific OER mechanisms have been proposed for the understanding of the OER process, such as the coupled oxygen evolution mechanism (COM) and oxide path mechanism (OPM) [25,26]. Understanding the OER mechanism can guide the development of more efficient OER electrocatalysts.

The HER occurs at the surface of the cathode, where the mechanism is associated with the adsorption/desorption of hydrogen intermediates on the catalyst surface, involving a two-electron transfer process: (1) the Volmer step; (2) the Heyrovsky or Tafel step (depending on the reaction pathway) [27–29]. The reaction intermediates and pathways vary depending on the pH:

In acidic electrolytes:

$$H^+ + * + e^- \rightarrow H^* \text{ (Volmer)}$$

$$H^* + H^+ + e^- \rightarrow H_2 + * \text{ (Heyrovsky) or } 2H^* \rightarrow H_2 + 2^* \text{ (Tafel)}$$

In alkaline or neutral electrolytes:

$$H_2O + * + e^- \rightarrow H^* + OH^-$$

$$H_2O + H^* + e^- \rightarrow H_2 + * + OH^- \text{ (Heyrovsky) or } 2H^* \rightarrow H_2 + 2^* \text{ (Tafel)}$$

The rate-determining step of the HER process can be identified using the Tafel slope (b) from the polarization curve. In general, if $b < 30$ mV dec^{-1}, the Tafel step is the rate-determining step; if 40 mV dec^{-1} $< b <$ 120 mV dec^{-1}, the Heyrovsky step is the rate-determining step; and if $b >$ 120 mV dec^{-1}, the Volmer step is the rate-determining step. More importantly, in alkaline or neutral electrolytes, the energy barrier of water dissociation is a prerequisite in the HER process, resulting in higher activation barriers and slower kinetics. Therefore, a larger overpotential is needed to drive the reaction in the alkaline or neutral electrolytes. In addition, alkaline HER is the simplest cathodic reaction, involving only water dissociation and a two-electron transfer process. This makes it an important model reaction for investigating other complex reduction reactions, such as CO_2 reduction and N_2 reduction [30,31].

2.3. Performance Evaluation Parameters

The performance of water electrolysis is determined by both the surface kinetics and thermodynamics of the catalyst. To evaluate the catalytic activity of a given electrocatalyst, several key electrochemical parameters have been established to elucidate the overall performance of electrocatalysts, including the overpotential, Tafel slope, mass activity, specific activity, turnover frequency, faradaic efficiency, and stability [32]:

1. **Overpotential (η)**. The overpotential for water electrolysis is the extra applied voltage relative to the theoretical voltage under standard conditions, which is the commonly used parameter to assess the activity of the given electrocatalyst. Typically, the overpotential at j_{geo} (the geometric current density normalized to the electrode surface area) = 10 mA/cm^2 (corresponding to a solar-to-hydrogen efficiency of 12.3%) is used as a key parameter to evaluate the catalytic activity [33]. However, the overpotential of the electrocatalysts is strongly affected by the specific surface area and loading mass, especially for nanocatalysts with large specific surface areas, which hinders the use of the overpotential to reveal the intrinsic catalytic activity [34].
2. **Tafel Slope (b) and Exchange Current Density (j_0)**. The two parameters can be obtained from the Tafel equation, $\eta = a + b\log j$, where b is the Tafel slope, j is the current density, and η is the overpotential, respectively. Extrapolating the linear part of the Tafel plot to zero overpotential gives the exchange current density (j_0), which reflects the intrinsic catalytic activity of the catalyst. Generally, an excellent electrocatalyst should have low b and high j_0 values [35].
3. **Mass Activity (MA) and Specific Activity (SA)**. To further characterize the intrinsic activity of the catalyst, additional parameters, such as the mass activity (A/g) and specific activity, have been proposed based on the electrocatalyst mass loading, specific surface area, and electrochemically active surface area (ECSA). Mass activity is the current normalized by the current based on the catalyst mass loading. For electrocatalysts in equal mass, higher mass activity indicates greater catalytic efficiency, making it a useful parameter for assessing cost efficiency [36]. Specific activity is obtained by normalizing the current with the specific surface area or ECSA of the electrocatalysts,

providing a more accurate reflection of intrinsic catalytic differences and facilitating the understanding of structure–activity relationships [37].

4. **Turnover Frequency (TOF), Faradaic Efficiency (FE), and Stability**. Turnover frequency represents the number of product molecules generated per active site per unit of time, making it a key parameter for evaluating the intrinsic catalytic activity of the electrocatalysts [38]. Faradaic efficiency is defined as the ratio of the experimental to the theoretical product, indicating the efficiency and selectivity of the electrocatalysts [39]. In general, the FE of efficient electrocatalysts for water electrocatalysis is expected to be close to 100%. In addition to the above parameters, the stability of electrocatalysts, including both performance stability and structural stability, is the key indicator to assess the application potential of the electrocatalyst [40]. Performance stability is usually assessed by long-term operating electrocatalytic measurements, such as cyclic voltammetry (CV), chronopotentiometry (CP), and chronoamperometry (CA) tests. Structural stability requires in situ or post-reaction characterizations to evaluate the changes in the composition, structure, and morphology of the electrocatalysts.

3. Overview of Bulk Oxides for Water Electrolysis

3.1. Overview of Bulk Oxide Electrocatalysts for Oxygen Evolution Reaction

Bulk oxides with flexible crystal structures and compositions provide excellent templates for investigating the structure–activity correlations in the OER [16,41,42]. Based on the widely accepted OER catalytic mechanisms, Nørskov et al. proposed a standard hydrogen electrode (SHE) computational model to simulate free energy changes during the OER process, which has greatly enhanced the understanding of OER mechanisms [43]. Later, Nørskov et al. investigated the relationship between the adsorption energy of reaction intermediates and the OER activity of the electrocatalysts [44,45]. For an ideal OER electrocatalyst, the free energy change (ΔG) for each reaction step is 1.23 V, resulting in an overall ΔG of 4.92 V (Figure 2a). However, in the actual OER processes, the free energy changes for different reaction steps are not identical, where the slowest step determines the overall reaction rate. More theoretical and experimental results indicate the adsorption energy of reaction intermediates can reflect the OER activity of a catalyst, such as ΔG_{O^*} and ΔG_{O^*}–ΔG_{HO^*} (Figure 2b–d). Therein, the electrocatalyst with oxygen intermediates with a binding energy that is neither too strong nor too weak exhibits the optimal activity. Experimentally, RuO_2 with moderate adsorption energy exhibits the best OER activity. In addition, the volcano plot is also applicable for predicting the performance of other transition metal-based systems, making it one of the most widely accepted and mature descriptors to date.

For transition metal-based electrocatalysts, the *d* orbital electron distribution of transition metals significantly influences the electronic structure. Understanding the relationship between the *d* electron density and OER activity is crucial for designing high-performance transition metal oxide catalysts. For simpler systems, the number of *d* electrons directly reflects the catalytic activity. According to Sabatier's principle, the adsorption energy of reaction intermediates should neither be too strong nor too weak. Transition metal-based oxides, particularly Ni- and Co-based oxides, generally exhibit better OER catalytic activity [46]. Rossmeisl et al. systematically investigated the relationship between oxygen-based intermediates and the *d* electron count during the OER process (shown in Figure 3a,b) [47]. Theoretical calculations revealed that as the *d* electron count increases, the binding strength of intermediates also increases. A subsequent study indicated that the enhanced activity of transition metal oxides with different *d* electrons is related to increased occupancy of *d* antibonding orbitals and optimized binding strength of HO*, which endows Ni-based oxides with superior catalytic activity [48]. In addition to the OER activity, the chemical stability of oxide electrocatalysts also follows a volcano-like trend with the *d* electron count (shown in Figure 3c,d) [49]. The surface adsorption energetics and bulk thermochemistry depend similarly on the number of outer electrons of the transition metal in the oxide. In addition to the effect of the *d* electron on the catalytic activity, the interaction energies between metal centers and ligands play a crucial role. According to HSAB theory, hard acids prefer to bind with hard bases, while soft acids favor soft bases. Transition metals typically act as borderline acids, which can interact effectively with both hard and soft bases, depending on their electronic configuration and oxidation state [50]. The enhanced activity observed in certain transition metal oxides (e.g., Ru-based perovskites and pyrochlores) can be attributed to their ability to form strong interactions with oxygen ligands due to their intermediate hardness. These interactions facilitate optimal binding energies for catalytic reactions such as the HER and OER. Conversely, zinc-based complexes are generally considered hard acids, leading to weaker interactions with oxygen ligands, which are relatively harder bases. This mismatch results in less favorable binding energies, explaining why zinc-based complexes typically do not exhibit excellent catalytic activity in these reactions. In light of the key role of *d* orbital electrons in the optimization of reaction intermediate binding energies, these groundbreaking works provide significant guidance for the development of oxide electrocatalysts.

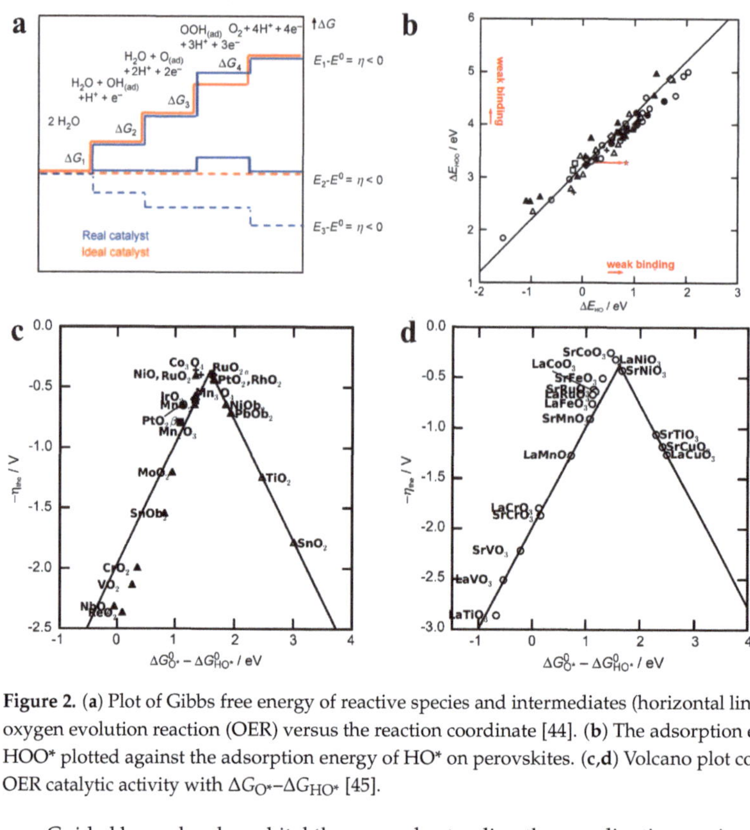

Figure 2. (**a**) Plot of Gibbs free energy of reactive species and intermediates (horizontal lines) of the oxygen evolution reaction (OER) versus the reaction coordinate [44]. (**b**) The adsorption energy of HOO* plotted against the adsorption energy of HO* on perovskites. (**c**,**d**) Volcano plot correlating OER catalytic activity with $\Delta G_{O^*}–\Delta G_{HO^*}$ [45].

Guided by molecular orbital theory, understanding the coordination environment of the metal–oxygen bond in the oxide lattice is crucial for tuning the electronic structure to regulate the surface binding energy. Taking first-row transition metals with octahedral coordination as examples (Figure 4a), the exposed *B*-sites have the coordination environment BO_5 at the surface, breaking symmetry and leading to the splitting of e_g and t_{2g} degenerate orbitals into different energy levels [14]. According to frontier orbitals theory, the adsorbate intermediates would interact with the vertically oriented e_g orbital with stronger overlap than that with the t_{2g} orbital, thus determining the adsorption/desorption energy of the reaction intermediates. It should be noted that the spin states presented in Figure 4a are based on the experimental measurements of bulk $LaMO_3$ (M = Cr, Mn, Fe, Co, Ni) materials at room temperature. Other configurations may also be observed under different conditions, such as varying temperatures or in thin-film geometries [51]. In fact, the spin states of the transition metal are largely dependent on the oxidation state and crystal structure (e.g., coordination environment, distortion), which should be determined by the magnetic measurements. Inspired by this point, Shao-Horn et al. systematically investigated the OER activity of a series of perovskite oxide electrocatalysts and proposed a significant e_g-filling descriptor [52,53]. Based on e_g orbital occupancy, the σ-e_g orbital bonding has a stronger overlap with the oxygen-based reaction intermediates than that of the π-t_{2g} orbital bonding, directly influencing the binding energy of the adsorbates and promoting charge transfer, thereby enhancing OER performance. Therefore, as shown in Figure 4b, the occupancy of e_g orbitals correlates with the OER activity of perovskites exhibits a volcano-like relationship, in which the $Ba_{0.5}Sr_{0.5}Co_{0.8}Fe_{0.2}O_{3-\delta}$ (BSCF) at the peak of the volcano shows the best OER activity with an e_g filling of ~1.2. The e_g-filling descriptor is also consistent with the Sabatier principle, where optimizing the binding strength between the reaction intermediates and the catalyst enhances the catalytic activity. Based on this theory, many outstanding OER electrocatalysts have been designed, such as $CaCu_3Ru_4O_{12}$, $Ca_2Mn_2O_5$, and $CaCu_3Fe_4O_{12}$ [54–56]. To design an excellent OER oxide electrocatalyst with optimal e_g orbital occupancy, the *d* orbital can be modulated by regulating the oxidation state and spin state.

Although the e_g-filling descriptor has been widely used in the electrocatalyst design, the available range of this descriptor is limited to explain the activity trends of all kinds of electrocatalysts. For example, the OER activity of the $LaBO_3$ (*B* = Cr, Mn, Fe, Co, Ni) system does not directly correlate with the e_g orbital occupancy [57]. Consequently, further theoretical and experimental studies have proposed more descriptors to establish catalytic-dependent models, such as metal–oxygen covalency and charge transfer energy [58,59]. For example, due to the sharing of electrons between the metal and oxygen atoms in highly covalent late-transition metal oxides, the metal and oxygen atoms can both be seen as active sites. To more precisely capture the mixed ionic–covalent character of transition metal oxides, metal–oxygen covalency derived from the bulk electronic structure has emerged as a powerful descriptor for OER theoretical studies. Specifically, the metal–oxygen covalency indicates the characters of metal *d*-orbitals and oxygen *p*-orbitals, dictating the stability and surface adsorption energetics [49,60,61]. Therefore, as shown in Figure 4c, the metal–oxygen covalency of oxides directly affects the oxygen binding energy, the oxygen vacancy formation energy, and the electron transfer barrier [14]. In detail, the enhanced metal–oxygen covalency would induce the O *p*-band center closer to the Fermi level, enhancing the charge transfer between the active site and reaction intermediates during the OER process, thus increasing the OER activity. Nevertheless, the stability of the oxide electrocatalysts would decrease with the increased metal–oxygen covalency; thus, the moderate metal–oxygen covalency would lead to both high activity and stability for the OER [58,59]. Therefore, metal–oxygen covalency can be used to predict and explain the OER activity of oxide electrocatalysts. For example, Xu et al. introduced Fe into $LaCoO_3$, observing a transition from low-spin to high-spin Co in $LaCo_{0.9}Fe_{0.1}O_3$, which increased the overlap between Co 3*d* and O 2*p* bands, enhancing the covalency and improving the OER performance [62]. Similar performance optimization can also be observed in spinel oxides such as $ZnFe_xCo_{2-x}O_4$ [63]. In practice, doping the metal anion with different electronegative atoms or tuning the oxidation state of the metal anion can regulate the Fermi level and O 2*p* states, which can alter the metal–oxygen covalency and hybridization, consequently optimizing the OER activity of oxide electrocatalysts [64,65].

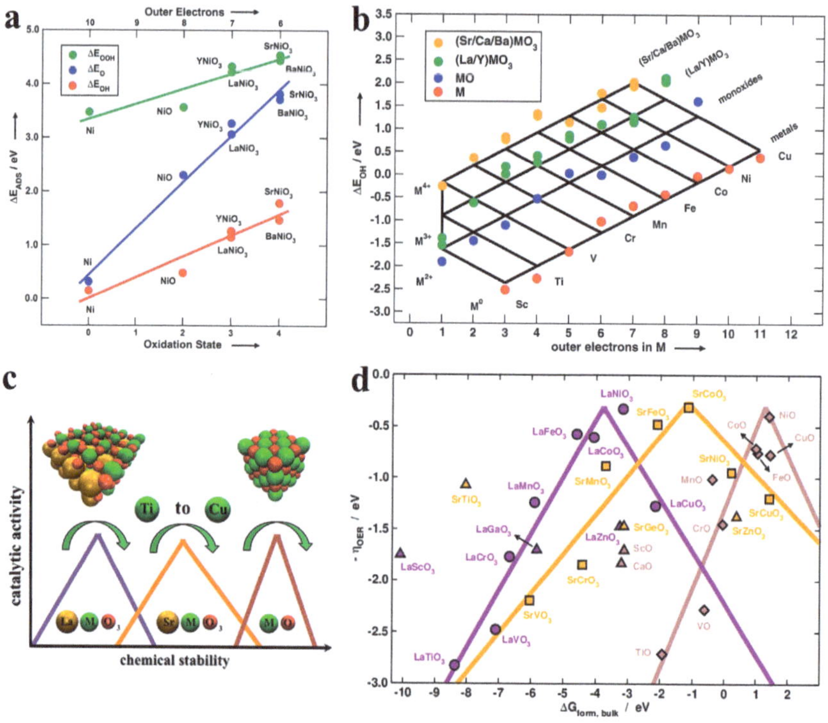

Figure 3. (**a**) Variations of the binding energy for different intermediates with the formula oxidation states in various compounds [47]. (**b**) The adsorption energy of *OH on various compounds with different outer electrons [47]. (**c**) The correlation of activity and stability in different compounds [49]. (**d**) Volcano-type activity plot for various compounds with different formation energy [49].

The discovery of new electrocatalysts is fundamental to investigating the structure–activity correlations, providing critical guidance for designing more efficient electrocatalysts and accelerating the development of practical applications. Based on advancements in theoretical and experimental studies, such as e_g orbital occupancy and metal–oxygen covalency, more novel oxide electrocatalysts have been developed for the OER. Goodenough et al. investigated the OER activity of the two isostructural $ACoO_3$ (A = Ca, Sr) oxides synthesized via high-pressure synthesis [66]. As shown in Figure 5a, lattice parameter changes caused by different *A*-site ions led to modifications in the electronic structure of the oxides, resulting in superior catalytic activity for $CaCoO_3$. Beyond simple perovskite oxides, quadruple perovskite has emerged as an efficient electrocatalytic system due to the complex electronic interaction between *A*- and *B*-site metal anions. For example, Long et al. synthesized a complex quadruple perovskite oxide $CaCu_3Ir_4O_{12}$ for OER (shown in Figure 5b), which exhibited a low overpotential of 252 mV at 10 mA cm^{-2} in 1 M KOH, along with excellent long-term stability, demonstrating great potential for commercial applications [67]. In situ analysis revealed that the enhanced OER activity can be ascribed to the strong *3d–2p–5d* orbital hybridization induced by the $A'-B$ intersite cooperation. In addition, Shao et al. reported a novel oxide catalyst, $Ba_4Sr_4(Co_{0.8}Fe_{0.2})_4O_{15}$, with a unique hexagonal structure. Both tetrahedral Co and octahedral O ions serve as OER active sites, resulting in exceptional OER activity (shown in Figure 5c) [68]. Additionally, many other outstanding oxide electrocatalysts (Table 1), such as Ba_2MIrO_6 (M = Y, La, Ce, Pr, Nd, and Tb) and $A_2Ru_2O_7$ (A = Y, Nd, Gd, Bi), have been developed based on the above descriptors, where the development of new material systems has significantly advanced catalysis research, offering more electrocatalyst options for the OER [69,70]. Typically, pyrochlore oxides $A_2Ru_2O_7$ (A = Y, Nd, Gd, Bi) are a class of promising OER catalysts with improved activity and stability compared to RuO_2 [70]. The catalytic performance is influenced by the A-site cation, where a longer Ru–O bond and weaker Ru *4d*–O *2p* hybridization enhance the initial OER activity. Under acidic conditions, both A-site and Ru dissolution occur, affecting long-term stability. DFT-based Pourbaix diagrams confirm the thermodynamic instability of these materials, while theoretical predictions and Bader charge analysis suggest that dissolution exposes highly oxidized Ru sites with enhanced activity. Despite degradation, these pyrochlores exhibit significantly higher stability than RuO_2, based on stability number metrics, offering a promising strategy for improving Ru-based catalysts in water electrolysis.

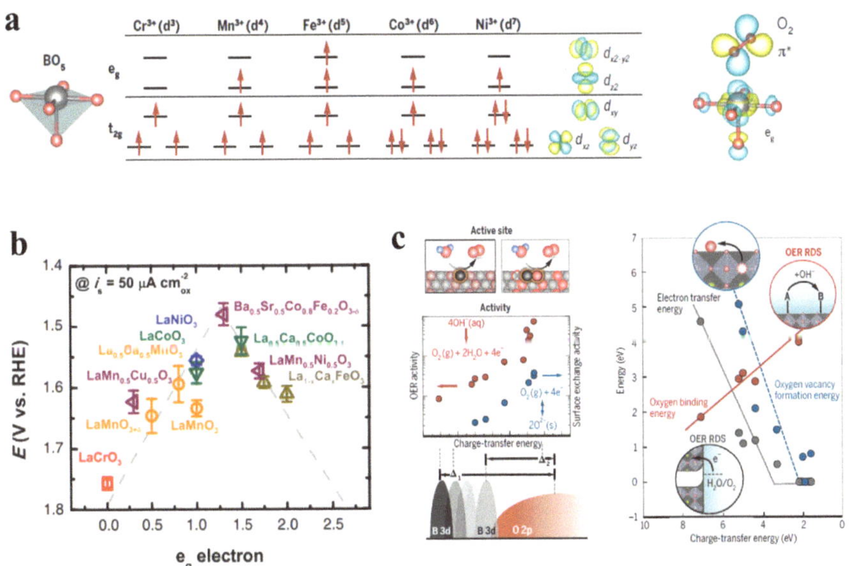

Figure 4. (**a**) Electronic configuration and relevant metal orbitals of first-row transition metals for a BO_5 configuration [14]. (**b**) Relationship between overpotential of perovskite oxides and e_g orbital occupancy [52]. (**c**) Correlations between charge transfer energy (Δ) and OER activity [14].

Table 1. Summary of representative bulk oxide electrocatalysts for OER.

Catalysts	Electrolyte	η at 10 mA cm^{-2} (mV)	Tafel Slope (mV dec^{-1})	Reference
Ba_2MIrO_6 (M = Y, La, Ce, Pr, Nd, Tb)	0.1 M $HClO_4$	>370	60–120	[69]
$Y_2Ir_2O_7$	0.1 M $HClO_4$	262	50	[71]
6*H*-$SrIrO_3$	0.5 M H_2SO_4	248	/	[72]
$CaCu_3Ru_4O_{12}$	0.5 M H_2SO_4	171	40	[54]
$Ba_3TiIr_2O_9$	0.1 M $HClO_4$	275	45.7	[73]
$Ba_4Sr_4(Co_{0.8}Fe_{0.2})_4O_{15}$	0.1 M KOH	340	47	[68]
Sr_2MIrO_6 (M = Ni, Co, Sc, Fe)	0.1 M $HClO_4$	295–420	48–90	[74]
$CaCu_3Ir_4O_{12}$	1 M KOH	252	47	[67]
$Sr_3Ir_2O_7$	0.5 M H_2SO_4	259	50	[75]
Cu_2IrO_3	1 M KOH	361	51	[35]
$SrIr_2O_6$	0.1 M $HClO_4$	303	44.2	[76]
Dy_2NiRuO_6	0.1 M $HClO_4$	277	58	[77]
$Li_2Mn_{0.85}Ru_{0.15}O_3$	0.1 M KOH	260	49.6	[78]

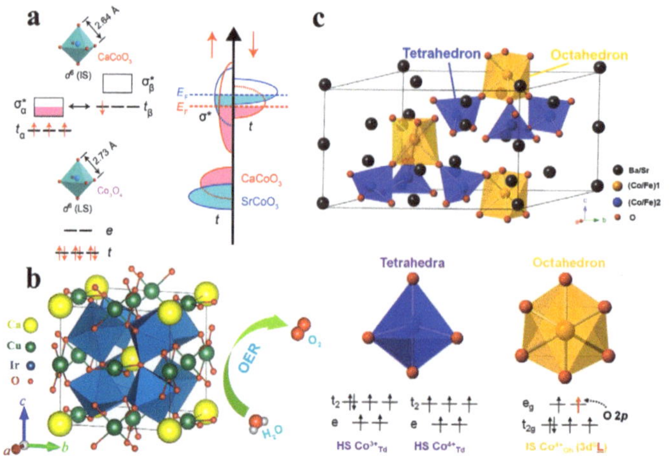

Figure 5. (**a**) Electronic structure changes in $CaCoO_3$ and $SrCoO_3$ [66]; (**b**) application of $CaCu_3Ir_4O_{12}$ in OER [67]; (**c**) novel oxide OER catalyst $Ba_4Sr_4(Co_{0.8}Fe_{0.2})_4O_{15}$ [68].

3.2. Overview of Bulk Oxide Electrocatalysts for Hydrogen Evolution Reaction

Due to the flexible composition and structures, bulk oxides have demonstrated significant performance advantages and application potential in the OER. In view of the excellent OER activity of oxides, it is of great demand to develop oxide electrocatalysts with highly efficient HER activity for practical applications of water electrolysis. However, the intrinsic disadvantages, such as poor electronic conductivity, inappropriate electronic structure for the HER, and limited active sites, have hindered the development of HER oxide electrocatalysts [79]. Therefore, due to the poor HER activity of bulk oxides, only a few oxides with high-performance HER electrocatalysts have been reported currently. To address this issue, substantial research has been dedicated to optimizing HER performance through different strategies, such as element doping, defect engineering, and morphology engineering [28,30,80]. To date, several bulk oxide systems have shown excellent HER activity that is even comparable to commercial Pt/C, indicating the promising potential for oxides in the HER. However, an ongoing debate still exists on the correlation between the intrinsic alkaline HER activity and electronic properties, which makes the design of bulk oxide catalysts for the HER largely a trial-and-error process.

As mentioned above, the intermediate H* is included in all the pathways for the HER. Therefore, the Gibbs free energy change in hydrogen adsorption (ΔG_{H^*}) can theoretically evaluate the interaction between the H* intermediate and the active site, thereby determining the HER activity. In general, ΔG_{H^*} for an ideal HER electrocatalyst should be close to zero value according to the Sabatier principle, which indicates the optimal absorption/desorption of H* on the catalyst surface. In addition, excellent HER electrocatalysts also enable a higher exchange current density. Therefore, by correlating experimental exchange current densities with theoretically calculated ΔG_{H^*} values, a volcano diagram has been proposed to predict the activity [81–83]. Therein, Pt is located at the peak of the volcano with ΔG_{H^*} close to 0, indicating that Pt is the most active HER electrocatalyst based on current research. An efficient HER electrocatalyst binds the reaction intermediate neither too strongly nor too weakly. Understanding how to regulate the binding energies of reactive intermediates on the catalyst surface is crucial for designing HER electrocatalysts with excellent activity. Based on the correlation between *d* electrons and catalytic performance, Nørskov proposed the *d*-band theory (also known as hydrogen binding energy, HBE), which analyzes the relationship between the *d*-orbital energy levels of transition metals and the adsorption energy of reaction intermediates [84–86]. The strength of the interaction between transition metals and reaction intermediates is determined by the filling of antibonding states, which is influenced by the energy of antibonding states relative to the Fermi level (shown in Figure 6a). Since antibonding states are always higher than *d* states, higher energy relative to the Fermi level results in stronger binding between intermediates and the catalyst. Thus, the energy of the *d*-state (relative to the Fermi level) can serve as an activity descriptor. The relative energy of the metal *d*-band and the antibonding hydrogen σ^* orbital directly determines their overlap, which governs the bond strength of H intermediates to the metal surface (ΔG_{H^*}), explaining why Pt-based catalysts remain the best HER electrocatalysts (shown in Figure 6b) [22,87]. The *d*-band theory has been widely used to explain and design more efficient HER oxide electrocatalysts. Lotsch et al. investigated and compared the changes in HER performance of a class of delafossite oxides $PdBO_2$ (B = Cr, Co) and $PtCoO_2$ under long-term electrocatalytic testing [88]. Through comprehensive analysis, the β-PdH_x layer formed in situ on the surface of $PdCoO_2$ inherently stretched the Pd lattice, leading to the broadening of the *d*-band, and thus optimizing the adsorption energy of H*, consequently achieving catalytic performance comparable to commercial Pt/C catalysts.

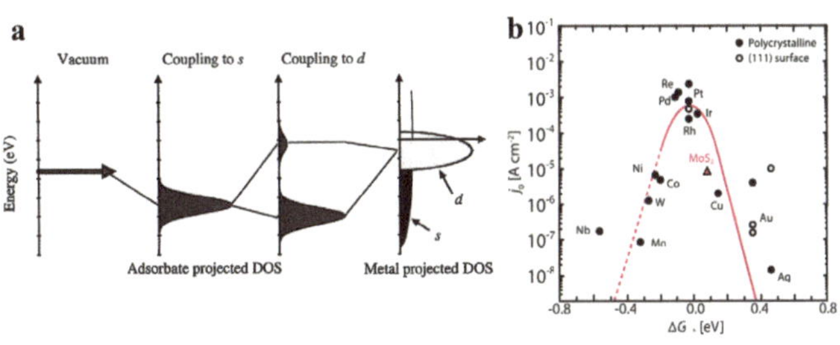

Figure 6. (**a**) The interaction between adsorbates and transition metal [85]. (**b**) Volcano plot of HER performance vs. ΔG_{H^*} [22].

Despite the *d*-band theory providing in-depth insight into the optimum catalyst for a given class of catalyst materials, there are additional factors that are needed to quantitatively determine the absolute reaction rates. Specifically, the kinetics barrier of the HER would change as a function of different pH conditions, leading to varying activity dependent on the pH values [89–91]. In acidic electrolytes, ΔG_{H^*} can be used as a powerful descriptor to evaluate the HER activity, while in the alkaline or neutral electrolytes, the activation barrier of water dissociation and the binding hydroxyl intermediate would also influence the catalytic activity of the catalyst. Therefore, in addition to the previously mentioned ΔG_{H^*}, the Gibbs free energy change in hydroxyl desorption (ΔG_{OH^*}) and the kinetic barrier of water dissociation (E_b) are jointly proposed as the descriptor for alkaline HER [92,93]. Water dissociation theory incorporates the kinetics of water dissociation and interfacial water/anion transfer, which can be ascribed to the H* originates from water dissociation under alkaline conditions [94]. Additionally, the adsorption of OH* competes with H* adsorption, affecting the utilization of active sites. Therefore, promoting OH* desorption or providing OH* adsorption sites can enhance H* generation and utilization, thereby improving the overall HER efficiency [95,96]. Recently, more specific HER descriptors have been proposed for the understanding of the HER process, such as zero-charge potential theory and a hydrogen bond network [97,98]. Zero-charge potential theory suggests that the surface double layer of the catalyst changes with the pH, controlling the binding of reaction intermediates H* and OH* on the catalyst surface, providing some explanation for the "dual-site" catalytic mechanism [99]. Nevertheless, no single descriptor can universally explain the experimental phenomena, and multiple theories are often needed to deepen the understanding of the structure–activity correlations for the HER. Therefore, further theoretical research is required to build effective explanatory models and guide the discovery and design of more efficient electrocatalysts.

Based on advancements in theoretical and experimental studies, such as *d*-band theory, more novel oxide electrocatalysts have been developed for the HER (Table 2). In view of the considerable OER activity of oxides, it is expected that oxide electrocatalysts may enable significant potential for water dissociation in the HER process. For example, Markovic et al. reported a $Ni(OH)_2$-Pt composite HER electrocatalyst with a factor of 8 activity increase in HER activity relative to state-of-the-art catalysts (Figure 7a) [94]. The enhanced HER performance can be attributed to the dissociation of water promoted by $Ni(OH)_2$ clusters. In addition, as shown in Figure 7b, Shao et al. systematically investigated the alkaline HER performance of *A*-site-ordered double perovskite oxides $RBaCo_2O_{5.5+\delta}$ (R = lanthanides) [100]. Therein, the synergistic effects of high-spin Co^{3+} tetrahedral oxygen vacancies and intermediate-spin Co^{4+} octahedral vacancies in $RBaCo_2O_{5.75}$ could enhance H_2O adsorption and dissociation accompanied by the optimization of H intermediate adsorption, ultimately improving the overall HER activity. Later, Shao et al. explored the influence of different *A*-site ions on the performance of $RBaCo_2O_{5.5+\delta}$, suggesting that A-site ion electronegativity can serve as a descriptor for HER activity prediction and evaluation [101]. Theoretical calculations of 13 different $RBaCo_2O_{5.5+\delta}$ compositions revealed that the *A*-site ion electronegativity plays a key role in determining the electronic structure of *B*-site ions, where $Gd_{0.5}La_{0.5}BaCo_2O_{5.5+\delta}$ (electronegativity~2.33) exhibits the best HER activity due to the optimal electronic structure. The flexible crystal structure and composition of perovskite oxides have facilitated the emergence of more oxide-based HER electrocatalysts, such as $SrRuO_3$, Sr_2RuO_4, $BaRuO_3$, and so on [102–104]. Notably, the Ruddlesden–Popper-type layered oxide Sr_2RuO_4 exhibits outstanding HER activity in alkaline media, comparable to the best electrocatalysts reported to date [104]. Theoretical calculations reveal that this performance stems from an unusual synergistic effect between its perovskite and rock-salt layers—the SrO-terminated (001) surface enables barrier-free water dissociation, while apical oxygen sites in the perovskite layer facilitate favorable hydrogen adsorption and evolution. Therefore, designing and identifying more high-performance oxide HER electrocatalysts through the combination of theory and experiments is crucial for designing and developing more efficient HER electrocatalysts.

Table 2. Summary of representative bulk oxide electrocatalysts for HER.

Catalysts	Electrolyte	η at 10 mA cm^{-2} (mV)	Tafel Slope (mV dec^{-1})	Reference
$_{.5}(Ba_{0.5}Sr_{0.5})_{0.5}Co_{0.8}Fe_{0.2}O_{3-\delta}$	1 M KOH	237	45	[105]
$Gd_{0.5}La_{0.5})BaCo_2O_{5.5+\delta}$	1 M KOH	210	27.6	[101]
$PrBaCo_2O_{5.5+\delta}$	0.1 M KOH	291	89	[106]
$SrRuO_3$	1 M KOH	101	67	[104]
Sr_2RuO_4	1 M KOH	61	51	[104]
$SrRu_{0.9}Co_{0.1}O_{3-x}$	1 M KOH	57.8	35	[102]
$Sr_4Ru_2O_9$	1 M KOH	28	55	[107]
$SrTi_{0.7}Ru_{0.3}O_{3-x}$	1 M KOH	46	40	[108]
$BaMoO_3$	1 M KOH	336	110	[109]
9*R*-$BaRuO_3$	1 M KOH	51	30	[7]
$La_2Sr_2PtO_{7+\delta}$	0.5 M H_2SO_4	13	22	[110]

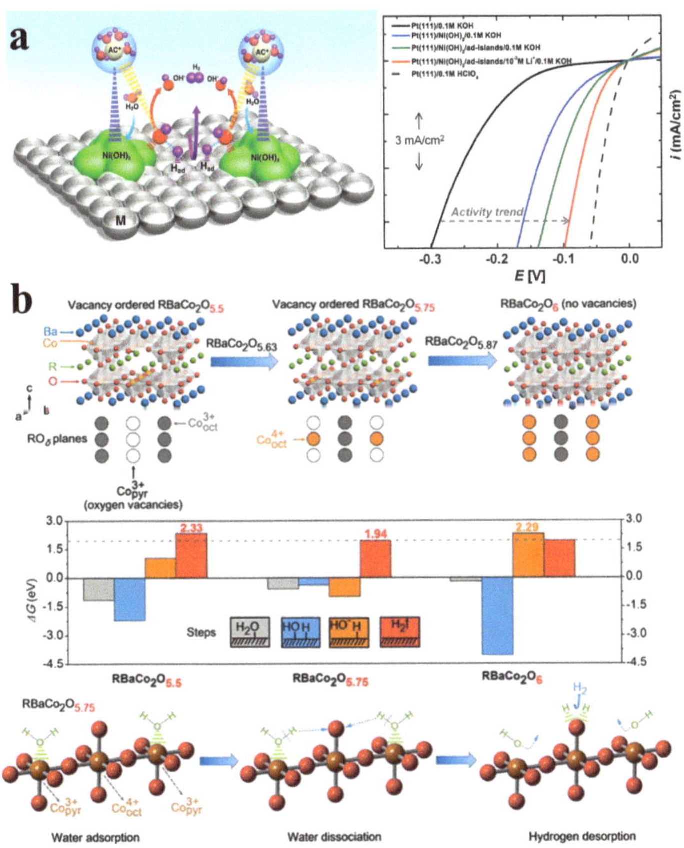

Figure 7. (**a**) Schematic of alkaline HER based on water dissociation theory [94]. (**b**) Relationship between HER performance of $RBaCo_2O_{5.5+\delta}$ and oxygen vacancies [100].

4. Design Strategy for Bulk Oxides with High Water Electrolysis Performance

The catalyst design has been the research focus on electrochemical reactions that promise energy conversions and storage. Great attention, for example, was dedicated to developing earth-abundant transition metal-based nanomaterials, aiming to achieve high catalytic performance at a low cost. Unfortunately, nanocatalysts typically suffer from unsatisfactory repeatability, especially the lack of a scaled-up synthesis strategy for industrial manufacture, thereby impeding their commercial potential. Therefore, research efforts on catalyst design are still urgently demanded for electrocatalysis. Bulk oxides address some critical aspects of water electrolysis (especially scaled-up synthesis using solid-state reactions), but they still have inherent drawbacks, such as unsatisfactory activity and severely limited performance by virtue of the finite active sites in the bulk materials. To further enhance the performance and selectivity of bulk oxide electrocatalysts, it is essential to design and develop more advanced electrocatalysts. Currently, there are generally two strategies to improve the activity of an electrocatalyst: increasing the number of active sites and enhancing the intrinsic activity of each site [22]. In fact, increasing the number of active sites and enhancing the intrinsic activity often occur simultaneously. Different design strategies can be tailored based on these principles to achieve superior catalytic activity.

4.1. Crystal Structure Engineering

Exploring a relatively simple and clear system to realize intrinsic activity regulation is urgent and significant for oxide catalysts. Structural phase transitions provide a treasure route to intensify the systematic understanding of the correlation between the electrocatalytic activity and the crystal, as well as the electronic structure of a material, such as noble metal and transition metal dichalcogenides [111–113]. Altering interatomic distances and bonding patterns without any composition change is expected to alter the electronic structure and thus adjust the catalytic performance. In our recent study, the model system barium ruthenate ($BaRuO_3$, BRO) with four distinct polymorphs is synthesized by using the solid-state reaction and high-pressure/high-temperature (HPHT) method to rationalize the impact of the crystal as well as electronic structure on the HER activity (Figure 8a) [7]. As a micron-sized oxide, the BRO in the 9*R* phase (a rhombohedral stacking variant with nine layers per unit cell) displays an exceptional alkaline HER activity comparable to commercial Pt/C (Figure 8b). The substantial electronic configuration is revealed by the combination of synchrotron X-ray analytical techniques, magnetic and electric measurements, and density function theory calculations. The direct *d*-*d* hopping in face-shared RuO_6 octahedra leads to strong orbital overlap with the H_2O molecule, which renders fast kinetics of water splitting and optimizes the binding energy with surface reactants, thus tremendously improving the HER efficiency of the bulk 9*R*-BRO. More importantly, 9*R*-BRO can be easily synthesized on a large scale using solid-state reactions, showing reproducible catalytic activity and superior transferability in commercial alkaline water electrolyzers, signifying the great potential of bulk oxide electrocatalysts in practical applications. This work highlights the influence of the crystal structure on the activity of oxide electrocatalysts, suggesting phase transition is a valid design principle for advanced electrocatalysts with rational activity enhancement.

In addition to the structural phase transition, the dimensionality control of the crystal structure can also boost the catalytic activity of the oxide electrocatalysts. The layered Ruddlesden–Popper (R-P) oxides have emerged as a more promising OER catalyst family due to the flexible perovskite layers in the lattice [16,114,115]. Previous studies have shown that the OER performances of R-P oxides are largely associated with the inherent dimensionality arising from the different perovskite layers. For example, the OER performances are investigated for a series of R-P oxides $La_nSrNi_nO_{3n+1}$ (n = 1, 2, 3 and ∞), showing the OER activity enhanced with the increasing perovskite layers n [116]. More theoretical and experimental results indicate that the optimized activity of the R-P nickelates can be ascribed to the insulator-to-metal transition and strengthened Ni-O hybridization induced by the increased dimensionality. It is worth noting that the trends of the OER activity are also associated with the composition. In our recent study (Figure 8c,d), we sought to intensify the systematic understanding of the intrinsic effect of the number of perovskite layers in the OER by investigating a prototypical family of R-P iridate oxides $Sr_{n+1}Ir_nO_{3n+1}$ (n = 1, 2, and ∞) [75]. Unveiled by experiments and theoretical calculations, both the conductivity and binding energy of oxygen intermediates on the metal site depend on the number of perovskite layers, resulting in the enhanced OER intrinsic activity of $Sr_3Ir_2O_7$. These studies provide enticing strategies for developing advanced electrocatalysts via unique crystal structure designs in renewable energy applications.

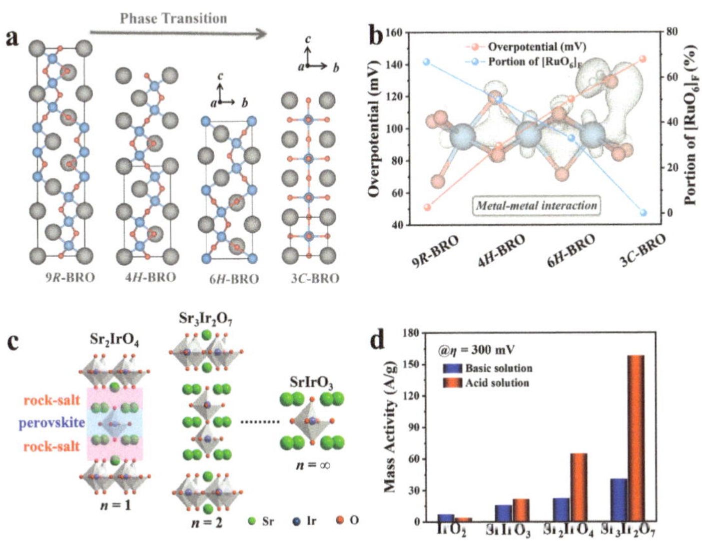

Figure 8. (**a**) Pressure-dependent structural transition of $BaRuO_3$ from 9*R*- to 4*H*-, 6*H*-, and 3*C*-polymorphs [7]. (**b**) Variation in overpotential at 10 mA cm^{-2} reflecting different proportions of face-shared octahedra in BRO polymorphs [7]. (**c**) Crystal structures of $Sr_{n+1}Ir_nO_{3n+1}$ (n = 1, 2, and ∞) [75]. (**d**) Mass activity of IrO_2 and $Sr_{n+1}Ir_nO_{3n+1}$ (n = 1, 2, and ∞) at the overpotential of η = 300 mV in acid and basic solutions [75].

4.2. Heteroatom Doping

The electronic structure plays a vital role in determining the catalytic activity of electrocatalysts, where heteroatom doping has been broadly used to modulate the electronic structure, like the *d*-band center, consequently altering the water electrolysis performance. Taking perovskite oxides as an example, as shown in Figure 9, over 90% of periodic table elements can be incorporated into the perovskite structures [117]. By replacing elements at different crystallographic positions in perovskite oxides (ABO_3), both the *A*- and *B*-site, as well as O-site, the electronic structure can be tuned to optimize the catalytic activity due to the different ionic radius and electronic structure from the heteroatom [118]. Typically, *A*-site ions are composed of alkali, alkaline earth, or rare-earth metals, which cannot directly participate in catalytic reactions. However, the heteroatom doping at the *A*-site would induce the change in the crystal structure, such as the contraction or expansion of the lattice parameters, structural distortions, and phase changes, and thus alter the electronic structure of *B*-site ions to affect the catalytic performance. For example, Müller et al. investigated the impact of different *A*-site ion doping on the OER performance of pyrochlore $Y_{1.8}M_{0.2}Ru_2O_{7-\delta}$ (M = Cu, Co, Ni, Fe, Y). Different *A*-site ions can influence the surface lattice oxygen defect concentration and alter the energy difference between the O 2*p* band and the Fermi level (EF), thereby modulating the intrinsic OER activity [119]. In addition, the introduction of heterovalent atoms can also alter the oxidation state of *B*-site ions, like the enhanced OER stability of $SrRuO_3$ by Na doping, which can be ascribed to the increased Ru oxidation states and the distortion of RuO_6 octahedra [120].

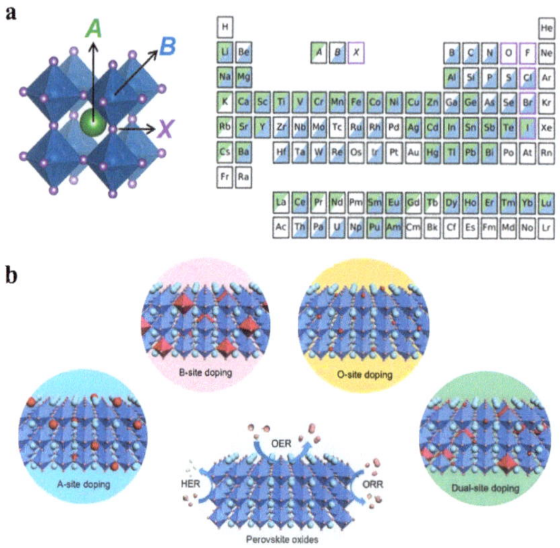

Figure 9. (**a**) The typical perovskite structure and composition [100]. (**b**) Element doping strategies at different sites of perovskite oxides [101].

B-site ion doping can directly modify the electronic structure of active sites, thus enhancing the electrocatalytic activity of electrocatalysts. For example, Lu et al. studied the effects of Co ion doping on the layered perovskite oxide $Bi_7Fe_3Ti_3O_{21}$ [121]. The elongation of oxygen octahedra in the layered structure can stabilize the intermediate-spin Co^{3+} state, providing an optimal electronic configuration for OER. Additionally, Co doping improved the conductivity of the oxide and optimized oxygen intermediate adsorption, resulting in efficient OER performance. More importantly, diluting the content of precious metal in the crystal structure is a sought-after strategy for designing efficient electrocatalysts. For instance, perovskite oxide $SrIrO_3$ has been demonstrated as a highly active electrocatalyst exceeding the benchmark OER catalysts such as IrO_2 and RuO_2 [72]. Further introduction of Zr or Ti at the Ir-site not only decreased the content of iridium, but also enhanced the intrinsic OER activity more than that of $SrIrO_3$, confirming the great advantages of the heteroatom doping strategy [122,123]. In addition to cation doping, anion doping is another effective strategy to improve the catalytic activity, such as the F, S, N, or halogen elements at O-sites. Due to the discrepancy of electronegativity and valency, these dopants can alter the oxygen vacancy formation energy, adjust the electronic structure, and induce surface structural changes, thereby influencing the overall catalytic performance. Therein, F atom doping is widely used in transition metal oxides due to the inherently most electronegative anion with a −1 charge. For example, F anion doping in $La_{0.5}Ba_{0.25}Sr_{0.25}CoO_{2.9-\delta}F_{0.1}$ could alter the CoO_6 coordination environment, which induced CoO_6 octahedra to transform into CoO_5F octahedra and CoO_5 pyramids, thus promoting Co e_g and t_{2g} orbital electron rearrangement [124]. Therefore, the introduction of the F anion resulted in the upshift of the O-*p* band center and activated the lattice O in the OER process, consequently boosting the OER activity for oxides.

4.3. Defect Engineering

In addition to heteroatom doping, defect engineering is also an effective strategy to alter the electronic structure without additional elements being introduced into the lattice, providing a relatively simple platform to uncover the activity–structure correlations [125]. Commonly, defects can be introduced into the cation and anion sublattices as vacancies or interstitials, respectively [126]. Thus, defect engineering can greatly tailor the physicochemical properties and electronic structure, consequently optimizing the performances. Anion defects have been extensively investigated in oxide electrocatalysts, which can be introduced through various methods, including a topochemical reduction reaction, heterovalent ion doping, etc. For example, Kim et al. synthesized single-phase $Ca_2Mn_2O_5$ by a reduction reaction under 5% H_2/Ar atmosphere, successfully introducing oxygen vacancies in the lattice and thus accelerating charge transfer during the OER [56]. Compared to the original $CaMnO_3$, $Ca_2Mn_2O_5$ exhibited a 100 mV lower onset potential and 4–6 times higher mass activity, which can be attributed to the regulation of Mn *d* orbital electron occupancy induced by oxygen vacancy and faster OH^- transfer in the OER process. In addition, theoretical studies also found that oxygen vacancy formation energy correlates with the energy difference between the O 2*p* band center and E_F, where the lower O 2*p* band center indicates the higher oxygen vacancy formation energies [64]. For example, Ciucci et al. synthesized layered $NdBaMn_2O_{5.5}$, where ordered oxygen vacancies occupied the apex of MnO_5 pyramids, alternating with MnO_6 octahedra along the *b* axis [127]. The introduction of oxygen vacancies optimized the Mn *d* orbital alignment and O 2*p* band center with the e_g orbital electron occupancy~1.04, consequently improving charge transfer and optimizing the adsorption energy between the catalysts and reaction intermediates. Therefore, at an applied voltage of 1.7 V, $NdBaMn_2O_{5.5}$ exhibited 24 times higher intrinsic activity and 2.5 times higher mass activity than the commercial RuO_2 catalyst for the OER.

In recent years, there has been growing interest in cation defect engineering to optimize oxide catalytic performance [128]. Cation defects can directly adjust the electronic structure of active sites, such as regulating the *d* orbital electron distribution of transition metals, thereby influencing the catalytic activity. Additionally, *A*-site defects can modify the oxidation state of B-site ions, which is accompanied by increased oxygen vacancies in the lattice. For example, Shao et al. synthesized a series of *A*-site-deficient $La_{1-x}FeO_{3-\delta}$, which exhibit enhanced OER and ORR activities than those in $LaFeO_3$ [129]. Structural characterization revealed that La vacancies promoted the formation of Fe_{4+} and surface oxygen vacancies, significantly improving the catalytic activity. In addition, *A*-site defects can also optimize the electronics of *B*-site ions, such as e_g orbital filling, further enhancing the OER activity. For example, Gong et al. synthesized *A*-site-deficient $Sr_xCo_{0.8}Fe_{0.2}O_{3-\delta}$ with enhanced OER performance under alkaline conditions, which can be ascribed to the optimization of e_g occupancy of active sites [130]. Nevertheless, undesired defects would also result in unexpected performance reduction. Thus, defect engineering requires precise control over the synthesis condition.

4.4. Morphology Engineering

In addition to enhancing the intrinsic activity mentioned in the above strategies, regulating the morphology of oxide electrocatalysts is an effective strategy to improve catalyst performance. Bulk oxides are typically synthesized via high-temperature solid-state reactions, resulting in micrometer-sized particles with low specific surface areas and limited active sites. Engineering the morphology of bulk oxides, such as the porous structure, nanostructure, and heterostructure, can increase the number of exposed active sites and thus boost the performance of electrocatalysts [27,131,132]. For example, Shao et al. synthesized a three-dimensional ordered microporous (3DOM) $LaFeO_3$ catalyst via the colloidal crystal template method [133]. As shown in Figure 10a,b, compared with bulk $LaFeO_3$, the 3DOM-$LaFeO_3$ catalyst exhibited an enhanced specific surface area and active site exposure, boosting the activity and stability markedly for the OER and HER. Additionally, Zou et al. fabricated porous $SrTi(Ir)O_3$ nanotubes by electrospinning methods (as shown in Figure 10c) [134]. Specifically, compared to $SrTi(Ir)O_3$ nanoparticles, the porous nanotubes exhibited a 2.7 times higher specific surface area and superior OER activity in acidic conditions. However, due to the limitation of high temperatures in synthesizing the mixed oxides, only limited oxides have been fabricated in the nanostructure. The design of well-defined nanostructures plays a crucial role in enhancing the electrocatalytic performance of oxide materials, as it can significantly increase the surface area, expose more active sites, and modulate the electronic structure. Alternatively, the sol–gel method offers a promising route for synthesizing oxide materials with well-controlled nanoparticle structures. For instance, Kim et al. successfully synthesized pyrochlore-type $Tl_2Ru_2O_7$ nanoparticles (~200 nm, Figure 10d) using a sol–gel approach, followed by surface functionalization with dihydrogen phosphate groups to yield P-$Tl_2Ru_2O_7$ nanoparticles [135]. The resulting P-$Tl_2Ru_2O_7$ catalyst exhibited excellent bifunctional activity toward both the oxygen evolution reaction (OER) and the oxygen reduction reaction (ORR), enabling high-performance air cathodes in hybrid Na–air batteries. The enhanced catalytic performance can be attributed to two key factors: (1) the increased Ru–O bond covalency induced by surface dihydrogen phosphate functionalization, and (2) the significantly enhanced specific surface area arising from the nanoparticulate structure.

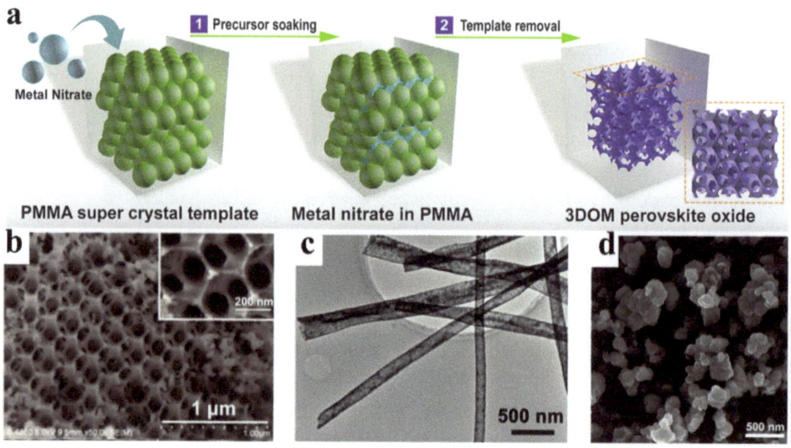

Figure 10. (**a**,**b**) The 3D porous $LaFeO_3$ synthesized via templating [133]; (**c**) $SrTi(Ir)O_3$ nanofibers [134]; (**d**) P-$Tl_2Ru_2O_7$ nanoparticles [135].

5. Conclusions and Perspective

Due to the advantages of flexible composition and structure, substantial development has been achieved in bulk oxide catalysts, providing a good platform for studying the correlation between the activity and structure. For instance, perovskite oxides such as 9*R*-$BaRuO_3$ demonstrate approachable scaled-up synthesis and efficient alkaline water electrolyzer performance [7]. Moreover, pyrochlore oxides $A_2Ru_2O_7$ (A = Y, Nd, Gd, Bi) represent a promising class of OER catalysts, showing not only enhanced catalytic activity but also improved durability compared to RuO_2, as evidenced by lower Tafel slopes and higher stability numbers [70]. In this review, we have introduced the recent breakthroughs in bulk oxide electrocatalysts for water electrolysis, especially for the development of main descriptors describing the activity dependence, such as the e_g orbital occupancy, *d*-band theory, etc. Although bulk oxides address the critical aspects of electrocatalysts (scaled-up synthesis using solid-state reactions), their inherent limitations, such as the finite active sites associated with the bulk nature, led to slightly inferior performance compared to nanocatalysts. To address the limitation of performance, several strategies are highlighted in this review, including optimizing the crystal/electronic structures and increasing the active sites. Although bulk oxides have demonstrated excellent water electrolysis performance in extensively researched studies, several limitations and challenges are still remaining in the current field. From our point of view, the bulk oxides for water electrolysis that need to be further investigated are schematically shown in the following aspects:

5.1. Developing New Electrocatalysts

While some bulk oxides have achieved significant developments in water electrolysis, the discovery of new bulk oxide electrocatalysts with better activity remains a hot spot in this field. To date, only a limited number of oxides have been explored for water electrolysis. Advances in solid-state chemistry have enabled the synthesis of a vast array of new oxide materials, with over 90% of the elements in the periodic table being potentially incorporated into their compositions and structures. Furthermore, the development of high-pressure synthesis techniques has significantly expanded the diversity of available bulk oxides. As a result, the extensive chemical space offered by these materials holds great promise for the design of novel electrocatalysts with superior performance. Up to now, only a few oxides have been investigated for water electrolysis, and thus, more corresponding research is invited to expand the variety of electrocatalysts. In addition, most bulk oxide electrocatalysts are typically micron-sized due to the limitations of widely used solid-state reactions (e.g., high temperature, long reaction time). Only limited classes can be synthesized in nanosized morphology using moderate strategies. Although the bulk oxides with micron-sized morphology have demonstrated stability under industrial plasma spraying for the electrode fabrication of AWE, developing nanostructured oxide electrocatalysts remains the focus of research of water electrolyzer techniques, like PEMWE. The bulk nature not only results in finite active sites for catalytic reactions, but also impedes the further preparation of a homogeneous catalyst layer in membrane-based water electrolyzer techniques. It is, therefore, a standing challenge to realize the synthesis of bulk oxide electrocatalysts in nanostructured morphology using moderate strategies.

5.2. Deepening the Understanding of Structure–Activity Correlations

The correlation between structure and activity is crucial for the design of efficient electrocatalysts. Although an in-depth understanding of water electrolysis has been achieved for bulk oxides, the intrinsic characteristic dominating activity and stability remains elusive. For example, the main reported descriptors, like the *d*-band center, are usually based on a linear scaling relationship (e.g., volcano plot), which has the upper limitation for optimizing performance. In addition, the available range of each descriptor is limited in predicting the activity trends of all kinds of electrocatalysts. Therefore, the development of more universal and comprehensive descriptors is vital for future research on structure–activity correlations. Moreover, identifying precisely the fundamental structures (crystal/electronic/surface structure) of the electrocatalysts is the premise for investigating the structure–activity correlations. To unravel the intrinsic activity dependence, combining integrated theoretical calculations and advanced characterization techniques (especially in situ analysis) would be an accessible way. Recently, emerging factors that may significantly influence the catalytic performance have attracted increasing attention. For example, $BaRuO_3$ exhibits four distinct crystalline phases with markedly different HER activities, suggesting that the connectivity modes of octahedra in perovskite structures—such as face-sharing, edge-sharing, or corner-sharing configurations—can play a critical role in determining the catalytic performance. Furthermore, recent studies indicate that noble metals such as Ir in higher oxidation states may exhibit enhanced catalytic activity. However, it remains to be determined whether this phenomenon is observed in just a few specific instances or represents a broader trend, and what underlying mechanisms are responsible.

5.3. Correlation of Atomic Properties of Transition Metals with Ligand Interactions

A profound understanding of the atomic characteristics of transition metals, including electronegativity, ionic radius, and *d*-orbital energy levels, is crucial for elucidating their potential interactions with ligands. These intrinsic properties significantly influence the binding affinity, electronic structure, and catalytic behavior of the metal centers. Despite numerous studies exploring transition metal complexes, most investigations tend to treat these atomic features separately or focus primarily on one aspect, which can overlook the complex, combined effects these properties have on coordination behavior and reactivity. Therefore, future research efforts should emphasize multi-parameter analyses that simultaneously consider these atomic factors, supported by both experimental data and theoretical modeling. Such integrated studies will fill a key gap in the field, enhancing our ability to tailor transition metal complexes for specific applications more effectively.

5.4. Practical Applications for Water Electrolysis

Despite decades of research, substantial advancements in efficient electrocatalyst design have been achieved on the laboratory scale, while these have not improved the urgent requirements of industrial water electrolyzer performance. Bridging the gap between advanced electrocatalysts and highly effective electrodes in the device is one of the main challenges in achieving cheap hydrogen. Therefore, bulk oxide electrocatalysts should be evaluated in real water electrolyzers in future research. In addition, comprehensive improvements in activity and stability under industrial conditions are another key target for practical applications (e.g., high current densities, long operating time, high temperature). Still, the advanced electrocatalysts in the laboratory fall short of meeting the demand for commercial devices. Therefore, it is imperative to develop robust electrodes with efficient electrocatalysts to ensure the long-term stability of water electrolyzers.

Author Contributions: C.Z. (Chuanhui Zhu): writing—original draft preparation, review, and editing; C.Z. (Changming Zhao): writing—review and editing; H.T.: writing—review and editing; S.-Y.T.: writing—review and editing, supervision, funding acquisition. All authors have read and agreed to the published version of the manuscript.

Funding: This work was funded by the National Science Foundation of China (Grant No. 12174326), the Program for Guangdong Introducing Innovative and Entrepreneurial Teams (2019ZT08L101), and Shenzhen Key Laboratory of Eco-materials and Renewable Energy (Grant No. ZDSYS20200922160400001).

Data Availability Statement: Source data are available from the corresponding authors upon reasonable request.

Conflicts of Interest: The authors declare no conflicts of interest.

:erences

Lagadec, M.F.; Grimaud, A. Water electrolysers with closed and open electrochemical systems. *Nat. Mater.* **2020**, *19*, 1140–1150. [CrossRef] [PubMed]

Gunathilake, C.; Soliman, I.; Panthi, D.; Tandler, P.; Fatani, O.; Ghulamullah, N.A.; Marasinghe, D.; Farhath, M.; Madhujith, T.; Conrad, K.; et al. A comprehensive review on hydrogen production, storage, and applications. *Chem. Soc. Rev.* **2024**, *53*, 10900–10969. [CrossRef] [PubMed]

Liu, R.T.; Xu, Z.L.; Li, F.M.; Chen, F.Y.; Yu, J.Y.; Yan, Y.; Chen, Y.; Xia, B.Y. Recent advances in proton exchange membrane water electrolysis. *Chem. Soc. Rev.* **2023**, *52*, 5652–5683. [CrossRef]

Wan, R.; Yuan, T.; Wang, L.; Li, B.; Liu, M.; Zhao, B. Earth-abundant electrocatalysts for acidic oxygen evolution. *Nat. Catal.* **2024**, *7*, 1288–1304. [CrossRef]

Yang, Y.; Peltier, C.R.; Zeng, R.; Schimmenti, R.; Li, Q.; Huang, X.; Yan, Z.; Potsi, G.; Selhorst, R.; Lu, X.; et al. Electrocatalysis in Alkaline Media and Alkaline Membrane-Based Energy Technologies. *Chem. Rev.* **2022**, *122*, 6117–6321. [CrossRef]

Song, Y.; Chen, H.; Wang, X.; Weng, C.; Zou, K.; Wang, C.; Yuan, Y.; Ma, Y.; Yang, X.; Lin, W. Engineering Ir-based catalysts for high current density applications in proton exchange membrane water electrolyzers. *Energy Environ. Sci.* **2025**, *18*, 130–154. [CrossRef]

Zhu, C.; Tian, H.; Tan, P.; Huang, B.; Zhao, S.; Cai, G.; Yuan, C.; Zhao, M.-H.; Cao, M.; Zhao, J.; et al. Ruthenate perovskite with face-sharing motifs for alkaline hydrogen evolution. *Chem Catal.* **2024**, *4*, 101132. [CrossRef]

Nguyen, L.T.; Cava, R.J. Hexagonal Perovskites as Quantum Materials. *Chem. Rev.* **2021**, *121*, 2935–2965. [CrossRef]

Zhu, Y.; Tang, Z.; Yuan, L.; Li, B.; Shao, Z.; Guo, W. Beyond conventional structures: Emerging complex metal oxides for efficient oxygen and hydrogen electrocatalysis. *Chem. Soc. Rev.* **2025**, *54*, 1027–1092. [CrossRef]

Humayun, M.; Li, Z.; Israr, M.; Khan, A.; Luo, W.; Wang, C.; Shao, Z. Perovskite Type ABO_3 Oxides in Photocatalysis, Electrocatalysis, and Solid Oxide Fuel Cells: State of the Art and Future Prospects. *Chem. Rev.* **2025**, *125*, 3165–3241. [CrossRef]

Wang, K.; Han, C.; Shao, Z.; Qiu, J.; Wang, S.; Liu, S. Perovskite Oxide Catalysts for Advanced Oxidation Reactions. *Adv. Funct. Mater.* **2021**, *31*, 2102089. [CrossRef]

Meharban, F.; Lin, C.; Wu, X.; Tan, L.; Wang, H.; Hu, W.; Zhou, D.; Li, X.; Luo, W. Scaling Up Stability: Navigating from Lab Insights to Robust Oxygen Evolution Electrocatalysts for Industrial Water Electrolysis. *Adv. Energy Mater.* **2021**, *11*, 2102886. [CrossRef]

Du, Y.; Liu, J.; Chen, J.; Wang, S.; Tang, Y.; Wang, A.L.; Fu, G.; Lu, X.F. Design Principle and Regulation Strategy of Noble Metal-Based Materials for Practical Proton Exchange Membrane Water Electrolyzer. *Adv. Energy Mater.* **2024**, *15*, 2404113. [CrossRef]

Hwang, J.; Rao, R.R.; Giordano, L.; Katayama, Y.; Yu, Y.; Shao-Horn, Y. Perovskites in catalysis and electrocatalysis. *Science* **2017**, *358*, 751–756. [CrossRef]

Zhao, Q.; Yan, Z.; Chen, C.; Chen, J. Spinels: Controlled Preparation, Oxygen Reduction/Evolution Reaction Application, and Beyond. *Chem. Rev.* **2017**, *117*, 10121–10211. [CrossRef]

Chen, D.; Chen, C.; Baiyee, Z.M.; Shao, Z.; Ciucci, F. Nonstoichiometric Oxides as Low-Cost and Highly-Efficient Oxygen Reduction/Evolution Catalysts for Low-Temperature Electrochemical Devices. *Chem. Rev.* **2015**, *115*, 9869–9921. [CrossRef]

Zou, X.; Zhang, Y. Noble metal-free hydrogen evolution catalysts for water splitting. *Chem. Soc. Rev.* **2015**, *44*, 5148–5180. [CrossRef]

Zhang, W.; Liu, M.; Gu, X.; Shi, Y.; Deng, Z.; Cai, N. Water Electrolysis toward Elevated Temperature: Advances, Challenges and Frontiers. *Chem. Rev.* **2023**, *123*, 7119–7192. [CrossRef]

Song, J.; Wei, C.; Huang, Z.-F.; Liu, C.; Zeng, L.; Wang, X.; Xu, Z.J. A review on fundamentals for designing oxygen evolution electrocatalysts. *Chem. Soc. Rev.* **2020**, *49*, 2196–2214. [CrossRef]

20. Xu, X.; Pan, Y.; Zhong, Y.; Shi, C.; Guan, D.; Ge, L.; Hu, Z.; Chin, Y.Y.; Lin, H.J.; Chen, C.T.; et al. New Undisputed Evidence Strategy for Enhanced Lattice-Oxygen Participation of Perovskite Electrocatalyst Through Cation Deficiency Manipulation. *Sci.* **2022**, *9*, e2200530. [CrossRef]
21. Pan, Y.; Xu, X.; Zhong, Y.; Ge, L.; Chen, Y.; Veder, J.M.; Guan, D.; O'Hayre, R.; Li, M.; Wang, G.; et al. Direct evidence of boo oxygen evolution over perovskite by enhanced lattice oxygen participation. *Nat. Commun.* **2020**, *11*, 2002. [CrossRef] [Pub
22. Seh, Z.W.; Kibsgaard, J.; Dickens, C.F.; Chorkendorff, I.; Norskov, J.K.; Jaramillo, T.F. Combining theory and experime electrocatalysis: Insights into materials design. *Science* **2017**, *355*, eaad4998. [CrossRef] [PubMed]
23. An, L.; Wei, C.; Lu, M.; Liu, H.; Chen, Y.; Scherer, G.G.; Fisher, A.C.; Xi, P.; Xu, Z.J.; Yan, C.H. Recent Development of Ox Evolution Electrocatalysts in Acidic Environment. *Adv. Mater.* **2021**, *33*, e2006328. [CrossRef]
24. Liu, Y.; Liang, X.; Chen, H.; Gao, R.; Shi, L.; Yang, L.; Zou, X. Iridium-containing water-oxidation catalysts in acidic electr *Chin. J. Catal.* **2021**, *42*, 1054–1077. [CrossRef]
25. Wang, X.; Xi, S.; Huang, P.; Du, Y.; Zhong, H.; Wang, Q.; Borgna, A.; Zhang, Y.W.; Wang, Z.; Wang, H.; et al. Pivotal ro reversible NiO_6 geometric conversion in oxygen evolution. *Nature* **2022**, *611*, 702–708. [CrossRef]
26. Lin, C.; Li, J.-L.; Li, X.; Yang, S.; Luo, W.; Zhang, Y.; Kim, S.-H.; Kim, D.-H.; Shinde, S.S.; Li, Y.-F.; et al. In-situ reconstructe atom array on α-MnO_2 with enhanced performance for acidic water oxidation. *Nat. Catal.* **2021**, *4*, 1012–1023. [CrossRef]
27. Fang, M.; Dong, G.; Wei, R.; Ho, J.C. Hierarchical Nanostructures: Design for Sustainable Water Splitting. *Adv. Energy M* **2017**, *7*, 1700559. [CrossRef]
28. Zhu, J.; Hu, L.; Zhao, P.; Lee, L.Y.S.; Wong, K.Y. Recent Advances in Electrocatalytic Hydrogen Evolution Using Nanopar *Chem. Rev.* **2020**, *120*, 851–918. [CrossRef]
29. Li, L.; Wang, P.; Shao, Q.; Huang, X. Metallic nanostructures with low dimensionality for electrochemical water splitting. C *Soc. Rev.* **2020**, *49*, 3072–3106. [CrossRef]
30. Wang, X.; Zheng, Y.; Sheng, W.; Xu, Z.J.; Jaroniec, M.; Qiao, S.-Z. Strategies for design of electrocatalysts for hydrogen evol under alkaline conditions. *Mater. Today* **2020**, *36*, 125–138. [CrossRef]
31. Ahsan, M.A.; He, T.; Noveron, J.C.; Reuter, K.; Puente-Santiago, A.R.; Luque, R. Low-dimensional heterostructures for adva electrocatalysis: An experimental and computational perspective. *Chem. Soc. Rev.* **2022**, *51*, 812–828. [CrossRef] [PubMed]
32. Jiao, Y.; Zheng, Y.; Jaroniec, M.; Qiao, S.Z. Design of electrocatalysts for oxygen- and hydrogen-involving energy conve reactions. *Chem. Soc. Rev.* **2015**, *44*, 2060–2086. [CrossRef] [PubMed]
33. Wei, C.; Xu, Z.J. The Comprehensive Understanding of 10 mA cm geo^{-2} as an Evaluation Parameter for Electrochemical W Splitting. *Small Methods* **2018**, *2*, 1800168. [CrossRef]
34. Montoya, J.H.; Seitz, L.C.; Chakthranont, P.; Vojvodic, A.; Jaramillo, T.F.; Norskov, J.K. Materials for solar fuels and chem *Nat. Mater.* **2016**, *16*, 70–81. [CrossRef]
35. Zhu, C.; Cai, G.-H.; Yuan, C.; Huang, B.; Li, G.; Croft, M.; Greenblatt, M.; Li, M.-R. Intersite Charge Transfer Enhanced Ox Evolution Reactivity on A_2IrO_3 (A = Li, Na, Cu) Delafossite Electrocatalysts. *J. Electrochem. Soc.* **2022**, *169*, 056523. [CrossR
36. Nong, H.N.; Reier, T.; Oh, H.-S.; Gliech, M.; Paciok, P.; Vu, T.H.T.; Teschner, D.; Heggen, M.; Petkov, V.; Schlögl, R.; et unique oxygen ligand environment facilitates water oxidation in hole-doped $IrNiO_x$ core–shell electrocatalysts. *Nat. Catal.* *1*, 841–851. [CrossRef]
37. Wei, C.; Sun, S.; Mandler, D.; Wang, X.; Qiao, S.Z.; Xu, Z.J. Approaches for measuring the surface areas of metal c electrocatalysts for determining their intrinsic electrocatalytic activity. *Chem. Soc. Rev.* **2019**, *48*, 2518–2534. [CrossRef]
38. Lettenmeier, P.; Wang, L.; Golla-Schindler, U.; Gazdzicki, P.; Canas, N.A.; Handl, M.; Hiesgen, R.; Hosseiny, S.S.; Gago, Friedrich, K.A. Nanosized IrO_x-Ir Catalyst with Relevant Activity for Anodes of Proton Exchange Membrane Electro Produced by a Cost-Effective Procedure. *Angew. Chem. Int. Ed.* **2016**, *55*, 742–746. [CrossRef]
39. Liu, Y.; Liang, X.; Gu, L.; Zhang, Y.; Li, G.D.; Zou, X.; Chen, J.S. Corrosion engineering towards efficient oxygen evol electrodes with stable catalytic activity for over 6000 hours. *Nat. Commun.* **2018**, *9*, 2609. [CrossRef]
40. Geiger, S.; Kasian, O.; Ledendecker, M.; Pizzutilo, E.; Mingers, A.M.; Fu, W.T.; Diaz-Morales, O.; Li, Z.; Oellers, T.; Frucht et al. The stability number as a metric for electrocatalyst stability benchmarking. *Nat. Catal.* **2018**, *1*, 508–515. [CrossRef]
41. Xu, X.; Wang, W.; Zhou, W.; Shao, Z. Recent Advances in Novel Nanostructuring Methods of Perovskite Electrocatalys Energy-Related Applications. *Small Methods* **2018**, *2*, 1800071. [CrossRef]
42. Zhou, D.; Li, P.; Lin, X.; McKinley, A.; Kuang, Y.; Liu, W.; Lin, W.F.; Sun, X.; Duan, X. Layered double hydroxide-b electrocatalysts for the oxygen evolution reaction: Identification and tailoring of active sites, and superaerophobic nano electrode assembly. *Chem. Soc. Rev.* **2021**, *50*, 8790–8817. [CrossRef] [PubMed]
43. Rossmeisl, J.; Logadottir, A.; Nørskov, J.K. Electrolysis of water on (oxidized) metal surfaces. *Chem. Phys.* **2005**, *319*, 178 [CrossRef]

Dau, H.; Limberg, C.; Reier, T.; Risch, M.; Roggan, S.; Strasser, P. The Mechanism of Water Oxidation: From Electrolysis via Homogeneous to Biological Catalysis. *ChemCatChem* **2010**, *2*, 724–761. [CrossRef]

Man, I.C.; Su, H.Y.; Calle-Vallejo, F.; Hansen, H.A.; Martínez, J.I.; Inoglu, N.G.; Kitchin, J.; Jaramillo, T.F.; Nørskov, J.K.; Rossmeisl, J. Universality in Oxygen Evolution Electrocatalysis on Oxide Surfaces. *ChemCatChem* **2011**, *3*, 1159–1165. [CrossRef]

Vojvodic, A.; Norskov, J.K. Chemistry. Optimizing perovskites for the water-splitting reaction. *Science* **2011**, *334*, 1355–1356. [CrossRef]

Calle-Vallejo, F.; Inoglu, N.G.; Su, H.-Y.; Martínez, J.I.; Man, I.C.; Koper, M.T.M.; Kitchin, J.R.; Rossmeisl, J. Number of outer electrons as descriptor for adsorption processes on transition metals and their oxides. *Chem. Sci.* **2013**, *4*, 1245–1249. [CrossRef]

Xu, X.; Pan, Y.; Zhong, Y.; Ran, R.; Shao, Z. Ruddlesden–Popper perovskites in electrocatalysis. *Mater. Horiz.* **2020**, *7*, 2519–2565. [CrossRef]

Calle-Vallejo, F.; Díaz-Morales, O.A.; Kolb, M.J.; Koper, M.T.M. Why Is Bulk Thermochemistry a Good Descriptor for the Electrocatalytic Activity of Transition Metal Oxides? *ACS Catal.* **2015**, *5*, 869–873. [CrossRef]

Pearson, R.G. Hard and Soft Acids and Bases. *J. Am. Chem. Soc.* **1963**, *85*, 3533–3539. [CrossRef]

Hong, W.T.; Risch, M.; Stoerzinger, K.A.; Grimaud, A.; Suntivich, J.; Shao-Horn, Y. Toward the rational design of non-precious transition metal oxides for oxygen electrocatalysis. *Energy Environ. Sci.* **2015**, *8*, 1404–1427. [CrossRef]

Suntivich, J.; May, K.J.; Gasteiger, H.A.; Goodenough, J.B.; Shao-Horn, Y. A perovskite oxide optimized for oxygen evolution catalysis from molecular orbital principles. *Science* **2011**, *334*, 1383–1385. [CrossRef] [PubMed]

Suntivich, J.; Gasteiger, H.A.; Yabuuchi, N.; Nakanishi, H.; Goodenough, J.B.; Shao-Horn, Y. Design principles for oxygen-reduction activity on perovskite oxide catalysts for fuel cells and metal-air batteries. *Nat. Chem.* **2011**, *3*, 546–550. [CrossRef]

Miao, X.; Zhang, L.; Wu, L.; Hu, Z.; Shi, L.; Zhou, S. Quadruple perovskite ruthenate as a highly efficient catalyst for acidic water oxidation. *Nat. Commun.* **2019**, *10*, 3809. [CrossRef]

Yagi, S.; Yamada, I.; Tsukasaki, H.; Seno, A.; Murakami, M.; Fujii, H.; Chen, H.; Umezawa, N.; Abe, H.; Nishiyama, N.; et al. Covalency-reinforced oxygen evolution reaction catalyst. *Nat. Commun.* **2015**, *6*, 8249. [CrossRef]

Kim, J.; Yin, X.; Tsao, K.C.; Fang, S.; Yang, H. $Ca_2Mn_2O_5$ as oxygen-deficient perovskite electrocatalyst for oxygen evolution reaction. *J. Am. Chem. Soc.* **2014**, *136*, 14646–14649. [CrossRef]

Ashok, A.; Kumar, A.; Bhosale, R.R.; Almomani, F.; Malik, S.S.; Suslov, S.; Tarlochan, F. Combustion synthesis of bifunctional $LaMO_3$ (*M* = Cr, Mn, Fe, Co, Ni) perovskites for oxygen reduction and oxygen evolution reaction in alkaline media. *J. Electroanal. Chem.* **2018**, *809*, 22–30. [CrossRef]

Hong, W.T.; Stoerzinger, K.A.; Lee, Y.-L.; Giordano, L.; Grimaud, A.; Johnson, A.M.; Hwang, J.; Crumlin, E.J.; Yang, W.; Shao-Horn, Y. Charge-transfer-energy-dependent oxygen evolution reaction mechanisms for perovskite oxides. *Energy Environ. Sci.* **2017**, *10*, 2190–2200. [CrossRef]

Sun, Y.; Liao, H.; Wang, J.; Chen, B.; Sun, S.; Ong, S.J.H.; Xi, S.; Diao, C.; Du, Y.; Wang, J.-O.; et al. Covalency competition dominates the water oxidation structure–activity relationship on spinel oxides. *Nat. Catal.* **2020**, *3*, 554–563. [CrossRef]

May, K.J.; Carlton, C.E.; Stoerzinger, K.A.; Risch, M.; Suntivich, J.; Lee, Y.-L.; Grimaud, A.; Shao-Horn, Y. Influence of Oxygen Evolution during Water Oxidation on the Surface of Perovskite Oxide Catalysts. *J. Phys. Chem. Lett.* **2012**, *3*, 3264–3270. [CrossRef]

Risch, M.; Grimaud, A.; May, K.J.; Stoerzinger, K.A.; Chen, T.J.; Mansour, A.N.; Shao-Horn, Y. Structural Changes of Cobalt-Based Perovskites upon Water Oxidation Investigated by EXAFS. *J. Phys. Chem. C* **2013**, *117*, 8628–8635. [CrossRef]

Duan, Y.; Sun, S.; Xi, S.; Ren, X.; Zhou, Y.; Zhang, G.; Yang, H.; Du, Y.; Xu, Z.J. Tailoring the Co 3*d*-O 2*p* Covalency in $LaCoO_3$ by Fe Substitution to Promote Oxygen Evolution Reaction. *Chem. Mater.* **2017**, *29*, 10534–10541. [CrossRef]

Zhou, Y.; Sun, S.; Song, J.; Xi, S.; Chen, B.; Du, Y.; Fisher, A.C.; Cheng, F.; Wang, X.; Zhang, H.; et al. Enlarged CoO Covalency in Octahedral Sites Leading to Highly Efficient Spinel Oxides for Oxygen Evolution Reaction. *Adv. Mater.* **2018**, *30*, e1802912. [CrossRef]

Mueller, D.N.; Machala, M.L.; Bluhm, H.; Chueh, W.C. Redox activity of surface oxygen anions in oxygen-deficient perovskite oxides during electrochemical reactions. *Nat. Commun.* **2015**, *6*, 6097. [CrossRef]

Suntivich, J.; Hong, W.T.; Lee, Y.-L.; Rondinelli, J.M.; Yang, W.; Goodenough, J.B.; Dabrowski, B.; Freeland, J.W.; Shao-Horn, Y. Estimating Hybridization of Transition Metal and Oxygen States in Perovskites from O K-Edge X-Ray Absorption Spectroscopy. *J. Phys. Chem. C* **2014**, *118*, 1856–1863. [CrossRef]

Li, X.; Wang, H.; Cui, Z.; Li, Y.; Xin, S.; Zhou, J.; Long, Y.; Jin, C.; Goodenough, J.B. Exceptional oxygen evolution reactivities on $CaCoO_3$ and $SrCoO_3$. *Sci. Adv.* **2019**, *5*, eaav6262. [CrossRef]

Ye, X.; Song, S.; Li, L.; Chang, Y.-C.; Qin, S.; Liu, Z.; Huang, Y.-C.; Zhou, J.; Zhang, L.-j.; Dong, C.-L.; et al. A′–B Intersite Cooperation-Enhanced Water Splitting in Quadruple Perovskite Oxide $CaCu_3Ir_4O_{12}$. *Chem. Mater.* **2021**, *33*, 9295–9305. [CrossRef]

Zhu, Y.; Tahini, H.A.; Hu, Z.; Chen, Z.G.; Zhou, W.; Komarek, A.C.; Lin, Q.; Lin, H.J.; Chen, C.T.; Zhong, Y.; et al. Boosting Oxygen Evolution Reaction by Creating Both Metal Ion and Lattice-Oxygen Active Sites in a Complex Oxide. *Adv. Mater.* **2020**, *32*, e1905025. [CrossRef]

69. Diaz-Morales, O.; Raaijman, S.; Kortlever, R.; Kooyman, P.J.; Wezendonk, T.; Gascon, J.; Fu, W.T.; Koper, M.T. Iridium-b double perovskites for efficient water oxidation in acid media. *Nat. Commun.* **2016**, *7*, 12363. [CrossRef]
70. Hubert, M.A.; Patel, A.M.; Gallo, A.; Liu, Y.; Valle, E.; Ben-Naim, M.; Sanchez, J.; Sokaras, D.; Sinclair, R.; Nørskov, J.K.; et al. A Oxygen Evolution Reaction Activity–Stability Relationships in Ru-Based Pyrochlores. *ACS Catal.* **2020**, *10*, 12182–12196. [Cross
71. Lebedev, D.; Povia, M.; Waltar, K.; Abdala, P.M.; Castelli, I.E.; Fabbri, E.; Blanco, M.V.; Fedorov, A.; Copéret, C.; Marzari, N.; Highly Active and Stable Iridium Pyrochlores for Oxygen Evolution Reaction. *Chem. Mater.* **2017**, *29*, 5182–5191. [CrossRef
72. Yang, L.; Yu, G.; Ai, X.; Yan, W.; Duan, H.; Chen, W.; Li, X.; Wang, T.; Zhang, C.; Huang, X.; et al. Efficient oxygen evol electrocatalysis in acid by a perovskite with face-sharing IrO_6 octahedral dimers. *Nat. Commun.* **2018**, *9*, 5236. [CrossRef] [Pub
73. Zhang, Q.; Liang, X.; Chen, H.; Yan, W.; Shi, L.; Liu, Y.; Li, J.; Zou, X. Identifying Key Structural Subunits and Their Synergis Low-Iridium Triple Perovskites for Oxygen Evolution in Acidic Media. *Chem. Mater.* **2020**, *32*, 3904–3910. [CrossRef]
74. Retuerto, M.; Pascual, L.; Piqué, O.; Kayser, P.; Salam, M.A.; Mokhtar, M.; Alonso, J.A.; Peña, M.; Calle-Vallejo, F.; Rojas, S. oxidation state and lattice distortion influence the oxygen evolution activity in acid of iridium double perovskites. *J. Mater. C A* **2021**, *9*, 2980–2990. [CrossRef]
75. Zhu, C.; Tian, H.; Huang, B.; Cai, G.; Yuan, C.; Zhang, Y.; Li, Y.; Li, G.; Xu, H.; Li, M.-R. Boosting oxygen evolution reactic enhanced intrinsic activity in Ruddlesden−Popper iridate oxides. *Chem. Eng. J.* **2021**, *423*, 130185. [CrossRef]
76. Wang, L.; Shi, L.; Liu, Q.; Huang, Y.; Yan, W.; Liang, X.; Zhao, X.; Chen, H.; Zou, X. Structurally Robust Honeycomb Lay Strontium Iridate as an Oxygen Evolution Electrocatalyst in Acid. *ACS Catal.* **2023**, *13*, 7322–7330. [CrossRef]
77. Rodríguez-García, I.; Gómez de la Fuente, J.L.; Galyamin, D.; Tolosana-Moranchel, Á.; Kayser, P.; Salam, M.A.; Alonso, Calle-Vallejo, F.; Rojas, S.; Retuerto, M. Dy_2NiRuO_6 perovskite with high activity and durability for the oxygen evolution rea in acidic electrolyte. *J. Mater. Chem. A* **2024**, *12*, 16854–16862. [CrossRef]
78. Zhong, X.; Sui, L.; Yang, M.; Koketsu, T.; Klingenhof, M.; Selve, S.; Reeves, K.G.; Ge, C.; Zhuang, L.; Kan, W.H.; et al. Stabiliz of layered lithium-rich manganese oxide for anion exchange membrane fuel cells and water electrolysers. *Nat. Catal.* **20** 546–559. [CrossRef]
79. Zhu, Y.; Lin, Q.; Zhong, Y.; Tahini, H.A.; Shao, Z.; Wang, H. Metal oxide-based materials as an emerging family of hydr evolution electrocatalysts. *Energy Environ. Sci.* **2020**, *13*, 3361–3392. [CrossRef]
80. Anantharaj, S.; Noda, S.; Jothi, V.R.; Yi, S.; Driess, M.; Menezes, P.W. Strategies and Perspectives to Catch the Missing Piec Energy-Efficient Hydrogen Evolution Reaction in Alkaline Media. *Angew. Chem. Int. Ed.* **2021**, *60*, 18981–19006. [CrossRef]
81. Parsons, R. The rate of electrolytic hydrogen evolution and the heat of adsorption of hydrogen. *Trans. Faraday Soc.* **195** 1053–1063. [CrossRef]
82. Conway, B.E.; Tilak, B.V. Interfacial processes involving electrocatalytic evolution and oxidation of H_2, and the role of chemisc H. *Electrochim. Acta* **2002**, *47*, 3571–3594. [CrossRef]
83. Nørskov, J.K.; Bligaard, T.; Logadottir, A.; Kitchin, J.R.; Chen, J.G.; Pandelov, S.; Stimming, U. Trends in the Exchange Curre Hydrogen Evolution. *J. Electrochem. Soc.* **2005**, *152*, J23–J26. [CrossRef]
84. Hammer, B.; Norskov, J.K. Why gold is the noblest of all the metals. *Nature* **1995**, *376*, 238–240. [CrossRef]
85. Nilsson, A.; Petterssonb, L.G.M.; Hammer, B.; Bligaard, T.; Christensen, C.H.; Norskov, J.K. The electronic structure eff heterogeneous catalysis. *Catal. Lett.* **2005**, *100*, 111–114. [CrossRef]
86. Greeley, J.; Norskov, J.K.; Kibler, L.A.; El-Aziz, A.M.; Kolb, D.M. Hydrogen evolution over bimetallic systems: Understandin trends. *Chemphyschem* **2006**, *7*, 1032–1035. [CrossRef]
87. Hansen, J.N.; Prats, H.; Toudahl, K.K.; Morch Secher, N.; Chan, K.; Kibsgaard, J.; Chorkendorff, I. Is There Anything Better Pt for HER? *ACS Energy Lett.* **2021**, *6*, 1175–1180. [CrossRef]
88. Podjaski, F.; Weber, D.; Zhang, S.; Diehl, L.; Eger, R.; Duppel, V.; Alarcón-Lladó, E.; Richter, G.; Haase, F.; Fontcuberta i Morra et al. Rational strain engineering in delafossite oxides for highly efficient hydrogen evolution catalysis in acidic media. *Nat.* **2019**, *3*, 55–63. [CrossRef]
89. Subbaraman, R.; Tripkovic, D.; Chang, K.C.; Strmcnik, D.; Paulikas, A.P.; Hirunsit, P.; Chan, M.; Greeley, J.; Stamenkov Markovic, N.M. Trends in activity for the water electrolyser reactions on 3*d M* (Ni, Co, Fe, Mn) hydr (oxy) oxide catalysts *Mater.* **2012**, *11*, 550–557. [CrossRef]
90. Zheng, J.; Yan, Y.; Xu, B. Correcting the Hydrogen Diffusion Limitation in Rotating Disk Electrode Measurements of Hydr Evolution Reaction Kinetics. *J. Electrochem. Soc.* **2015**, *162*, F1470–F1481. [CrossRef]
91. Liu, E.; Li, J.; Jiao, L.; Doan, H.T.T.; Liu, Z.; Zhao, Z.; Huang, Y.; Abraham, K.M.; Mukerjee, S.; Jia, Q. Unifying the Hydr Evolution and Oxidation Reactions Kinetics in Base by Identifying the Catalytic Roles of Hydroxyl-Water-Cation Adducts. *Chem. Soc.* **2019**, *141*, 3232–3239. [CrossRef] [PubMed]
92. Sheng, W.; Zhuang, Z.; Gao, M.; Zheng, J.; Chen, J.G.; Yan, Y. Correlating hydrogen oxidation and evolution activity on plat at different pH with measured hydrogen binding energy. *Nat. Commun.* **2015**, *6*, 6848. [CrossRef] [PubMed]

Zheng, J.; Sheng, W.; Zhuang, Z.; Xu, B.; Yan, Y. Universal dependence of hydrogen oxidation and evolution reaction activity of platinum-group metals on pH and hydrogen binding energy. *Sci. Adv.* **2016**, *2*, e1501602. [CrossRef]
Subbaraman, R.; Tripkovic, D.; Strmcnik, D.; Chang, K.C.; Uchimura, M.; Paulikas, A.P.; Stamenkovic, V.; Markovic, N.M. Enhancing hydrogen evolution activity in water splitting by tailoring Li^+-$Ni(OH)_2$-Pt interfaces. *Science* **2011**, *334*, 1256–1260. [CrossRef]
Strmcnik, D.; Uchimura, M.; Wang, C.; Subbaraman, R.; Danilovic, N.; van der Vliet, D.; Paulikas, A.P.; Stamenkovic, V.R.; Markovic, N.M. Improving the hydrogen oxidation reaction rate by promotion of hydroxyl adsorption. *Nat. Chem.* **2013**, *5*, 300–306. [CrossRef]
Mao, B.; Sun, P.; Jiang, Y.; Meng, T.; Guo, D.; Qin, J.; Cao, M. Identifying the Transfer Kinetics of Adsorbed Hydroxyl as a Descriptor of Alkaline Hydrogen Evolution Reaction. *Angew. Chem. Int. Ed.* **2020**, *59*, 15232–15237. [CrossRef]
Zhang, L.; Hu, H.; Sun, C.; Xiao, D.; Wang, H.T.; Xiao, Y.; Zhao, S.; Chen, K.H.; Lin, W.X.; Shao, Y.C.; et al. Bimetallic nanoalloys planted on super-hydrophilic carbon nanocages featuring tip-intensified hydrogen evolution electrocatalysis. *Nat. Commun.* **2024**, *15*, 7179. [CrossRef]
Chen, X.H.; Li, X.L.; Li, T.; Jia, J.H.; Lei, J.L.; Li, N.B.; Luo, H.Q. Enhancing neutral hydrogen production by disrupting the rigid hydrogen bond network on Ru nanoclusters through Nb_2O_5-mediated water reorientation. *Energy Environ. Sci.* **2024**, *17*, 5091–5101. [CrossRef]
Ledezma-Yanez, I.; Wallace, W.D.Z.; Sebastián-Pascual, P.; Climent, V.; Feliu, J.M.; Koper, M.T.M. Interfacial water reorganization as a pH-dependent descriptor of the hydrogen evolution rate on platinum electrodes. *Nat. Energy* **2017**, *2*, 17031. [CrossRef]
Guan, D.; Zhou, J.; Hu, Z.; Zhou, W.; Xu, X.; Zhong, Y.; Liu, B.; Chen, Y.; Xu, M.; Lin, H.J.; et al. Searching General Sufficient and Necessary Conditions for Ultrafast Hydrogen Evolving Electrocatalysis. *Adv. Funct. Mater.* **2019**, *29*, 1900704. [CrossRef]
Guan, D.; Zhou, J.; Huang, Y.C.; Dong, C.L.; Wang, J.Q.; Zhou, W.; Shao, Z. Screening highly active perovskites for hydrogen-evolving reaction via unifying ionic electronegativity descriptor. *Nat. Commun.* **2019**, *10*, 3755. [CrossRef] [PubMed]
Pan, S.; Yang, X.; Sun, J.; Wang, X.; Zhu, J.; Fu, Y. Competitive Adsorption Mechanism of Defect-Induced *d*-Orbital Single Electrons in $SrRuO_3$ for Alkaline Hydrogen Evolution Reaction. *Adv. Energy Mater.* **2023**, *13*, 2301779. [CrossRef]
Huang, C.; Zhu, C.; Yang, J.; Han, T.; Liang, M.; Zhao, M.-H.; Li, G.; Li, M.-R. Boost alkaline hydrogen evolution reaction by intermetallic charge transfer in perovskite ruthenate encapsulated in ZIF-67 derived cobalt sulfide nanoparticles. *Appl. Surf. Sci.* **2024**, *663*, 160128. [CrossRef]
Zhu, Y.; Tahini, H.A.; Hu, Z.; Dai, J.; Chen, Y.; Sun, H.; Zhou, W.; Liu, M.; Smith, S.C.; Wang, H.; et al. Unusual synergistic effect in layered Ruddlesden-Popper oxide enables ultrafast hydrogen evolution. *Nat. Commun.* **2019**, *10*, 149. [CrossRef]
Xu, X.; Chen, Y.; Zhou, W.; Zhu, Z.; Su, C.; Liu, M.; Shao, Z. A Perovskite Electrocatalyst for Efficient Hydrogen Evolution Reaction. *Adv. Mater.* **2016**, *28*, 6442–6448. [CrossRef]
Sun, Q.; Dai, Z.; Zhang, Z.; Chen, Z.; Lin, H.; Gao, Y.; Chen, D. Double perovskite $PrBaCo_2O_{5.5}$: An efficient and stable electrocatalyst for hydrogen evolution reaction. *J. Power Sources* **2019**, *427*, 194–200. [CrossRef]
Zhang, F.; Hao, S.; Zheng, G.; Lei, L.; Zhang, X. Enhanced hydrogen evolution from the face-sharing [RuO_6] octahedral motif. *J. Energy Chem.* **2021**, *56*, 276–282. [CrossRef]
Dai, J.; Zhu, Y.; Tahini, H.A.; Lin, Q.; Chen, Y.; Guan, D.; Zhou, C.; Hu, Z.; Lin, H.J.; Chan, T.S.; et al. Single-phase perovskite oxide with super-exchange induced atomic-scale synergistic active centers enables ultrafast hydrogen evolution. *Nat. Commun.* **2020**, *11*, 5657. [CrossRef]
Xu, X.; Pan, Y.; Zhong, Y.; Ge, L.; Jiang, S.P.; Shao, Z. From scheelite $BaMoO_4$ to perovskite $BaMoO_3$: Enhanced electrocatalysis toward the hydrogen evolution in alkaline media. *Compos. Part B Eng.* **2020**, *198*, 108214. [CrossRef]
Dai, J.; Zhu, Y.; Chen, Y.; Wen, X.; Long, M.; Wu, X.; Hu, Z.; Guan, D.; Wang, X.; Zhou, C.; et al. Hydrogen spillover in complex oxide multifunctional sites improves acidic hydrogen evolution electrocatalysis. *Nat. Commun.* **2022**, *13*, 1189. [CrossRef]
Voiry, D.; Mohite, A.; Chhowalla, M. Phase engineering of transition metal dichalcogenides. *Chem. Soc. Rev.* **2015**, *44*, 2702–2712. [CrossRef] [PubMed]
Zhao, M.; Xia, Y. Crystal-phase and surface-structure engineering of ruthenium nanocrystals. *Nat. Rev. Mater.* **2020**, *5*, 440–459. [CrossRef]
Chen, Y.; Lai, Z.; Zhang, X.; Fan, Z.; He, Q.; Tan, C.; Zhang, H. Phase engineering of nanomaterials. *Nat. Rev. Chem.* **2020**, *4*, 243–256. [CrossRef]
Jung, K.N.; Jung, J.H.; Im, W.B.; Yoon, S.; Shin, K.H.; Lee, J.W. Doped lanthanum nickelates with a layered perovskite structure as bifunctional cathode catalysts for rechargeable metal-air batteries. *ACS Appl. Mater. Interfaces* **2013**, *5*, 9902–9907. [CrossRef]
Takeguchi, T.; Yamanaka, T.; Takahashi, H.; Watanabe, H.; Kuroki, T.; Nakanishi, H.; Orikasa, Y.; Uchimoto, Y.; Takano, H.; Ohguri, N.; et al. Layered perovskite oxide: A reversible air electrode for oxygen evolution/reduction in rechargeable metal-air batteries. *J. Am. Chem. Soc.* **2013**, *135*, 11125–11130. [CrossRef]

116. Cao, C.; Shang, C.; Li, X.; Wang, Y.; Liu, C.; Wang, X.; Zhou, S.; Zeng, J. Dimensionality Control of Electrocatalytic Activi Perovskite Nickelates. *Nano Lett.* **2020**, *20*, 2837–2842. [CrossRef]
117. Bartel, C.J.; Sutton, C.; Goldsmith, B.R.; Ouyang, R.; Musgrave, C.B.; Ghiringhelli, L.M.; Scheffler, M. New tolerance fact predict the stability of perovskite oxides and halides. *Sci. Adv.* **2019**, *5*, aav0693. [CrossRef]
118. Liu, Y.; Huang, H.; Xue, L.; Sun, J.; Wang, X.; Xiong, P.; Zhu, J. Recent advances in the heteroatom doping of perovskite oxide efficient electrocatalytic reactions. *Nanoscale* **2021**, *13*, 19840–19856. [CrossRef]
119. Kuznetsov, D.A.; Naeem, M.A.; Kumar, P.V.; Abdala, P.M.; Fedorov, A.; Muller, C.R. Tailoring Lattice Oxygen Bindir Ruthenium Pyrochlores to Enhance Oxygen Evolution Activity. *J. Am. Chem. Soc.* **2020**, *142*, 7883–7888. [CrossRef]
120. Retuerto, M.; Pascual, L.; Calle-Vallejo, F.; Ferrer, P.; Gianolio, D.; Pereira, A.G.; Garcia, A.; Torrero, J.; Fernandez-Diaz, Bencok, P.; et al. Na-doped ruthenium perovskite electrocatalysts with improved oxygen evolution activity and durabili acidic media. *Nat. Commun.* **2019**, *10*, 2041. [CrossRef]
121. Li, X.; Sun, Y.; Wu, Q.; Liu, H.; Gu, W.; Wang, X.; Cheng, Z.; Fu, Z.; Lu, Y. Optimized Electronic Configuration to Improv Surface Absorption and Bulk Conductivity for Enhanced Oxygen Evolution Reaction. *J. Am. Chem. Soc.* **2019**, *141*, 3121– [CrossRef] [PubMed]
122. Liang, X.; Shi, L.; Liu, Y.; Chen, H.; Si, R.; Yan, W.; Zhang, Q.; Li, G.D.; Yang, L.; Zou, X. Activating Inert, Nonprecious Perovs with Iridium Dopants for Efficient Oxygen Evolution Reaction under Acidic Conditions. *Angew. Chem. Int. Ed.* **201** 7631–7635. [CrossRef] [PubMed]
123. Liang, X.; Shi, L.; Cao, R.; Wan, G.; Yan, W.; Chen, H.; Liu, Y.; Zou, X. Perovskite-Type Solid Solution Nano-Electrocatalysts E Simultaneously Enhanced Activity and Stability for Oxygen Evolution. *Adv. Mater.* **2020**, *32*, e2001430. [CrossRef]
124. Hua, B.; Li, M.; Pang, W.; Tang, W.; Zhao, S.; Jin, Z.; Zeng, Y.; Shalchi Amirkhiz, B.; Luo, J.-L. Activating p-Blocking Cente Perovskite for Efficient Water Splitting. *Chem* **2018**, *12*, 2902–2916. [CrossRef]
125. Su, C.; Wang, W.; Shao, Z. Cation-Deficient Perovskites for Clean Energy Conversion. *Acc. Mater. Res.* **2021**, *2*, 477–488. [Cros
126. Arandiyan, H.; Mofarah, S.S.; Sorrell, C.C.; Doustkhah, E.; Sajjadi, B.; Hao, D.; Wang, Y.; Sun, H.; Ni, B.J.; Rezaei, M.; et al. D engineering of oxide perovskites for catalysis and energy storage: Synthesis of chemistry and materials science. *Chem. Soc.* **2021**, *50*, 10116–10211. [CrossRef]
127. Wang, J.; Gao, Y.; Chen, D.; Liu, J.; Zhang, Z.; Shao, Z.; Ciucci, F. Water Splitting with an Enhanced Bifunctional Double Perov *ACS Catal.* **2017**, *8*, 364–371. [CrossRef]
128. Liu, D.; Zhou, P.; Bai, H.; Ai, H.; Du, X.; Chen, M.; Liu, D.; Ip, W.F.; Lo, K.H.; Kwok, C.T.; et al. Development of Perov Oxide-Based Electrocatalysts for Oxygen Evolution Reaction. *Small* **2021**, *17*, e2101605. [CrossRef]
129. Zhu, Y.; Zhou, W.; Yu, J.; Chen, Y.; Liu, M.; Shao, Z. Enhancing Electrocatalytic Activity of Perovskite Oxides by Tuning C Deficiency for Oxygen Reduction and Evolution Reactions. *Chem. Mater.* **2016**, *28*, 1691–1697. [CrossRef]
130. Da, Y.; Zeng, L.; Wang, C.; Gong, C.; Cui, L. A simple approach to tailor OER activity of $Sr_xCo_{0.8}Fe_{0.2}O_3$ perovskite cata *Electrochim. Acta* **2019**, *300*, 85–92. [CrossRef]
131. Chen, G.; Zhu, Y.; Chen, H.M.; Hu, Z.; Hung, S.F.; Ma, N.; Dai, J.; Lin, H.J.; Chen, C.T.; Zhou, W.; et al. An Amorphous N Iron-Based Electrocatalyst with Unusual Local Structures for Ultrafast Oxygen Evolution Reaction. *Adv. Mater.* **2019**, *31*, e190 [CrossRef] [PubMed]
132. Zhang, H.; Guan, D.; Gao, X.; Yu, J.; Chen, G.; Zhou, W.; Shao, Z. Morphology, crystal structure and electronic state one co-tuning strategy towards developing superior perovskite electrocatalysts for water oxidation. *J. Mater. Chem. A* **20** 19228–19233. [CrossRef]
133. Dai, J.; Zhu, Y.; Zhong, Y.; Miao, J.; Lin, B.; Zhou, W.; Shao, Z. Enabling High and Stable Electrocatalytic Activity of Iron-B Perovskite Oxides for Water Splitting by Combined Bulk Doping and Morphology Designing. *Adv. Mater. Interfaces* **2018**, *6*, 180 [CrossRef]
134. Chen, H.; Shi, L.; Liang, X.; Wang, L.; Asefa, T.; Zou, X. Optimization of Active Sites via Crystal Phase, Composition, Morphology for Efficient Low-Iridium Oxygen Evolution Catalysts. *Angew. Chem. Int. Ed.* **2020**, *59*, 19654–19658. [CrossRe
135. Kim, M.; Ju, H.; Kim, J. Dihydrogen phosphate ion functionalized nanocrystalline thallium ruthenium oxide pyrochlore bifunctional electrocatalyst for aqueous Na-air batteries. *Appl. Catal. B Environ.* **2019**, *245*, 29–39. [CrossRef]

molecules

MDPI

ew

ɒre Engineering in Biomass-Derived Carbon Materials for ɑhanced Energy, Catalysis, and Environmental Applications

Vang [1,2,†], Bolong Luo [1,2,†], Zhaoyu Wang [1,†], Yao Hu [1,*] and Mingliang Du [1,*]

1 Key Laboratory of Synthetic and Biological Colloids, Ministry of Education, School of Chemical and Material Engineering, Jiangnan University, Wuxi 214122, China; 17730127768@163.com (Q.W.); 15399706779@163.com (B.L.); 18769411559@163.com (Z.W.)

2 School of Environmental and Ecology, Jiangnan University, Wuxi 214122, China

* Correspondence: huyao15880@gmail.com (Y.H.); du@jiangnan.edu.cn (M.D.)

† These authors contributed equally to this work.

Abstract: Biomass-derived carbon materials (BDCs) are highly regarded for their renewability, environmental friendliness, and broad potential for application. A significant advantage of these materials lies in the high degree of customization of their physical and chemical properties, especially in terms of pore structure. Pore engineering is a key strategy to enhance the performance of BDCs in critical areas, such as energy storage, catalysis, and environmental remediation. This review focuses on pore engineering, exploring the definition, classification, and adjustment techniques of pore structures, as well as how these factors affect the application performance of BDCs in energy, catalysis, and environmental remediation. Our aim is to provide a solid theoretical foundation and practical guidance for the pore engineering of BDCs to facilitate the rapid transition of these materials from the laboratory to industrial applications.

Keywords: biomass-derived carbon; pore engineering; energy; catalysis; environmental

1. Introduction

In recent years, BDCs have attracted significant attention due to their renewability, environmental friendliness, and broad application potential [1–4]. These materials are produced from renewable resources, such as agricultural waste, forestry by-products, and other organic materials, making them a sustainable alternative to traditional carbon materials, like activated carbon or graphene [5,6]. The renewable nature of BDCs aligns closely with current green technology trends, contributing not only to the reduction of carbon footprints but also to the promotion of a circular economy [7]. Additionally, the diverse range of biomass sources allows for remarkable tunability in BDCs, enabling them to adapt to various application scenarios [8].

One of the key advantages of BDCs lies in their highly customizable physical and chemical properties, particularly in terms of pore structure [9]. Through careful design, BDCs can form complex porous structures with large surface areas, allowing precise control over pore size distribution while optimizing surface chemical properties and functional groups [10]. This tunability significantly enhances their performance in areas such as energy storage, catalysis, and environmental remediation [11–13].

Pore engineering is a crucial strategy for significantly enhancing the performance of BDCs across a wide range of applications, particularly in energy storage, catalysis, and environmental remediation [14–18]. The precise control of pore structure, including parameters such as pore size, distribution, and interconnectivity, plays a vital role in determining the effectiveness of BDCs in these fields [19]. A well-tailored pore architecture not only affects the physical properties of the material but also influences its chemical reactivity, surface interactions, and overall functionality [20]. As a result, optimizing pore characteristics can lead to substantial improvements in the efficiency and adaptability of BDCs, making them more suitable for advanced technological applications [21]. By leveraging the unique properties of BDCs and employing pore engineering techniques, these materials can be fine-tuned to meet specific technological needs, further expanding their potential in the fields of energy, catalysis, and environmental protection [14–16,22].

There have already been numerous excellent reviews that delve into the properties, sustainability, and environmental benefits of biochar, covering its broad applications in fields such as water treatment, catalysis, soil remediation, energy storage, and construction materials [17,18,23–29]. However, despite the rapid development of BDCs, existing research often overlooks the critical role of pore engineering in enhancing material performance. Precise control of pore structure is essential for optimizing the performance of BDCs in various applications, and therefore, more systematic and in-depth studies are needed to fully unlock their potential in energy, catalysis, and environmental remediation [30–33].

As shown in Figure 1, Our review will focus on pore engineering, examining the definition, classification, and regulation methods of pore structures, and exploring their impact on material performance, addressing gaps in current research. The article will systematically introduce various pore regulation techniques, including physical activation, chemical activation, self-activation, and templating methods, with detailed analysis of how these methods are applied in different carbonization processes, such as hydrothermal carbonization, pyrolysis, microwave carbonization, and laser carbonization. Additionally, the review will explore the broad applications of pore structures in energy, catalysis, and environmental remediation, highlighting their potential to enhance material performance.

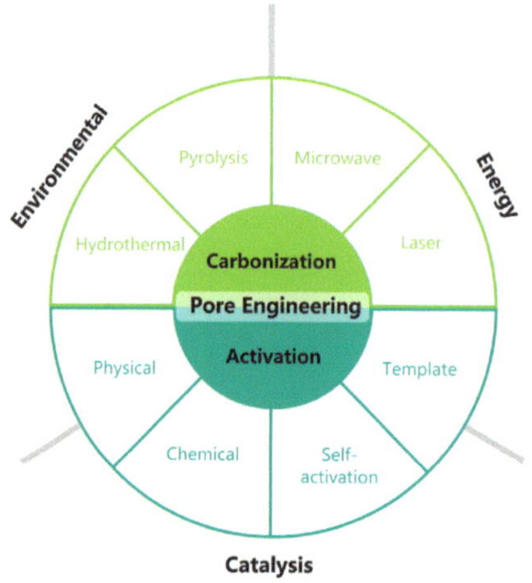

Figure 1. Overview of BDC materials with Pore Structure, Pore Engineering, and wide applications.

2. Concepts and Classification of Pore Structure

2.1. Definition and Classification of Pores

Owing to the unique properties of BDCs, the design and construction of their porous structures in materials are of great significance during their preparation process [34]. As shown in the Figure 2, pores in materials can be categorized into three types: micropores (pore size less than 2 nm), mesopores (pore size of 2–50 nm) and macropores (pore size more than 50 nm). Microporous materials are renowned for their extremely high surface areas and a pore structure composed of minuscule channels and cavities, endowing the materials with unique interactive capabilities at the molecular level. Mesoporous materials, on the other hand, feature an orderly pore structure along with relatively larger diameters, offering considerable flexibility in pore size adjustment; their pore walls may be continuous or discontinuous, and sometimes they include regularly arranged channels. Macroporous materials are distinguished by their spacious pore structures and larger diameters, often manifesting a more open pore network that permits the free movement of large molecules or particles; thus, macroporous materials typically exhibit greater permeability.

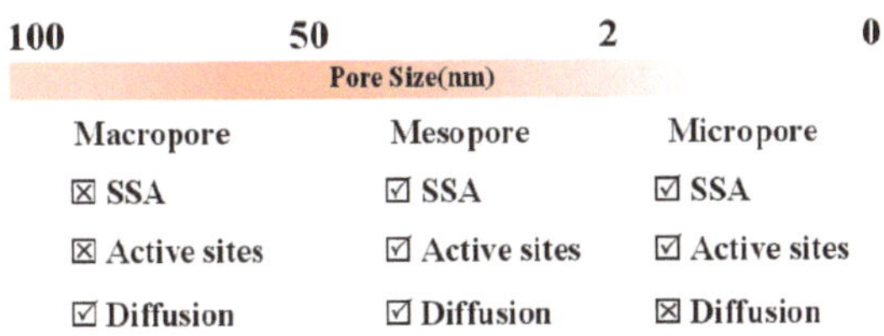

Figure 2. Particle size distribution and characteristics of various types of pores.

2.2. Impact of Pore Structure on Performance

Micropores play a crucial role in increasing the specific surface area of materials, as they provide abundant inner surface areas. When it comes to further increasing the specific surface area, mesopores play a positive role, not only increasing the additional surface area, but also promoting the transport and diffusion of gas molecules within the material. As for macropores, their main function is to affect the macroscopic structure and pore network of materials, and they play a crucial role in improving the quality transfer efficiency of materials [35,36]. Overall, the synergistic effect of micropores, mesopores, and macropores determines the specific surface area and pore characteristics of materials, thereby affecting their performance in various applications. In catalytic reactions, the high surface area of microporous materials provides numerous active sites for catalytic reactions and gas adsorption [37,38], which are specific regions on the catalyst surface that can interact with reactants and participate in the reaction. The active site plays a crucial role in reducing the activation energy and accelerating the reaction rate [39]. However, the small pore size of micropores may limit the diffusion path of substances, affecting their efficiency in applications that require rapid material transport [40]. In contrast, although mesoporous and macroporous materials may have slightly inferior surface area, their larger pore size and ordered pore structure provide advantages for the accessibility of active sites and material transport, which is particularly important in heterogeneous catalytic reactions. In the field of energy storage, the porosity and pore size distribution of porous carbon materials as electrodes for supercapacitors play a crucial role in ion adsorption and transport rates [41, 42]. Micropores help increase the surface area and storage capacity but may sacrifice transmission rate. Meanwhile, mesopores and macropores can enhance the ion transfer rate, thereby increasing the power density of supercapacitors [42]. Environmental management also benefits from the pore size and structure of porous materials, which determine their ability to adsorb pollutants [43]. Microporous materials perform better in adsorbing small molecule pollutants, while mesoporous and macroporous materials are more suitable for capturing large molecule or particle pollutants [41,44]. Therefore, the pore structure of porous materials not only determines their surface area and distribution of active sites, but also directly affects the efficiency of material transport and application related properties.

3. Main Methods for Regulating Pore Structure

3.1. Physical and Chemical Activation

Activation is a process during which different kinds of gas or activators can be used to act with carbonaceous materials [45]. The activation process comprises two steps: carbonization and activation. In the carbonization stage, biomass undergoes carbonization under high temperature and anaerobic conditions to form carbon materials. At the same time, non-carbon elements are expelled in the form of gas, resulting in the formation of pores [46]. Activation is a crucial step in controlling the morphology of carbon materials, and hierarchical porosity is continuously introduced on the surface of the carbon material in this process, which increases its specific surface areas (SSAs), enriches its pore structure, and beneficially improves the overall physicochemical properties of the carbon material. Activation is a key step in controlling the morphology of carbon materials. Hence, it is extremely important to design activation process parameters reasonably (e.g., activation temperature and the activator) [47]. Activation can be divided into four types: physical activation, chemical activation, self-activation and the template method.

In the physical activation process, biomass will go through two stages. The first stage is the pyrolysis of biomass precursors in an inert gas atmosphere, which will lead to partial carbonization and the elimination of non-carbon elements. The second stage is activation of the carbonized material with an oxidizing gas at elevated temperatures (800–1100 °C) to adjust the pore structure, with the entire activation process taking place in an oxidizing gas medium [48].

Activation time is a crucial parameter in physical activation. Zgrzebnicki et al. [49]. prepared N-doped activated carbon derived from furfuryl alcohol through CO_2 activation. Different activation times of 15, 60, 120, and 240 min were set. The variation in the pore structure of the material with activation time is presented in Figure 3a. Apparently, the first three materials can all be classified as microporous materials. The latter two materials are porous materials with the coexistence of micropores and mesopores. Carbonization heating rates are also an important factor. Jing et al. [50]. reported a simple and scalable strategy for preparing activated carbon, aiming at producing activated carbon with a developed pore structure in high yield through rapid pyrolysis physical activation. Compared with slow pyrolysis semi-coke (M) of the same particle size, fast pyrolysis semi-coke (K) has a lower nitrogen adsorption capacity, indicating that a high heating rate is not conducive to the formation of micropores. The "hysteresis loop" of K shows a higher degree of separation, indicating that high heating rates promote the formation of mesopores and macropores (Figure 3b). Physical activation is an effective technique for improving the specific surface area and porosity of biomass-derived carbon materials, which is crucial for enhancing their performance in adsorption [51], catalysis, and energy storage [52,53]. This method is environmentally friendly, especially when using water vapor or carbon dioxide as activators, which can reduce the use of chemical additives [54]. However, precise regulation of pore structure to meet specific application requirements still faces challenges. In addition, the high temperature and prolonged processing during the physical activation process may result in high energy consumption, which to some extent limits its application on an industrial scale [55]. Future research and development work needs to focus on optimizing activation conditions to achieve improved production efficiency and material properties, while maintaining environmental sustainability.

Unlike physical activation, in chemical activation, activators are added. Here, carbonization and activation take place simultaneously in a single step [56]. During activation, the activator and precursor become involved in dehydration, cross-linking, or condensation reactions, with carbon atoms etched and separated from the carbon skeleton to form a porous structure. In addition, diffusion of the generated gas molecules (CO_2, CO, and H_2O) leads to the formation of many pores. Common chemical activators are currently divided into three categories: alkaline (e.g., KOH and NaOH), acidic (e.g., H_3PO_4 and H_2SO_4), and molten salt (e.g., $ZnCl_2$, Na_2CO_3, and K_2CO_3) [57].

Treating pinecones with different chemical activators could yield porous carbon (PC) of different morphologies (Figure 3c). The presence of KOH and $ZnCl_2$ produces micropores, which are different from those of NaOH and K_2CO_3 activated porous carbon. Among them, the sample activated by K_2CO_3 has an ordered, dense, and honeycomb-like porous structure [58]. The amount of activator used is also crucial. Yang et al. [59] prepared porous carbon CS-H3 using dry wood sawdust and KOH as an activator. The pore size distribution of carbon microstructure was regulated by the content of KOH and the addition of heteroatoms, as in the illustration for Figure 3d. At a lower KOH content in MS (CS-H1), the limited oxidizing action of KOH limits the compatibility of MS and carbon, and thus reduces the template action of MS to form pores. The MS deeply penetrates the carbon phase with a larger amount of KOH used (CS-H3 and CS-H5), making the formation of an open pore structure and even sheets. Compared with chemical activation, physical activation demands a longer activation time, a higher activation temperature, and consumes a substantial amount of energy [60]. Chemical activation is more effective than physical activation in improving the specific surface area and pore structure of biomass-derived carbon materials, especially in the generation of micropores [61]. By precisely adjusting the preparation conditions, chemical activation can customize the control of pore size and pore distribution to meet diverse application needs. Another major advantage of this method is its wide applicability to raw materials, which can convert biomass materials, such as agricultural waste and forestry by-products, into high-value porous carbon materials, achieving resource recycling [62]. Although chemical activators, such as KOH, have relatively low costs, some activators may be expensive and may have potential negative impacts on the environment, especially those that are difficult to recycle and may increase environmental burdens [54]. In addition, chemical activation requires precise control of reaction conditions, which increases the complexity of the operation and places higher demands on the accuracy of the production process. Therefore, future research and development should focus on reducing the cost of chemical activation, minimizing environmental impact, while improving material performance and production efficiency to achieve more sustainable and economically efficient production processes.

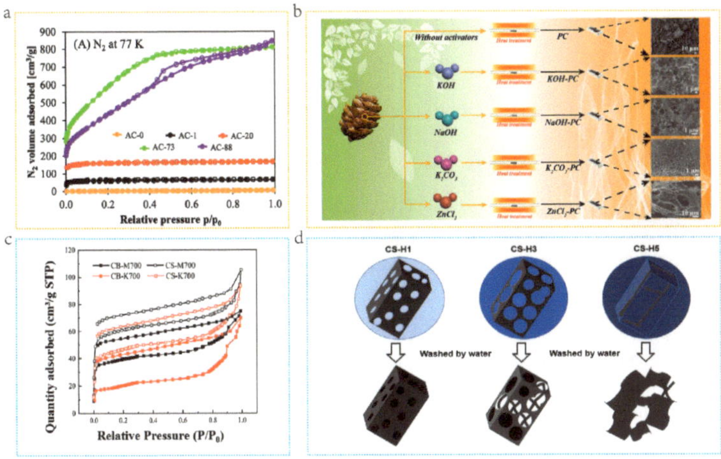

Figure 3. (**a**) Adsorption/desorption isotherms of nitrogen at 77 K (the AC-0 sample was heated under an inert atmosphere at 1273 K. All the other samples were physically activated with carbon dioxide at 1273 K. Burn-offs of 1%, 20%, 73%, and 88% were confirmed for materials activated for of 15, 60, 120 and 240 min, respectively. Samples were named according to measured burn-off, i.e., AC-1, AC-20, AC-73, and AC–88, respectively.) [49]. Copyright 2022, Elsevier. (**b**) Adsorption/desorption isotherms of different heating rates [50]. Copyright 2022, Elsevier. (**c**) Schematic illustration of the preparation of porous carbonaceous treated by different activators [58]. Copyright 2020, Elsevier. (**d**) Carbon formation with different amounts of KOH [59]. Copyright 2020, Elsevier.

3.2. Self-Activation

Self-activation is a strategy that does not rely on the addition of external activators. Its principle is that biomass precursors are abundant in inherent inorganic compounds, which can activate them in situ and convert the precursors into porous carbon [63]. Currently, some studies have reported the utilization of organic acid salts of Na, K, and Ca as carbon precursors and self-activators [64]. The principle is shown in Figure 4a. Certain metal nanoparticles and corresponding oxides are generated during the pyrolysis process. These can act as in situ templates and can be removed by thermal or etching methods. Subsequently, the originally occupied area is released and transformed into pores, thus achieving permeation into the carbon matrix at the molecular level to regulate the pore structure. Biomass is typically composed of lignin, cellulose, hemicellulose, or polysaccharides. These components produce H_2, CO_2, CO, H_2O, and CH_4 during the activation stage [65]. Therefore, these pyrolysis gases can be used as self-activators. The potential activation mechanism is shown in Figure 4b. During the activation process, these gases can laterally transfer or vertically penetrate carbon nanosheets, leading to the formation of pores. Jing et al. [66]. prepared porous carbon (PC) with high defect density by directly carbonizing bovine bone containing hydroxyapatite (HA), without the need for any additional activators or templates. Through in situ self-activation of HA, PC with a hierarchical porous structure dominated by mesopores was obtained (Figure 4c). Self-activation technology stands out in the field of BDC preparation due to its environmental and economic benefits. This technology not only reduces the generation of chemical waste by omitting additional chemical activators [61], but also simplifies the traditional two-step activation process by combining carbonization and activation processes into one step, effectively reducing production costs. However, the unevenness of gas distribution during self-activation may lead to inconsistencies in the pore structure of carbon materials. In addition, compared to BDCs prepared using chemical activators, BDCs obtained by self-activation methods usually have smaller SSAs and pore size [54]. With advancement in technology, these challenges may gradually be overcome, making self-activation a mainstream method of biomass carbonization.

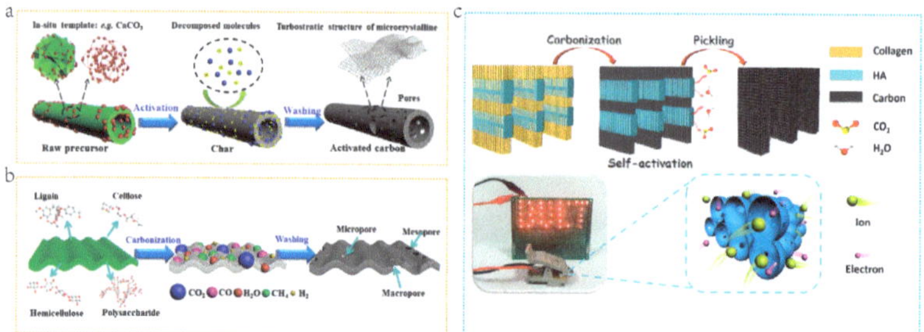

Figure 4. (**a**) Activation mechanism of self-activating agent via in situ template [54]. Copyright 2020, Elsevier. (**b**) Activation mechanism of self-activating agent via pyrolysis gases [54]. Copyright 2020, Elsevier. (**c**) Preparation of mesoporous dominant porous carbon by direct carbonization of bovine bone [66]. Copyright 2017, American Chemical Society.

3.3. Templating Methods

During this period, the template is impregnated into the biomass precursor. After carbonization and removal of the template with strong acid or alkali, uniformly sized pores will be generated [67]. The most prominent feature of this method is that the type, structure, spatial size, and shape of the template can control and adjust the structure of the material [68].

The hard template method, also known as the "nano casting method", makes use of the excellent spatial control offered by the template to achieve ordered micro-size and three-dimensional structure arrangement in carbon materials [67]. Typical hard templates include MgO, ZnO, and SiO_2 [69]. The main steps of the hard template method are as follows: (1) Select or synthesize the required materials as the hard template. (2) Thoroughly and effectively mix carbon sources and hard templates. (3) Conduct pyrolysis under certain conditions at high temperatures. 4) Rinse and remove the hard template with chemical reagents [70]. Hu et al. [71] took lotus seed shells as carbon precursors and sodium phytate as hard template precursors to prepare natural biomass-derived porous carbon with a hierarchical porous structure through in situ template combined with NaOH activation. As shown in Figure 5a, the lotus seed shell and sodium phytate aqueous solution are fully mixed at 60 °C to form a mixed gel. During carbonization, soluble sodium phytate undergoes pyrolysis to form nano $Na_5P_3O_{10}$, which then reacts with NaOH to transform into nano Na_2CO_3 and nano Na_3PO_4 particles. These particles are uniformly dispersed in the carbon substrate and leave large mesopores/macropores after washing treatment. A well-developed hierarchical porous carbon with a hollow nest structure is obtained by combining the micropores generated by NaOH activation. Compared with the hard template method, the soft template method does not require the use of a corrosive agent to removal the template, which improves synthesis efficiency [72].The soft template method employs organic compounds as templates, including surfactants, block copolymers, and ionic micelle [73]. Selecting appropriate carbon precursors and template materials is crucial for ensuring the use of soft templates for directional synthesis. The soft template method typically involves ordered aggregates formed by amphiphilic molecules, such as block copolymers and surfactants. When the concentration exceeds the critical concentration, these molecules will self-assemble into micelles or vesicles and interact with carbon precursors through hydrogen bonding, hydrophobic or hydrophilic interactions, and electrostatic interactions to form coatings on the precursors [74]. During the subsequent carbonization process, the micelles decompose to form a unique pore structure [75]. Tu et al. [72] prepared catalyst CNTs from lignin by employing a soft template method (Figure 5b). They controlled the size and dispersion of catalyst particles by loading the catalyst onto nano micelles. The hard template method has shown excellent performance in preparing highly ordered porous carbon materials, which can precisely control the pore size and pore shape. However, this advantage comes with the high cost and environmental risks of using strong acids or bases to remove the template [76]. Compared with this, the soft template method is favored for its simple operation process and environmental friendliness, and the template is easy to remove during high-temperature carbonization [77]. However, the soft template method may require more expensive template agents during the synthesis process and, due to insufficient filling of raw materials, it may lead to discontinuities and structural defects in the product pores [78]. In addition, mesoporous carbon materials prepared by this method typically have lower surface area and pore volume [79]. Therefore, to combine the advantages of both, a dual template method was introduced. Lu et al. [80]. synthesized FeOx embedded Fe-N doped carbon (Hemin/NPC) via carbonization by employing acorn shell and $Mg_5(OH)_2(CO_3)_4$ and NaCl as dual templates. The removal of template by acid washing generates abundant mesopores, while the melted NaCl nano-droplets were immersed into carbon framework to produce porous structure (Figure 5c).

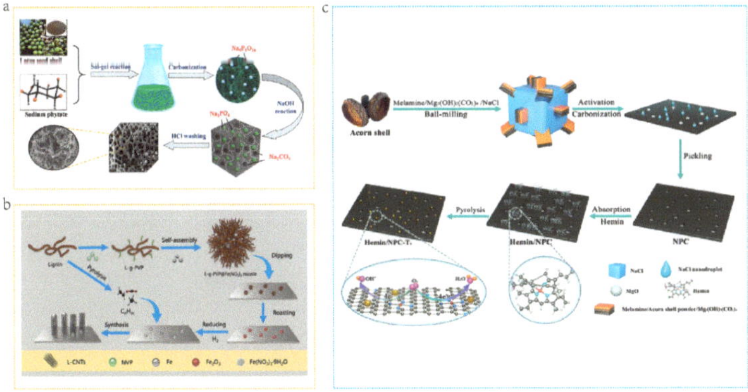

Figure 5. (**a**) Schematic illustration of the preparation of natural biomass-derived hierarchical porous carbons [71]. Copyright 2023, Elsevier. (**b**) Preparation of L-CNTs [72]. Copyright 2018, American Chemical Society. (**c**) Schematic illustration of the proposed procedure for fabricating Hemin/NPC catalysts [80]. Copyright 2021, Wiley Online Library.

4. Impact of Carbonization Methods on Pore Structure

The above activation methods can regulate the structure of BDC. However, high-temperature carbonization is still required to convert biomass into BCM. This process mainly includes decomposition, polymerization, aromatization, carbonization, and graphitization. At the same time, different carbonization methods also affect the structure of BDC, such as hydrothermal carbonization (HTC), pyrolysis, and other methods.

4.1. Hydrothermal Carbonization

Hydrothermal carbonization (HTC) is a green method for preparing carbon containing materials using biomass, such as cellulose, lignin, and hemicellulose, as raw materials. It is usually carried out in a temperature range of 100–250 °C and a pressure range of 2–6 MPa, with an aqueous solution as the medium [81]. During the HTC process, as the temperature increases, the water vapor pressure increases sharply, reducing surface tension. By simulating the natural coalification process, carbon materials with unique structures are produced [75]. Several reactions occur during the HTC process, including condensation, polymerization, hydrolysis, decarboxylation, dehydration, and aromatization.

The preparation conditions of the HTC method, such as HTC temperature, residence time, catalyst usage, and water-to-biomass ratio, have a significant impact on the structure of the obtained BDCs. Romero et al. [82] prepared carbon spheres by using sucrose, glucose, and cellulose as carbon sources via the hydrothermal method. As shown in Figure 6a after hydrothermal treatment, all three types of biomasses formed spherical carbon structures under the same preparation conditions. However, the diameter of the carbon spherical varied with changes in carbon source. At the same time, the temperature and reaction time of hydrothermal carbonization can lead to differences in the diameter of carbon spheres. Taking sucrose spherical carbon as an example, as the temperature increases, the spherical structure starts to develop from 180 °C, and the maximum size of spherical carbon appears at 200 °C. As the temperature continues to rise, the size of the sphere begins to decrease, which can be ascribed to the complete decomposition of the raw materials (Figure 6b). Regarding the HTC residence time, it can be observed that, after 12 h, spherical carbon is not completely formed. After a certain period, spherical carbon with a definite morphology is obtained (Figure 6b). Additionally, the continuous increase in time has little effect on the structure of spherical carbon. Residence time will also affect the porosity of BDCs, and longer residence time is beneficial to the formation of defined structure porosity, pore volume, and high SSAs. Due to the different thermal decomposition behaviors of biomass components, different biomass precursors will have an impact on the structure of BDCs prepared by the HTC method. For example, BDCs with a higher lignin content produce more biochar than those rich in hemicellulose or cellulose, as lignin degradation is difficult [83]. Chen et al. [84]. prepared hydrothermal carbon from glucose, lignin, and cellulose. As shown in Figure 6c, for HTC-G (glucose), its spherical primary particle size is approximately 200 nm and the surface is smooth. HTC-L (lignin) and HTC-C (cellulose) also display similar primary particle shapes, but with average sizes of 4.2 and 6.0 μm respectively. Additionally, the activated carbon AC-70 obtained by direct high-temperature pyrolysis of glucose shows irregular block-shaped particles with a size of over ten microns. This once again demonstrates that hydrothermal treatment is more conducive to the formation of spherical BDCs compared to direct pyrolysis. The mild temperature of the hydrothermal method helps to cultivate crystals with almost no defects and good orientation [85]. In addition, the hydrothermal method is highly favored for its green and environmentally friendly characteristics. It can not only synthesize small single crystals of a single component, but also prepare various complex compound powders, including composite oxides [86]. However, the demand for high-temperature and high-pressure environments in hydrothermal methods has increased their dependence on specific production equipment. The expansion of fluid volume in a closed reaction vessel poses safety hazards, which to some extent limit the widespread application of hydrothermal methods [87]. To promote the further development of hydrothermal methods, future research should focus on developing safer and more efficient synthesis technologies, reducing dependence on equipment, and fully leveraging its unique advantages in the field of material synthesis.

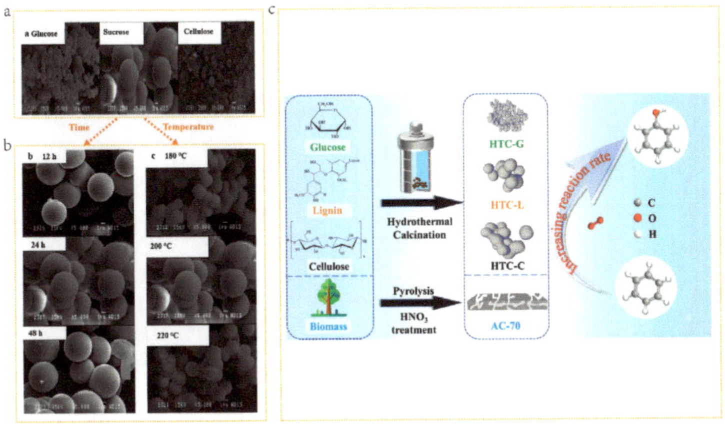

Figure 6. (**a**) The influence of biomass type [82]. Copyright 2014, Elsevier. (**b**) The influence of temperature and reaction time [82]. Copyright 2014, Elsevier. (**c**) Preparation of BDCs from different types of biomass through hydrothermal carbonization [84]. Copyright 2021, American Chemical Society.

4.2. Pyrolysis Carbonization

Pyrolysis carbonization is another practical approach for preparing BDCs. It entails the heating and decomposing of organic biomass under an inert gas atmosphere. In this process, volatile substances and non-carbon elements are eliminated, leading to the decomposition of biomass into solid carbon and gaseous products. The structure and porosity of pores formed within carbon materials can be effectively adjusted by controlling carbonization parameters, like heating rate, carbonization temperature, and annealing time [88].

Heating rates are one of the crucial operating parameters that exert a significant influence on the structure of BDCs. Research indicates that rapid heating facilitates the formation of three-dimensional structures in carbon materials. Conversely, slow heating results in high porosity and high SSAs [89,90]. Zhang et al. [90]. prepared porous carbon materials by using the macromolecular components of biomass pyrolysis steam and calcium citrate as carbon sources and templates, respectively. They also investigated the effect of heating rate on the structure of BDCs. As depicted in Figure 7a, with acceleration in the heating rate, the specific surface area and micropore volume of porous carbon initially increase and then decrease, reaching their peak at 10 K/min. When the heating rate increases, the pyrolysis vapor and calcium citrate can be fully mixed, which is beneficial for the formation of a pore structure dominated by mesopores and micropores in the subsequent carbonization process. However, when the heating rate is overly fast, excessive steam is generated and the amount of calcium citrate is insufficient, leading to the collapse of the pore structure and a significant reduction in SSAs. The carbonization temperature is another crucial preparation condition that impacts the carbon structure. Pariyar et al. [91] compared biomass materials obtained from five raw materials, pine sawdust (PD), rice husk (RH), food waste (FW), poultry waste (PL), and papermaking sludge (PS), at different pyrolysis temperatures (350, 450, 550, and 650 °C) and found that the pyrolysis temperature also affects the SSAs of BDCs. As shown in Figure 7b, the SSAs of the five types of BDCs increase with the rise in carbonization temperature. On the other hand, the degree of graphitization of carbon materials is influenced by the carbonization temperature. At a lower temperature (such as 600 °C), the graphitization degree of BDCs is relatively low. However, a high degree of graphitization can be attained at a high carbonization temperature (such as 1100 °C) [92]. Cho et al. [93]. systematically studied the structural and chemical changes of silk protein at temperatures above the onset of thermal degradation, using silk protein as a carbon source. Silk protein undergoes a structural transformation into a sp^2 hybridized carbon hexagonal structure by simple heating to 350 °C. When heated to 2800 °C degrees Celsius, a highly developed graphite structure is formed, as shown in Figure 7c.

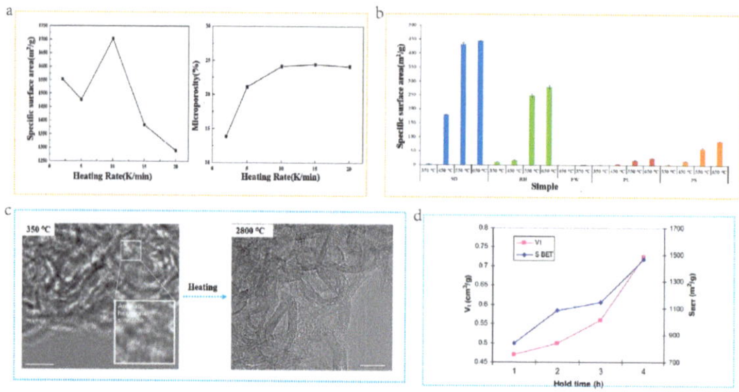

Figure 7. (**a**) The influence of heating rate on SSA and on porosity [90]. Copyright 2020, Elsevier. (**b**) The influence of carbonization temperature on SSA [91]. Copyright 2020, Elsevier. (**c**) SEM images of ilk protein at different temperatures [93]. Copyright 2015, Nature. (**d**) The effect of insulation time on SSA and pore volume [94]. Copyright 2012, Elsevier.

During the pyrolysis process, holding time is extremely important. Bouchelta et al. [94]. investigated the effect of pyrolysis conditions (holding time) on the preparation of BDCs from jujube seeds. As shown in Figure 7d, holding time has a significant impact on the comparison of surface area and pore volume. Increasing the holding time is beneficial for improving pore volume and SSAs. This is because the longer the holding time, the greater the mass of volatile compounds released, resulting in higher SSAs and porous volume. In addition, studies have shown that longer insulation times lead to BDCs having fewer heteroatoms and a higher degree of graphitization [95]. The pyrolysis method is carried out under moderately low temperature and an inert atmosphere, which is conducive to finely controlling the graphitization degree and surface functional groups of carbon materials, thereby precisely adjusting their physical and chemical properties [96]. However, the carbon materials prepared by this method may face the problem of insufficient carbonization degree. In addition, uneven heating during pyrolysis may lead to inconsistent quality and performance of carbon materials [97]. Meanwhile, potential by-products generated during the process, such as tar and gas, need to be effectively treated to prevent environmental pollution [96]. Therefore, although pyrolysis has shown great potential in converting biomass into carbon materials, further optimization of reaction conditions and post-treatment steps is still needed to enhance the performance of carbon materials and ensure their environmental sustainability.

4.3. *Microwave and Laser Carbonization*

Microwave carbonization (MAP) is a novel carbonization approach. The MAP mechanism involves converting electromagnetic energy into heat within the irradiated material [98]. MAP employs electromagnetic waves (EMWs), having wavelengths ranging from 1 mm to 1 m and frequencies ranging from 300 MHz to 300 GHz. In comparison with traditional techniques, MAP is a rapid, efficient, and comprehensive process of heating from within the material and has high safety. Owing to the swift carbonization process of MAP, it is difficult to control heating by adjusting the heating rate. The irradiation time and energy density are controllable parameters. Microwaves can open the closed pores in carbon materials during the carbonization process, thereby increasing their specific surface area. Nevertheless, the MAP time cannot be prolonged indefinitely [99]. Jimenez et al. [100] reported on BDC through microwave carbonization of walnut shells. They observed that, when the energy absorbed by the material surpasses a certain value, the micropores on the material surface will collapse due to excessive microwave heating, leading to decreased SSAs (Figure 8a). MAP has advantages of high heating efficiency, easy operation, and environmental friendliness in the preparation of biomass derived carbon materials [101], but there are also challenges, such as uneven energy absorption and high equipment costs. For example, when the microwave energy absorbed by a material exceeds a certain value, it may lead to a loss of specific surface area of the material [102], as further microwave heating may cause the collapse of micropores.

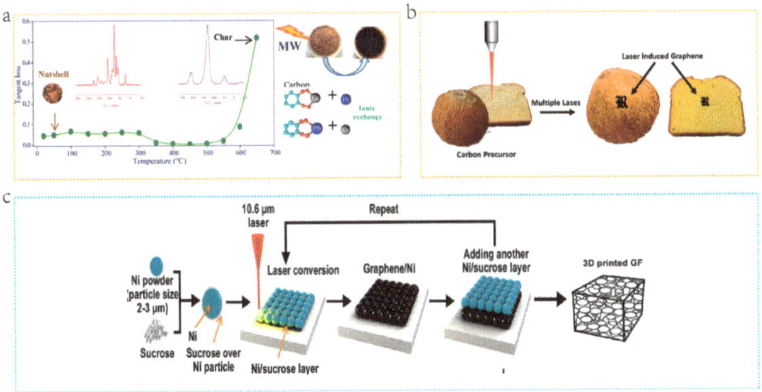

Figure 8. (**a**) Microwave pyrolysis of biomass [100]. Copyright 2017, Elsevier. (**b**) Schematic diagram of selecting specific areas of biomass to produce conductive graphite carbon [103]. Copyright 2017, American Chemical Society. (**c**) Schematic of in situ synthesis of 3D GF using a simulated 3D printing process [103]. Copyright 2017, American Chemical Society.

Laser-induced graphitization (LIG) is a method for converting biomass into carbon materials. Lasers can generate extremely high temperatures in a short time, enabling them to be used to induce the carbonization of biomass. The most prominent feature of this method is that it can select specific regions of biomass to produce conductive graphite carbon, as shown in Figure 8b. Additionally, metal catalysts and gases are not required. Tour et al. [103] prepared freestanding three-dimensional graphene foam (GF) by laser writing a Ni/sucrose mixture in a hydrogen atmosphere (as shown in Figure 8c). During the laser-induced carbonization (LIC) process, nickel powder and sucrose can absorb laser energy and be heated to high temperatures. This causes the nickel powder to sinter at the laser point and form a stable and loose nickel scaffold. Meanwhile, the nickel scaffolds act as templates for GFs and catalyze the carbonization of sucrose and its transformation into graphene. LIG allows direct fabrication of predesigned LIG patterns on various carbon materials, and precisely control their microstructure, conductivity, doping of heteroatoms, etc. [104]. This technology is selective and does not rely on chemicals or masks, significantly reducing the consumption of raw materials and environmental impact [105]. However, LIG technology still faces some challenges in research and application promotion, mainly including how to accurately manipulate the internal microstructure and atomic arrangement of LIG, and how to effectively apply it to more fields. In addition, the non-uniformity of energy absorption may also lead to some side reactions, which require further research and resolution.

5. Role of Pore Structure Regulation in Energy Applications

With the rapid development of renewable energy sources, the demand for efficient energy storage devices is growing. Supercapacitors, as a new type of energy storage device, have attracted widespread attention due to their fast charging and discharging, long lifespan, and high-power density. BDCs, as a new type of carbon material, have shown great potential in the application of supercapacitor electrode materials due to their unique porous structure and excellent conductivity.

BDC is a material prepared from biomass resources through thermochemical methods. Its hierarchical porous structure endows it with a large specific surface area, high porosity, rich pore structure, and excellent conductivity. These characteristics have a significant impact on the ion transport, specific capacitance, and charging and discharging performance of supercapacitors. The effects of different pore sizes on the charge storage and electrolyte ion diffusion of capacitors vary, and the synergistic action of micropores, mesopores, and macropores is key to improving the performance of supercapacitors.

Mostafa et al. [106], according to the steps in Figure 9a, used coconut shells as precursors to prepare two activated carbon samples: SUSCAP-01 (high micropore ratio) and SUSCAP-02 (high mesopore ratio) through carbonization and activation processes. These samples demonstrated a hierarchical porous structure of micropores, mesopores, and macropores, where micropores increased the specific surface area and served as adsorption sites for electrolyte ions, mesopores facilitated rapid ion diffusion, and macropores acted as ion buffers, reducing the ion diffusion distance and effectively enhancing the specific capacitance of supercapacitors.

BDC, with its high SSAs, abundant pore structure, and rich surface functional groups, has become a preferred material for supercapacitor electrodes. Researchers, such as Ravi et al. [107], have utilized spruce bark and activation agents, like KOH and $ZnCl_2$, to prepare efficient carbon electrodes through a one-step pyrolysis method. The pore structure is crucial for charge storage, as it affects the internal resistance of the electrode and the diffusion conditions of the charge. Biochar activated with KOH has higher SSAs, while biochar activated with $ZnCl_2$ has more mesopores, which help reduce resistance and enhance the phase angle, thereby strengthening charge storage capacity. $ZnCl_2$-activated biochar exhibits lower internal resistance and a higher phase angle due to its high mesopore ratio and residual Zn, which is beneficial for charge storage. The pore structure also provides transmission channels for electrolyte ions, promoting charge separation. KOH-activated biochar has a rough and porous surface, while $ZnCl_2$-activated biochar features denser cavities and pores, facilitating electrolyte penetration. The $ZnCl_2$-activated biochar electrode demonstrates superior supercapacitor performance, with an area capacitance as high as 342 mF/cm^2, thanks to its hydrophilic interactions and electrolyte penetration. In summary, the pore structure plays a decisive role in charge storage for supercapacitors, affecting internal resistance, charge accumulation, and the transmission efficiency of electrolyte ions. By optimizing the pore structure, the performance of supercapacitors can be significantly improved, expanding their application prospects in energy storage.

Kang et al. [108] further confirmed the impact of pore structure on the charging and discharging rate of supercapacitors. By following the steps in Figure 9b, starch was prepared into BDCs and, by adjusting the pore structure of the starch-derived BDCs, especially the development of mesopores and macropores, the diffusion of electrolyte ions can be significantly affected, thereby enhancing the charging and discharging rate of supercapacitors. The HT-900 sample demonstrated the smallest charge transfer resistance and excellent charging and discharging performance due to its larger pore size and high degree of graphitization.

The role of micropores in supercapacitors should not be overlooked. Ewelina et al. [109] prepared nitrogen-containing nano porous carbon materials (NCNMs) through direct pyrolysis, which showed a high degree of chemical composition uniformity and developed micropore–mesopore structure. According to the data in Figure 9c, this structure helps to improve the energy performance of supercapacitors, mainly due to its large internal surface area and more active sites.

As a new type of carbon material, BDC is gradually replacing traditional battery cathode materials due to its advantages, such as high specific surface area and porous structure, chemical stability, wide range of raw materials, and preparation methods. Lee et al. [110] prepared BDCs according to the steps in Figure 9d and, by adjusting the pore structure of microporous carbon, especially the pore size and porosity, the storage capacity and ion transport rate of lithium-ion batteries can be significantly improved.

Daniel et al. [111] further confirmed the importance of optimizing the pore structure and chemical composition to improve the electrochemical stability and reversible capacity of supercapacitor electrode materials. The porous carbon material prepared by KOH activation of cellulose has a high specific surface area and total porosity. According to the data in Figure 9e, the best electrochemical performance was obtained by optimizing the KOH/cellulose ratio.

In summary, the hierarchical porous structure and excellent conductivity of BDC give it great potential in the application of supercapacitor electrode materials. By optimizing the pore structure, especially with the development of appropriately sized mesopores and macropores, the diffusion rate of electrolyte ions can be significantly improved, thereby enhancing the charging and discharging performance of supercapacitors. In addition, the chemical stability and wide range of raw materials for BDCs make them sustainable and environmentally friendly supercapacitor electrode materials. Future research should continue to explore the preparation methods and optimization of pore structures of BDCs to achieve more efficient and stable supercapacitor electrode materials.

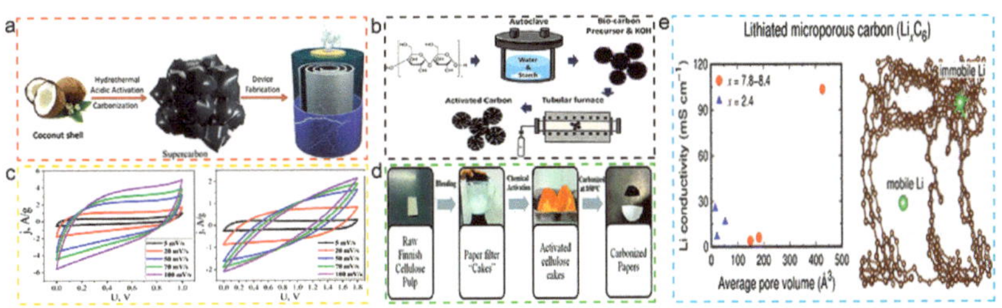

Figure 9. (**a**) Schematic illustration of the preparation of activated carbon for supercapacitor [106]. Copyright 2024, Royal Society of Chemistry. (**b**) Schematic of the preparation of activated biocarbon [108]. Copyright 2024, Maney Publishing. (**c**) Cyclic voltammograms of supercapacitors made from bio-carbon material NCNM with electrolyte 30% KOH and 1 M Na_2SO_4 [109]. Copyright 2024, Multidisciplinary Digital Publishing Institute. (**d**) Porous Characteristics of Microporous Carbon and Structure of Litigated Microporous Carbon [110]. Copyright 2018, Royal Society of Chemistry. (**e**) Schematic illustration of the synthetic procedure of activation and carbonization of precursor cellulose materials [111]. Copyright 2022, Royal Society of Chemistry.

6. Role of Pore Structure Regulation in Catalysis Applications

BDC exhibits a broad potential for applications in the field of catalysis due to its unique porous structure and chemical properties [112]. As a type of porous carbon material, the regulation of BDC's pore structure is of great significance for enhancing catalytic performance. By adjusting the pore structure of BDC, its catalytic activity, selectivity, and stability can be significantly improved. The following discussion will explore the mechanisms by which the regulation of BDC's pore structure enhances its catalytic activity, selectivity, and stability in hydrogen evolution reactions, redox reactions, and photocatalytic reactions.

The regulation of BDC's pore structure is achieved by altering the size, distribution, shape, and number of pores. This regulation can significantly affect the catalytic activity, selectivity, and stability of BDCs. For instance, BDCs with a high specific surface area can enhance hydrogen storage capacity, thereby strengthening the catalytic role in hydrogen evolution reactions [113].

Yang et al. [114] evaluated the impact of different engineered biochar-based catalysts on the production of synthesis gas during biomass pyrolysis and catalytic reforming without additional steam input. It was found that the BDC catalyst, activated physically for 2 h in Figure 10a, induced the highest H_2/CO ratio (1.15), and activated BDCs as catalysts produced comparable yields of synthesis gas. In the studies mentioned in the article, BDCs underwent various engineering treatments, such as physical and chemical activation, nitrogen doping, and nickel doping, to enhance catalytic performance. These treatments increased BDC's specific surface area and porosity, introduced new active sites, and thus enhanced its ability to catalyze the cracking of tar and the reforming of volatiles, improving the yield and quality of synthetic hydrogen. Therefore, a high specific surface area and hierarchical porous structure are beneficial to produce synthesis gas and hydrogen, and these characteristics of BDCs demonstrate good catalytic action in hydrogen evolution reactions.

In addition to playing a significant role in hydrogen evolution reactions, the regulation of BDCs' pore structure also plays a certain role in redox reactions. Wu et al. [115] studied the redox properties of wheat straw-derived nanoscale biochar (NBC), particularly exploring the relationship between its pore structure and redox reactions. The typical current–time (i–t) response of NBC upon continuous addition is shown in Figure 10b. By sequentially adding different concentrations (i.e., 0.2, 0.4, 0.6, 0.8, and 1.0 mg NBC) of the sample, the electron acceptance and electron-donating reactions of NBC on the GC electrode were observed. In redox or oxidation reactions, the quantity of transferred electrons (Q) is proportional to the addition of NBC mass. As shown in Figure 10b, EAC is generally greater than EDC, indicating that most of the redox groups in BDCs are composed of oxidized groups. The results show that NBC has significant redox activity and can participate in redox reactions as an electron acceptor and donor. NBC-700, produced by high-temperature pyrolysis, exhibits higher redox activity than NBC-400 due to its higher specific surface area and porosity. This indicates that the pore structure of BDCs has a significant impact on redox characteristics. The development of the pore structure may increase the active sites on the BDC's surface, thereby enhancing its ability to participate in redox reactions.

BDCs are often used as a catalyst, and BDC catalysts can significantly alter the distribution and properties of biomass pyrolysis products, promoting the production of bio-oil and gas through the interaction of volatiles with BDCs [116]. The relationship between the pore structure of BDCs and their performance as a catalyst in pyrolysis and catalytic reforming reactions is very close. As mentioned above, Yang et al. [117] evaluated the performance of various engineered BDC catalysts in the production of synthesis gas during biomass pyrolysis and catalytic reforming and found that the high specific surface area and hierarchical porous structure of BDCs are beneficial for the production of synthesis gas and hydrogen. Since the reactions in the study belong to thermal catalytic reactions, the pore structure of BDCs plays a role in thermal catalytic reactions. The pore structure of BDCs improves the efficiency of biomass pyrolysis and catalytic reforming processes by providing more active sites, regulating the diffusion of reactants and products, affecting the physical and chemical stability of the catalyst, and influencing the formation of coke.

The photocatalytic properties of BDCs are crucial for understanding and controlling the transformation and cycling of environmental pollutants. The photocatalytic reactions of BDCs mainly involve free radical and non-free radical redox reactions, triggered by quinone and phenolic functional groups, porous structure, and persistent free radicals in BDCs [118]. Mian et al. [119] discussed the research progress of biochar-supported photocatalysts (BSPs) and show in Figure 10c the ways to reduce the bandgap energy of titanium dioxide photocatalysts loaded on BDCs by sensitization, forming intermediate bandgap energy levels, and forming local trap states. This indicates that, by combining with BDCs, the light response range of photocatalysts can be extended to the visible light region, improving photocatalytic efficiency. From Figure 10d proposed by Colmenares et al. [120], it was observed that photocatalysts doped with different BDCs showed significant differences in the degradation efficiency of phenol, from which it can be concluded that the introduction of BDCs enhanced the performance of photocatalysts due to providing more active sites, enhancing light absorption, or promoting the separation of electron-hole pairs.

Therefore, the regulation of BDCs' pore structure is a key means to improve their catalytic performance. By changing the pore structure, not only can the catalytic activity and selectivity of BDCs be enhanced, but also their stability and durability in various practical applications can be strengthened. Moreover, the regulation of the BDC pore structure also plays an important role in photocatalytic reactions, providing a new perspective for understanding and controlling the transformation and cycling of environmental pollutants.

Zhang et al. [121] investigated the application of sludge-derived biochar in the electrocatalytic oxidation of azo dyes (such as methyl orange), highlighting the significant impact of its porous structure on electrocatalytic performance. Firstly, the porous structure of the biochar provides more active sites, enabling effective contact with pollutants and thus significantly enhancing the efficiency of electrocatalytic reactions. Secondly, the porous structure helps to improve the mass transfer efficiency near the electrode, facilitating the exchange of materials between pollutants and the electrode, which is crucial for accelerating the rate of electrocatalytic reactions. Furthermore, as the pyrolysis temperature increases, the conductivity of the biochar is enhanced, which is closely related to the formation of the porous structure, thereby improving the transfer of electrons. Lastly, the study also points out that biochar-modified electrodes at high pyrolysis temperatures have a higher oxygen evolution potential, which can reduce side effects in the electrocatalytic process, such as the production of oxygen, thereby further enhancing electrocatalytic efficiency. In summary, the porous structure of biochar has a direct and significant impact on the enhancement of its electrocatalytic performance.

Solid waste generated by the food industry, despite its diversity, contains the potential to be transformed into a wide range of industrial application products, among which biomass-based catalysts, industrial enzymes, and biofuels are particularly prominent sustainable products. For instance, Mohd et al. [122] prepared biochar from the high-temperature calcination of green coconut waste shells, which was used in solid-state fermentation to produce microbial β-glucosidase (BGL enzyme). Its porous structure provides an abundant reaction site for the enzyme, increasing the chances of contact between the enzyme and substrate, thereby enhancing the efficiency of enzyme-catalyzed reactions. Moreover, the large specific surface area of the biochar helps to adsorb enzyme molecules, reducing their deactivation, while also promoting the transfer of substrates and products, reducing mass transfer resistance, which is crucial for improving reaction rates and efficiency. The porous structure of the biochar also facilitates the recovery of enzymes, helping to reduce costs and improve economic viability, while optimizing their catalytic performance by altering the enzyme's microenvironment.

Furthermore, utilizing renewable resources, especially low-cost forestry waste, to produce bioenergy and materials is of great significance for achieving green and sustainable chemical processes. Xiang et al. [123] prepared biochar from sawdust through an H_3PO_4-assisted hydrothermal carbonization method, and then prepared a zirconium-coordinated biochar catalyst Zr-SDBC-P through a self-assembly method, which showed high activity in the catalytic transfer hydrogenation reaction of converting ethyl succinate to γ-valerolactone. Data show that the biochar treated with H_3PO_4 has a larger specific surface area and porous structure, which helps to improve the production efficiency of enzymes, and has good high-temperature stability, which is conducive to maintaining the stability of the catalyst during the reaction process. The recycling experiment results of the Zr-SDBC-P-4 catalyst also confirmed its good reusability, indicating that the porous structure helps to maintain the active center of the catalyst and, even after multiple uses, it can maintain its coordination catalytic performance. Therefore, the porous structure of biochar plays a key role in improving catalytic efficiency, reducing production costs, and achieving green chemical processes.

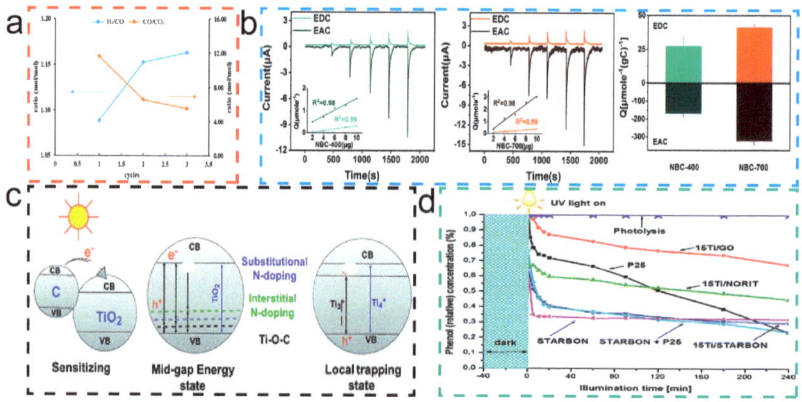

Figure 10. (**a**) H_2/CO and CO/CO_2 molar ratios based on the syngas composition versus feeding times in the combined pyrolysis and in-line catalytic reforming process using a NiBC-30 catalyst. [117]. Copyright 2023, American Chemical Society. (**b**) Reductive and oxidative current responses of NBC-400 (inset: linear relationship between the number of electrons and the added amounts of NBC-400); reductive and oxidative current responses of NBC-700 (inset: linear relationship between the electron numbers and the added amounts of NBC-700); electron transfer capacity of NBC-400 and NBC-700. [115]. Copyright 2022, Royal Society of Chemistry. (**c**) Reduction of the TiO_2–BSP band gap energy via sensitizing, forming a mid-gap energy state and forming a local trapping state. [119] Copyright 2018, Royal Society of Chemistry. (**d**) Photocatalyst activities in the aqueous phase degradation of phenol (reaction conditions: 150 mL of mother solution, 150 mg of photocatalyst, Cphenol = 50 ppm, temperature 30 °C, reaction pressure 1 bar). [120] Copyright 2013, Royal Society of Chemistry.

7. Role of Pore Structure Regulation in Environmental Applications

Firstly, the pore structure of BDCs can be designed and prepared through specific processes to achieve molecular size selective separation. This size selectivity is achieved by controlling the size and distribution of pores, allowing only molecules of specific sizes to enter and be adsorbed within the pores. For instance, Mirtha et al. [124] prepared sponge-like BDC adsorbents through freeze-drying and pyrolysis, which enhance the adsorption of CO_2 from CO_2/CH_4 and CO_2/N_2 gas mixtures. As shown in Figure 11a, by altering the pyrolysis temperature, the porosity and specific surface area of BDCs can be adjusted, providing more adsorption sites and enhancing the gas adsorption capacity. This finding indicates that, by finely tuning the pore structure of BDCs, their separation performance in gas mixtures can be significantly improved. In addition, the pyrolysis temperature is able to alter the porosity of biochar, according to Reis et al. [125] When exploring the synthesis methods of sustainable activated biochar as carbon-free anode materials for lithium-ion batteries (LIBs) and sodium-ion batteries (NIBs), researchers have found that there are significant differences in the pore structures of biochar synthesized from Norwegian spruce bark using chemical activation methods (with $ZnCl_2$ and KOH). Specifically, the biochar activated by $ZnCl_2$ (Biochar-1) formed highly mesoporous carbon, with 96.1% of its structure being mesopores, while the biochar activated by KOH (Biochar-2) showed a lower degree of graphitization and a more disordered, defective carbon structures, with a mesoporosity content of only 56.1%. Furthermore, Biochar-1 demonstrated the formation of more ordered graphene layers in its structure, whereas Biochar-2 exhibited a greater degree of disorder. In terms of surface functionality, although both biochars showed high functionality, Biochar-1 had a better electrochemical response due to the presence of pyridinic nitrogen functional groups. Consequently, the article concludes that, owing to its highly developed mesoporous structure and ordered carbon layer structure, Biochar-1 outperforms Biochar-2 in electrochemical properties, making it an excellent biomass anode material for LIB and NIB applications. Therefore, the article concludes that, due to Biochar-1's highly developed mesoporous structure and ordered carbon layer structure, its electrochemical performance is superior to Biochar-2, making it an outstanding biomass anode material for LIB and NIB applications.

Secondly, the hierarchical pore structure of BDCs provides a large surface area, which increases the contact opportunities between molecules and separation materials, thereby enhancing adsorption capacity and selectivity. This high specific surface area characteristic makes BDCs an efficient adsorbent, especially when dealing with low-concentration gases. Additionally, the hierarchical pore structure also helps to reduce light reflection and enhance light scattering effects, thereby improving solar energy absorption rates, which is particularly important for solar-driven water purification technologies.

Furthermore, the hierarchical pore structure of BDCs also shows great potential in water purification applications. As shown in Figure 11b, Zhang et al. [126] used BDCs derived from sorghum stalks, prepared through a simple carbonization process, to create a solar evaporator that performed excellently in the purification of various water sources, including desalination of seawater, treatment of acidic and alkaline wastewater, and treatment of organic polluted water. The success of these applications proves that BDCs have broad application prospects not only in the field of gas separation but also in liquid separation.

In summary, the hierarchical pore structure of BDCs plays a crucial role in both gas and liquid separation fields. By finely tuning the pore structure, not only can the adsorption efficiency and selectivity be improved, but the rate and efficiency of catalytic reactions can also be enhanced. Therefore, BDCs, as a sustainable and environmentally friendly material, have broad application prospects in separation science and technology. Future research should continue to explore the relationship between the pore structure of BDCs and separation performance to develop more efficient and economical separation materials, contributing to environmental protection and sustainable development.

When discussing the role of BDCs in environmental management, the research by Chen et al. [127] provides a powerful perspective. They found, as shown in Figure 11c, that BDC modification significantly increased the specific surface area and pore structure of Co_3O_4, which is generally conducive to the occurrence of catalytic reactions. The biochar-modified Co_3O_4 catalyst showed a significant promotional effect in the heterogeneous activation process of peroxy-monosulfate (PMS). This finding not only confirms the extraordinary activity of Co_3O_4 synthesized with BDC assistance in PMS activation but also highlights its potential application in the efficient degradation of ofloxacin, further emphasizing the promising role of BDCs as a catalyst for the removal of organic compounds.

The research by Jin et al. [128] further deepens our understanding of the catalytic and adsorption performance of BDCs in organic compounds. They pointed out that BDCs prepared under high-temperature pyrolysis conditions (BC900) have excellent adsorption capacity for trichloroethylene (TCE) and show excellent catalytic activity in the activation of persulfate (PS) for the degradation of sulfamethazine (SMT). This finding not only confirms the adsorption and catalytic performance of BDCs in water treatment but also reveals the significant impact of BDC particle size and pyrolysis temperature on performance. It is particularly noteworthy that BC900 with a smaller particle size (0–75 μm) is 19.5–62.3% more efficient in TCE adsorption than larger BDC particles. Moreover, the removal rate of SMT using the BC900/PS system increased significantly with the decrease in BDC particle size, from 24.2% to 98.3%. This trend indicates that the particle size and pyrolysis temperature of BDCs are key factors affecting adsorption and catalytic performance in water treatment.

In summary, the particle size and pyrolysis temperature of BDCs have a significant impact on their adsorption and catalytic performance in water treatment. BDCs with smaller particle size, after high-temperature pyrolysis, show excellent TCE adsorption and SMT degradation performance due to their larger specific surface area and rich microporous structure. These findings provide new strategies for the application of BDCs in environmental management and open new directions for future research and application.

In addition to organic pollutants, in the environmental field, BDCs also play a role in the management of exhaust gases. The research by Díaz-Maroto et al. [129] provides in-depth insights into the NO removal capacity of BDCs. They found, as shown in Figure 11d, that when the amount of BDCs is halved (from 3 g to 1.5 g), the NO removal capacity of both BDCs decreases. For an initial concentration of 5 ppmv of NO, the removal capacity of both activated carbons decreased by about 36% and 40%, respectively. Further analysis indicates that the specific surface area (SBET) of BDCs is directly related to NO removal capacity. A larger specific surface area means more active sites, which not only helps in the adsorption of NO molecules but also promotes the subsequent oxidation process. Moreover, the total pore volume and pore size distribution of BDCs are also important factors affecting NO removal capacity. BDCs with a larger micropore volume can provide more contact opportunities for NO and O_2, thus promoting the oxidation reaction. The role of BDC pore structure in exhaust gas treatment cannot be ignored, as it directly affects the adsorption of NO molecules, contact with oxygen, and subsequent catalytic oxidation reactions. These findings emphasize the potential of BDCs' pore characteristics and surface chemical properties in improving their performance in practical exhaust gas treatment applications. By optimizing these characteristics, we can expect significant improvements in the performance of BDCs in the field of exhaust gas treatment.

BDCs, as a carbonaceous material derived from biomass pyrolysis, play an important role in the treatment and recovery of heavy metals. Cao et al. [130] developed a new type of BDC by using dehydrated algal waste and combining it with KOH and $FeCl_3$ for co-activation treatment, preparing a highly efficient adsorbent specifically for elemental mercury in coal combustion flue gas. This method of preparing BDCs not only achieves the resource utilization of waste but also significantly increases the material's specific surface area and porosity, thereby enhancing its adsorption capacity for mercury. Algal waste-derived BDCs show excellent performance in removing elemental mercury, far exceeding that of unmodified BDCs. This high removal capacity is attributed to the Fe and Cl elements on the surface of the BDCs, which are retained during the KOH activation process and enhance the BDCs' magnetic properties by forming Fe-Cl complexes, thus simplifying the recovery process of the adsorbent. In the management of heavy metal pollution, the application of BDCs is mainly based on their high specific surface area and abundant surface functional groups, which enable BDCs to provide many active sites to effectively adsorb heavy metal ions in the solution, such as lead (Pb), cadmium (Cd), mercury (Hg), and chromium (Cr), etc. [131]. In summary, the application prospect of BDCs in heavy metal pollution management is broad, and their low cost, environmentally friendly, and adjustable characteristics make them an effective tool for heavy metal removal. With in-depth research, BDCs are expected to play a more significant role in the management and control of heavy metal pollution.

In summary, we have compiled a table (Table 1) to illustrate the adsorption efficiency of various biochars for different environmental pollutants based on the readings of some articles and the adsorption of pollutants by biochars in the environment.

Table 1. The adsorption efficiency of various biochars for different pollutants in the environment.

Types of Biochar	Adsorbed Contaminants	The Size of the Contaminant	Separation Efficiency	Refere
Carbonized sorghum straw	Seawater, acidic/alkaline wastewater, organic wastewater		Energy conversion efficiency 100%, Evaporation rate 3.173 kg m^{-2} h^{-1}	[126
Coconut shell activated carbon	Oils and organic solvents		The oil–water separation efficiency is maintained at more than 98%.	[13

Table 1. *Cont.*

Types of Biochar	Adsorbed Contaminants	The Size of the Contaminant	Separation Efficiency	Reference
Biochar esidue/ethylcellulose ixed matrix membrane after lignin depolymerization	CO_2/CH_4 and CO_2/N_2	Molecular level	CO_2/CH_4's selectivity is 9.97, CO_2/N_2's electivity is 20.01	[133]
ıgnetic biochar (M-RSB and M-SSB)	Cd^{2+} in solution	Ionic level	The maximum adsorption capacities are as follows: 42.48 mg/g (M-RSB) and 4.64 mg/g (M-SSB)	[134]
Biochar derived from wheat straw (SB)	Cd^{2+} and Co^{2+} in solution	Ionic level	The maximum adsorption capacities are as follows: 193 μmol g^{-1} (Cd^{2+}) and 89.7 μmol g^{-1} (Co^{2+})	[135]
'ine needles magnetic char (pncm and pncom)	Cu (II)	Ionic level	pncm: 0.4 mmol/g, pncom: 1.0 mmol/g	[136]
ınut Shell Biochar (PSB) l Modified PSB (MPSB)	As (III) and As (V)	Ionic level	MPSB: As (III) 86, As (V) 91.26%	[137]
Distiller's grains are lerived from biochar	Pb^{2+}	Ionic level	79.12 mg/g	[138]
ıb shell-derived biochar (CSAB)	Diesel Oil		93.9 mg/g	[139]
heep Manure Biochar 1B) and Robinia Biochar (RPB)	Mixed heavy metals (Pb^{2+}, Cu^{2+}, Cd^{2+})	Ionic level	SMB3: Pb^{2+} 20.2 mg/g, Cu^{2+} 13.9 mg/g, Cd^{2+} 3.2 mg/g RPB8: Pb^{2+} 7.4 mg/g, Cu^{2+} 10.5 mg/g	[140]

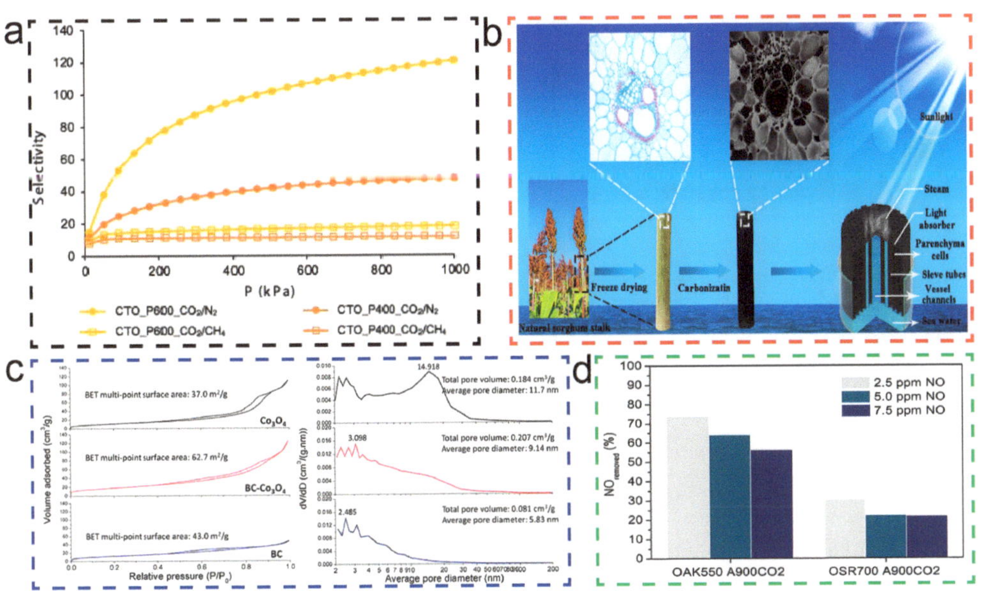

Figure 11. (**a**) Comparative mean selectivity, at 25 °C, of CTO_P400 (orange) and CTO_P600 (yellow), for the CO_2/CH_4 separation (square symbols) and for the CO_2/N_2, separation (circle symbols). (For

interpretation of the references by color in this figure legend, the reader is referred to the web version of this article.) [124].Copyright 2023, Elsevier. (**b**) Construction and schematic process of carbonized sorghum straw for solar evaporation [126]. Copyright 2021, Multidisciplinary Digital Publishing Institute. (**c**) N_2 adsorption/desorption isotherm and pore size distribution of the three samples [127]. Copyright 2018, Elsevier. (**d**) NO removal capacity of activated carbons as a function of the NO concentration in the air stream. Experimental conditions: Temperature: 25 °C; carbon bed: 1.5 g [129]. Copyright 2023, Elsevier.

8. Conclusions and Outlook

Pore engineering in BDCs has garnered increasing attention due to its significant impact on enhancing the performance of these materials across a wide range of applications, such as energy storage, catalysis, and environmental remediation. The ability to tailor the pore structure, including pore size, distribution, and interconnectivity, allows for the optimization of specific physical and chemical properties that directly influence the material's efficiency and adaptability in diverse technological applications.

In energy storage, the hierarchical porous structure of BDCs has shown immense potential in improving the performance of devices like supercapacitors, lithium-ion batteries, and sodium-ion batteries. The combination of micropores, mesopores, and macropores within BDCs enhances both ion transport and charge storage capabilities. Micropores offer a high surface area for ion adsorption, which increases the energy storage density, while mesopores and macropores reduce ion diffusion pathways and enhance electrolyte transport, improving power density and charge–discharge rates. As such, optimizing the pore architecture is critical for achieving high-performance energy storage devices. Furthermore, the tunability of pore size distribution in BDCs is essential for balancing energy density with power density in energy storage applications.

In catalysis, pore-engineered BDCs are highly effective due to the increased accessibility of active sites and the improvement of mass transport. The hierarchical porous structures facilitate the diffusion of reactants and products, which is especially critical in heterogeneous catalysis where the interplay between surface area and pore architecture determines the catalytic efficiency. The ability to create a finely tuned pore network not only boosts catalytic activity but also enhances selectivity and stability, particularly in reactions such as hydrogen evolution, oxygen reduction, and carbon dioxide reduction. BDCs with well-defined pore structures are increasingly being used as metal-free catalysts or as support for single-atom or nanoparticle catalysts, owing to their high specific surface area and chemical stability.

Environmental applications of BDCs, particularly in gas separation, water purification, and pollutant removal, benefit significantly from precise pore structure control. The hierarchical porosity of BDCs facilitates the adsorption of gases, such as CO_2, NOx, and volatile organic compounds (VOCs), through molecular sieve effects and enhanced surface interactions. In water treatment, the combination of micro- and mesopores enables efficient adsorption of organic and inorganic pollutants, while macropores improve fluid flow and regeneration capacity. The adaptability of BDCs in environmental remediation is further expanded by chemical modifications, such as doping with heteroatoms, which improve their affinity for specific contaminants and enhance photocatalytic or redox capabilities. Moreover, recent advancements in pore engineering have enabled the design of BDCs with tunable surface chemistry, further expanding their applicability in environmental cleanup.

Looking forward, further advancements in pore engineering techniques are essential to unlock the full potential of BDCs. Novel strategies such as dual-template synthesis, self-activation methods, and advanced templating techniques, which combine both hard and soft templates, are promising pathways to achieve precise control over pore architectures and enable the development of hierarchical structures with optimized performance. The integration of these strategies with emerging carbonization techniques, such as microwave and laser carbonization, presents new opportunities to fine-tune the pore structure at the nanoscale, leading to enhanced performance in energy, catalysis, and environmental applications.

Moreover, the combination of BDCs with other functional materials, such as metal oxides, metal-organic frameworks (MOFs), and conductive polymers, can create hybrid materials with synergistic properties, further expanding their functionality. The development of BDCs-based composites or heterostructures could lead to breakthroughs in applications such as electrochemical catalysis, gas sensing, and energy harvesting. For instance, incorporating transition metals into the porous carbon framework can generate active sites for electrocatalysis, while maintaining the high conductivity and structural integrity of the carbon matrix.

In conclusion, pore engineering is a pivotal strategy for maximizing the performance of BDCs. By refining control over pore structure, enhancing surface chemistry, and exploring new synthesis routes, BDCs are poised to play an increasingly critical role in addressing global challenges related to renewable energy, environmental sustainability, and green technology. Future research should focus on bridging the gap between laboratory-scale synthesis and large-scale production of BDCs, ensuring that these materials can be manufactured in an economically viable and environmentally sustainable manner. Additionally, further exploration of the relationships between pore structure, surface functionality, and material performance will be key to developing next-generation BDCs with tailored properties for specific applications. By continuing to innovate in pore engineering, BDCs can become a cornerstone of future technologies that contribute to a sustainable and circular economy.

Author Contributions: Writing—original draft preparation, Q.W., B.L. and Z.W.; writing—review and editing, Y.H.; supervision, Y.H.; project administration, Y.H. and M.D.; funding acquisition, M.D. All authors have read and agreed to the published version of the manuscript.

Funding: This research received no external funding.

Conflicts of Interest: The authors declare no conflicts of interest.

rences

Belluati, M.; Tabasso, S.; Calcio Gaudino, E.; Cravotto, G.; Manzoli, M. Biomass-Derived Carbon-Based Catalysts for Lignocellulosic Biomass and Waste Valorisation: A Circular Approach. *Green Chem.* **2024**, *26*, 8642–8668. [CrossRef]

Liu, T.; Yabu, H. Biomass-Derived Electrocatalysts: Low-Cost, Robust Materials for Sustainable Electrochemical Energy Conversion. *Adv. Energy Sustain. Res.* **2024**, *5*, 2300168. [CrossRef]

Mondal, S.; Ruidas, S.; Chongdar, S.; Saha, B.; Bhaumik, A. Sustainable Porous Heterogeneous Catalysts for the Conversion of Biomass into Renewable Energy Products. *ACS Sustain. Resour. Manag.* **2024**, *1*, 1672–1704. [CrossRef]

Wang, R.; Feng, Y.; Li, D.; Li, K.; Yan, Y. Towards the Sustainable Production of Biomass-Derived Materials with Smart Functionality: A Tutorial Review. *Green Chem.* **2024**, *26*, 9075–9103. [CrossRef]

Anvari, S.; Aguado, R.; Jurado, F.; Fendri, M.; Zaier, H.; Larbi, A.; Vera, D. Analysis of Agricultural Waste/Byproduct Biomass Potential for Bioenergy: The Case of Tunisia. *Energy Sustain. Dev.* **2024**, *78*, 101367. [CrossRef]

Blasi, A.; Verardi, A.; Lopresto, C.G.; Siciliano, S.; Sangiorgio, P. Lignocellulosic Agricultural Waste Valorization to Obtain Valuable Products: An Overview. *Recycling* **2023**, *8*, 61. [CrossRef]

Anttiroiko, A.-V. Smart Circular Cities: Governing the Relationality, Spatiality, and Digitality in the Promotion of Circular Economy in an Urban Region. *Sustainability* **2023**, *15*, 12680. [CrossRef]

Basumatary, S.F.; Patir, K.; Das, B.; Saikia, P.; Brahma, S.; Basumatary, B.; Nath, B.; Basumatary, B.; Basumatary, S. Production of Renewable Biodiesel Using Metal Organic Frameworks Based Materials as Efficient Heterogeneous Catalysts. *J. Clean. Prod.* **2022**, *358*, 131955. [CrossRef]

Lin, R.-B.; Zhang, Z.; Chen, B. Achieving High Performance Metal–Organic Framework Materials through Pore Engineering. *Acc. Chem. Res.* **2021**, *54*, 3362–3376. [CrossRef]

Kazemi Shariat Panahi, H.; Dehhaghi, M.; Ok, Y.S.; Nizami, A.-S.; Khoshnevisan, B.; Mussatto, S.I.; Aghbashlo, M.; Tabatabaei, M.; Lam, S.S. A Comprehensive Review of Engineered Biochar: Production, Characteristics, and Environmental Applications. *J. Clean. Prod.* **2020**, *270*, 122462. [CrossRef]

Shu, R.; Jiang, H.; Xie, L.; Liu, X.; Yin, T.; Tian, Z.; Wang, C.; Chen, Y. Efficient Hydrodeoxygenation of Lignin-Derived Phenolic Compounds by Using Ru-Based Biochar Catalyst Coupled with Silicotungstic Acid. *Renew. Energy* **2023**, *202*, 1160–1168. [CrossRef]

Wang, Z.; Li, Z.; Zhang, Y.; Fu, P. Carbon-Based Catalysts for Hydrodeoxygenation of Biomass Derivatives: A Review of Heteroatom Doping Strategies. *Mol. Catal.* **2024**, *569*, 114539. [CrossRef]

13. Xu, Y.; Fan, Z.; Li, X.; Yang, S.; Wang, J.; Zheng, A.; Shu, R. Cooperative Production of Monophenolic Chemicals and Ca Adsorption Materials from Cascade Pyrolysis of Acid Hydrolysis Lignin. *Bioresour. Technol.* **2024**, *399*, 130557. [Cros [PubMed]
14. Xia, H.; Hu, Y.; Li, Z.; Lan, H.; Zhang, J. Electron Spin Polarization in Rechargeable Batteries: Theoretical Foundation and Pra Applications. *Adv. Funct. Mater.* **2024**, 2413491. [CrossRef]
15. Hu, Y.; Gong, J.; Zeb, H.; Lan, H.; Asif, M.; Xia, H.; Du, M. Mechanistic Insights into C–C Coupling in Electrocatalytic Reduction Reaction. *Chem. Commun.* **2024**, *60*, 10618–10628. [CrossRef]
16. Hu, Y.; Lan, H.; He, J.; Fang, W.; Zhang, W.-D.; Lu, S.; Duan, F.; Du, M. Entropy-Engineered Middle-In Synthesis of Single-Atom Compounds for Nitrate Reduction Reaction. *ACS Nano* **2024**, *18*, 23168–23180. [CrossRef]
17. Yameen, M.Z.; Naqvi, S.R.; Juchelková, D.; Khan, M.N.A. Harnessing the Power of Functionalized Biochar: Progress, Challe and Future Perspectives in Energy, Water Treatment, and Environmental Sustainability. *Biochar* **2024**, *6*, 25. [CrossRef]
18. Luo, L.; Wang, J.; Lv, J.; Liu, Z.; Sun, T.; Yang, Y.; Zhu, Y.-G. Carbon Sequestration Strategies in Soil Using Biochar: Adv Challenges, and Opportunities. *Environ. Sci. Technol.* **2023**, *57*, 11357–11372. [CrossRef]
19. Leng, L.; Xiong, Q.; Yang, L.; Li, H.; Zhou, Y.; Zhang, W.; Jiang, S.; Li, H.; Huang, H. An Overview on Engineering the Su Area and Porosity of Biochar. *Sci. Total Environ.* **2021**, *763*, 144204. [CrossRef]
20. Soler-Illia, G.J.A.A.; Azzaroni, O. Multifunctional Hybrids by Combining Ordered Mesoporous Materials and Macromole Building Blocks. *Chem. Soc. Rev.* **2011**, *40*, 1107–1150. [CrossRef]
21. Hwang, J.; Ejsmont, A.; Freund, R.; Goscianska, J.; Schmidt, B.V.K.J.; Wuttke, S. Controlling the Morphology of Metal–Or Frameworks and Porous Carbon Materials: Metal Oxides as Primary Architecture-Directing Agents. *Chem. Soc. Rev.* **202** 3348–3422. [CrossRef] [PubMed]
22. Sun, W.; Hu, Y.; Gong, J.; Zhang, L.; Li, Z.; Fu, Q.; Lan, H.; Wang, D.; Yan, M.; Xia, H.; et al. Anomalous Catalytic Performar Non-Active-Site Carbon Substrate. *ChemSusChem* **2024**, e202401301. [CrossRef]
23. Zhao, X.; Elcin, E.; He, L.; Vithanage, M.; Zhang, X.; Wang, J.; Wang, S.; Deng, Y.; Niazi, N.K.; Shaheen, S.M.; et al. Using Bi for the Treatment of Continuous Cropping Obstacle of Herbal Remedies: A Review. *Appl. Soil Ecol.* **2024**, *193*, 105127. [Cros
24. Gao, G.; Yan, L.; Tong, K.; Yu, H.; Lu, M.; Wang, L.; Niu, Y. The Potential and Prospects of Modified Biochar for Comprehe Management of Salt-Affected Soils and Plants: A Critical Review. *Sci. Total Environ.* **2024**, *912*, 169618. [CrossRef] [PubMe
25. Liu, Z.; Tan, C.; Zhao, Y.; Song, C.; Lai, J.; Song, M. Singlet Oxygen in Biochar-Based Catalysts-Activated Persulfate Process: Generation to Detection and Selectivity Removing Emerging Contaminants. *Chem. Eng. J.* **2024**, *485*, 149724. [CrossRef]
26. Yuan, Y.; Liu, Q.; Zheng, H.; Li, M.; Liu, Y.; Wang, X.; Peng, Y.; Luo, X.; Li, F.; Li, X.; et al. Biochar as a Sustainable To Improving the Health of Salt-Affected Soils. *Soil Environ. Health* **2023**, *1*, 100033. [CrossRef]
27. Qi, S.; Degen, A.; Wang, W.; Huang, M.; Li, D.; Luo, B.; Xu, J.; Dang, Z.; Guo, R.; Shang, Z. Systemic Review for the Use of Bi to Mitigate Soil Degradation. *GCB Bioenergy* **2024**, *16*, e13147. [CrossRef]
28. Xiao, X.; Chen, B.; Chen, Z.; Zhu, L.; Schnoor, J.L. Insight into Multiple and Multilevel Structures of Biochars and Their Pot Environmental Applications: A Critical Review. *Environ. Sci. Technol.* **2018**, *52*, 5027–5047. [CrossRef] [PubMed]
29. Wang, Y.; Zhang, M.; Shen, X.; Wang, H.; Wang, H.; Xia, K.; Yin, Z.; Zhang, Y. Biomass-Derived Carbon Materials: Contro Preparation and Versatile Applications. *Small* **2021**, *17*, 2008079. [CrossRef]
30. Hu, Q.; Zhang, Z.; He, D.; Wu, J.; Ding, J.; Chen, Q.; Jiao, X.; Xie, Y. Progress and Perspective for "Green" Strategies of Cat Plastics Conversion into Fuels by Regulating Half-Reactions. *J. Am. Chem. Soc.* **2024**, *146*, 16950–16962. [CrossRef]
31. Ding, J.; He, D.; Du, P.; Wu, J.; Hu, Q.; Chen, Q.; Jiao, X. Design Photocatalysts to Boost Carrier Dynamics in Plastics Photoc sion into Fuels. *ACS Appl. Mater. Interfaces* **2024**, *16*, 35865–35873. [CrossRef] [PubMed]
32. Li, M.; Han, Z.; Hu, Q.; Fan, W.; Hu, Q.; He, D.; Chen, Q.; Jiao, X.; Xie, Y. Recent Progress in Solar-Driven CO_2 Reduct Multicarbon Products. *Chem. Soc. Rev.* **2024**, *53*, 9964–9975. [CrossRef] [PubMed]
33. Tan, L.; Sun, M.; Wang, H.; Wang, J.; Kim, J.; Lee, M. Enantiocontrolled Macrocyclization by Encapsulation of Substrates in Capsules. *Nat. Synth.* **2023**, *2*, 1222–1231. [CrossRef]
34. Lee, D.-W.; Jin, M.-H.; Park, J.-H.; Lee, Y.-J.; Choi, Y.-C. Flexible Synthetic Strategies for Lignin-Derived Hierarchically P Carbon Materials. *ACS Sustain. Chem. Eng.* **2018**, *6*, 10454–10462. [CrossRef]
35. Phan, H.H.; Phan, A.N. Carbon Nanostructures with the Ultra-High Surface Area and Porosity Derived from Bioma *Biomass-Based Functional Carbon Nanostructures for Supercapacitors*; Green Energy and Technology; Tiwari, S.K., Bystrzejews Kumar, V., Eds.; Springer: Berlin/Heidelberg, Germany, 2023.
36. Rouquerol, J.; Baron, G.; Denoyel, R.; Giesche, H.; Groen, J.; Klobes, P.; Levitz, P.; Neimark, A.V.; Rigby, S.; Skudas, R.; et al. L Intrusion and Alternative Methods for the Characterization of Macroporous Materials (IUPAC Technical Report). *Pure Chem.* **2012**, *84*, 107–136. [CrossRef]
37. Zhou, Q.; Guo, Y.; Zhu, Y. Photocatalytic Sacrificial H2 Evolution Dominated by Micropore-Confined Exciton Trans Hydrogen-Bonded Organic Frameworks. *Nat. Catal.* **2023**, *6*, 574–584. [CrossRef]
38. Zhang, Z.; Chen, Y.; Zhou, L.; Chen, C.; Han, Z.; Zhang, B.; Wu, Q.; Yang, L.; Du, L.; Bu, Y.; et al. The Simplest Construct Single-Site Catalysts by the Synergism of Micropore Trapping and Nitrogen Anchoring. *Nat. Commun.* **2019**, *10*, 1657. [Cro
39. Bennett, T.D.; Coudert, F.X.; James, S.L.; Cooper, A.I. The Changing State of Porous Materials. *Nat. Mater.* **2021**, *20*, 1179 [CrossRef] [PubMed]

Chen, K.; Mousavi, S.H.; Singh, R.; Snurr, R.Q.; Li, G.; Webley, P.A. Gating Effect for Gas Adsorption in Microporous Materials–Mechanisms and Applications. *Chem. Soc. Rev.* **2022**, *51*, 1139–1166. [CrossRef]
Yang, K.; Fan, Q.; Song, C.; Zhang, Y.; Sun, Y.; Jiang, W.; Fu, P. Enhanced Functional Properties of Porous Carbon Materials as High-Performance Electrode Materials for Supercapacitors. *Green Energy Resour.* **2023**, *1*, 100030. [CrossRef]
Liu, M.; Waugh, J.; Babu, S.K.; Spendelow, J.S.; Kang, Q. Numerical Modeling of Ion Transport and Adsorption in Porous Media: A Pore-Scale Study for Capacitive Deionization Desalination. *Desalination* **2022**, *526*, 115520. [CrossRef]
Gangadhar, J.; Tirumuruhan, B.; Sujith, R. Applications and Future Trends in Mesoporous Materials. In *Advanced Functional Porous Materials*; Engineering Materials; Uthaman, A., Thomas, S., Li, T., Maria, H., Eds.; Springer: Berlin/Heidelberg, Germany, 2022.
Peng, X.; Chen, L.; Li, Y. Ordered Macroporous MOF-Based Materials for Catalysis. *Mol. Catal.* **2022**, *529*, 112568. [CrossRef]
Huang, F.C.; Lee, C.K.; Han, Y.L.; Chao, W.C.; Chao, H.P. Preparation of Activated Carbon Using Micro-Nano Carbon Spheres through Chemical Activation. *J. Taiwan Inst. Chem. Eng.* **2014**, *45*, 2805–2812. [CrossRef]
Canales-Flores, R.A.; Prieto-García, F. Activation Methods of Carbonaceous Materials Obtained from Agricultural Waste. *Chem. Biodivers.* **2016**, *13*, 261–268. [CrossRef] [PubMed]
Suhas; Carrott, P.J.M.; Ribeiro Carrott, M.M.L. Lignin—From Natural Adsorbent to Activated Carbon: A Review. *Bioresour. Technol.* **2007**, *98*, 2301–2312. [CrossRef]
Sajjadi, B.; Chen, W.Y.; Egiebor, N.O. A Comprehensive Review on Physical Activation of Biochar for Energy and Environmental Applications. *Rev. Chem. Eng.* **2019**, *35*, 735–776. [CrossRef]
Zgrzebnicki, M.; Nair, V.; Mitra, S.; Kałamaga, A.; Przepiórski, J.; Wrobel, R.J. N-Doped Activated Carbon Derived from Furfuryl Alcohol–Development of Porosity, Properties, and Adsorption of Carbon Dioxide and Ethene. *Chem. Eng. J.* **2022**, *427*, 131709. [CrossRef]
Alazmi, A. Synergistic Effect of Hydrothermal and Physical Activation Approaches to Fabricate Activated Carbon for Energy Storage Applications. *Ceram. Int.* **2022**, *48*, 22131–22140. [CrossRef]
Serafin, J.; Dziejarski, B. Activated Carbons—Preparation, Characterization and Their Application in CO_2 Capture: A Review. *Environ. Sci. Pollut. Res.* **2024**, *31*, 40008–40062. [CrossRef]
Guo, Z.; Han, X.; Zhang, C.; He, S.; Liu, K.; Hu, J.; Yang, W.; Jian, S.; Jiang, S.; Duan, G. Activation of Biomass-Derived Porous Carbon for Supercapacitors: A Review. *Chin. Chem. Lett.* **2024**, *35*, 109007. [CrossRef]
Yuan, C.; Xu, H.; El-khodary, S.A.; Ni, G.; Esakkimuthu, S.; Zhong, S.; Wang, S. Recent Advances and Challenges in Biomass-Derived Carbon Materials for Supercapacitors: A Review. *Fuel* **2024**, *362*, 130795. [CrossRef]
Gao, Y.; Yue, Q.; Gao, B.; Li, A. Insight into Activated Carbon from Different Kinds of Chemical Activating Agents: A Review. *Sci. Total Environ.* **2020**, *746*, 141094. [CrossRef] [PubMed]
Shahcheragh, S.K.; Bagheri Mohagheghi, M.M.; Shirpay, A. Effect of Physical and Chemical Activation Methods on the Structure, Optical Absorbance, Band Gap and Urbach Energy of Porous Activated Carbon. *SN Appl. Sci.* **2023**, *5*, 313. [CrossRef]
Zhang, G.; Liu, X.; Wang, L.; Fu, H. Recent Advances of Biomass Derived Carbon-Based Materials for Efficient Electrochemical Energy Devices. *J. Mater. Chem.* **2022**, *10*, 9277–9307. [CrossRef]
Liu, X.; Antonietti, M. Molten Salt Activation for Synthesis of Porous Carbon Nanostructures and Carbon Sheets. *Carbon* **2014**, *69*, 460–466. [CrossRef]
Zhou, X.; Jia, Z.; Feng, A.; Wang, K.; Liu, X.; Chen, L.; Cao, H.; Wu, G. Dependency of Tunable Electromagnetic Wave Absorption Performance on Morphology-Controlled 3D Porous Carbon Fabricated by Biomass. *Compos. Commun.* **2020**, *21*, 100404. [CrossRef]
Yang, L.; Qiu, J.; Wang, Y.; Guo, S.; Feng, Y.; Dong, D.; Yao, J. Molten Salt Synthesis of Hierarchical Porous Carbon from Wood Sawdust for Supercapacitors. *J. Electroanal. Chem.* **2020**, *856*, 113673. [CrossRef]
Hu, H.; Yan, M.; Jiang, J.; Huang, A.; Cai, S.; Lan, L.; Ye, K.; Chen, D.; Tang, K.; Zuo, Q.; et al. A State-of-the-Art Review on Biomass-Derived Carbon Materials for Supercapacitor Applications: From Precursor Selection to Design Optimization. *Sci. Total Environ.* **2024**, *912*, 169141. [CrossRef]
He, H.; Zhang, R.; Zhang, P.; Wang, P.; Chen, N.; Qian, B.; Zhang, L.; Yu, J.; Dai, B. Functional Carbon from Nature: Biomass-Derived Carbon Materials and the Recent Progress of Their Applications. *Adv. Sci.* **2023**, *10*, 2205557. [CrossRef]
Heidarinejad, Z.; Dehghani, M.H.; Heidari, M.; Javedan, G.; Ali, I.; Sillanpää, M. Methods for Preparation and Activation of Activated Carbon: A Review. *Environ. Chem. Lett.* **2020**, *18*, 393–415. [CrossRef]
Kleszyk, P.; Ratajczak, P.; Skowron, P.; Jagiello, J.; Abbas, Q.; Frąckowiak, E.; Béguin, F. Carbons with Narrow Pore Size Distribution Prepared by Simultaneous Carbonization and Self-Activation of Tobacco Stems and Their Application to Supercapacitors. *Carbon* **2015**, *81*, 148–157. [CrossRef]
Fuertes, A.B.; Ferrero, G.A.; Sevilla, M. One-Pot Synthesis of Microporous Carbons Highly Enriched in Nitrogen and Their Electrochemical Performance. *J. Mater. Chem. A* **2014**, *2*, 14439–14448. [CrossRef]
Sun, K.; Leng, C.Y.; Jiang, J.C.; Bu, Q.; Lin, G.F.; Lu, X.C.; Zhu, G.Z. Microporous Activated Carbons from Coconut Shells Produced by Self-Activation Using the Pyrolysis Gases Produced from Them, That Have an Excellent Electric Double Layer Performance. *New Carbon Mater.* **2017**, *32*, 451–459. [CrossRef]
Niu, J.; Shao, R.; Liang, J.; Dou, M.; Li, Z.; Huang, Y.; Wang, F. Biomass-Derived Mesopore-Dominant Porous Carbons with Large Specific Surface Area and High Defect Density as High Performance Electrode Materials for Li-Ion Batteries and Supercapacitors. *Nano Energy* **2017**, *36*, 322–330. [CrossRef]

67. Wang, L.; Wang, T.; Hao, R.; Wang, Y. Synthesis and Applications of Biomass-Derived Porous Carbon Materials in Er Utilization and Environmental Remediation. *Chemosphere* **2023**, *339*, 139635. [CrossRef] [PubMed]
68. Deng, C.; Ding, F.; Li, X.; Guo, Y.; Ni, W.; Yan, H.; Sun, K.; Yan, Y.-M. Templated-Preparation of a Three-Dimensional Molybde Phosphide Sponge as a High Performance Electrode for Hydrogen Evolution. *J. Mater. Chem.* **2016**, *4*, 59–66. [CrossRef]
69. Wei, J.; Sun, Z.; Luo, W.; Li, Y.; Elzatahry, A.A.; Al-Enizi, A.M.; Deng, Y.; Zhao, D. New Insight into the Synthesis of Large Ordered Mesoporous Materials. *J. Am. Chem. Soc.* **2017**, *139*, 1706–1713. [CrossRef]
70. Xiao, Q.; Wang, K.; Wang, X.; Huang, S.; Cai, N.; Li, N. Solvent-Free Template Synthesis of SnO2/C Hybrid Hollow Sphere Superior Lithium-Sulfur Batteries. *Mater. Chem. Phys.* **2020**, *239*, 122070. [CrossRef]
71. Hu, L.; Zhu, Q.; Wu, Q.; Li, D.; An, Z.; Xu, B. Natural Biomass-Derived Hierarchical Porous Carbon Synthesized by an *ir* Hard Template Coupled with NaOH Activation for Ultrahigh Rate Supercapacitors. *ACS Sustain. Chem. Eng.* **2018**, *6*, 13949–1 [CrossRef]
72. Zheng, K.; Wu, Y.; Hu, Z.; Wang, S.; Jiao, X.; Zhu, J.; Sun, Y.; Xie, Y. Progress and Perspective for Conversion of Plastic Waste Valuable Chemicals. *Chem. Soc. Rev.* **2023**, *52*, 8–29. [CrossRef]
73. Asif, M.; Sidra Bibi, S.; Ahmed, S.; Irshad, M.; Shakir Hussain, M.; Zeb, H.; Kashif Khan, M.; Kim, J. Recent Advances in C Hydrogen Production, Storage and Commercial-Scale Use via Catalytic Ammonia Cracking. *Chem. Eng. J.* **2023**, *473*, 14 [CrossRef]
74. Wu, Y.; Hu, Q.; Chen, Q.; Jiao, X.; Xie, Y. Fundamentals and Challenges of Engineering Charge Polarized Active Sites for Photoreduction toward C_2 Products. *Acc. Chem. Res.* **2023**, *56*, 2500–2513. [CrossRef] [PubMed]
75. Zhang, W.; Cheng, R.; Bi, H.; Lu, Y.; Ma, L.; He, X. A Review of Porous Carbons Produced by Template Methods for Supercap Applications. *New Carbon Mater.* **2021**, *36*, 69–81. [CrossRef]
76. Chauhan, S. Synthesis of Ordered Mesoporous Carbon by Soft Template Method. *Mater. Today Proc.* **2021**, *81*, 842–847. [Cros
77. Pavlenko, V.; Khosravi, H.S.; Żółtowska, S.; Haruna, A.B.; Zahid, M.; Mansurov, Z.; Supiyeva, Z.; Galal, A.; Ozoemena Abbas, Q.; et al. A Comprehensive Review of Template-Assisted Porous Carbons: Modern Preparation Methods and Adva Applications. *Mater. Sci. Eng. R Rep.* **2022**, *149*, 100682. [CrossRef]
78. Lan, T.; Yalavarthi, R.; Shen, Y.; Gao, M.; Wang, F.; Hu, Q.; Hu, P.; Beladi-Mousavi, M.; Chen, X.; Hu, X.; et al. Polyoxometa Mediated Selectivity in Pt Single-Atoms on Ceria for Environmental Catalysis. *Angew. Chem. Int. Ed.* **2024**, e2024 [CrossRef]
79. Tang, Z.-E.; Lim, S.; Pang, Y.-L.; Ong, H.-C.; Lee, K.-T. Synthesis of Biomass as Heterogeneous Catalyst for Application in Bio Production: State of the Art and Fundamental Review. *Renew. Sustain. Energy Rev.* **2018**, *92*, 235–253. [CrossRef]
80. Lu, Z.; Chen, J.; Wang, W.; Li, W.; Sun, M.; Wang, Y.; Wang, X.; Ye, J.; Rao, H. Electrocatalytic, Kinetic, and Mechanism Ins into the Oxygen-Reduction Catalyzed Based on the Biomass-Derived FeO_x@N-Doped Porous Carbon Composites. *Small* **20** 2007326. [CrossRef] [PubMed]
81. Shrestha, A.; Acharya, B.; Farooque, A.A. Study of Hydrochar and Process Water from Hydrothermal Carbonization c Lettuce. *Renew. Energy* **2021**, *163*, 589–598. [CrossRef]
82. Romero-Anaya, A.J.; Ouzzine, M.; Lillo-Ródenas, M.A.; Linares-Solano, A. Spherical Carbons: Synthesis, Characterizatio Activation Processes. *Carbon* **2014**, *68*, 296–307. [CrossRef]
83. Yang, H.; Yan, R.; Chen, H.; Lee, D.H.; Zheng, C. Characteristics of Hemicellulose, Cellulose and Lignin Pyrolysis. *Fuel* **20** 1781–1788. [CrossRef]
84. Chen, T.; Ye, T.; Zhu, J.; Li, G.; Chen, Z.; Xie, M.; Zhou, Y.; Wang, J. Small-Sized Biomass-Derived Hydrothermal Carbon Enriched Oxygen Groups Quickens Benzene Hydroxylation to Phenol with Dioxygen. *Appl. Catal. A Gen.* **2021**, *626*, 1 [CrossRef]
85. Tripathi, V.S. Hydrothermal Method for Synthesis of Materials. In *Handbook on Synthesis Strategies for Advanced Materials*; I Institute of Metals Series; Tyagi, A.K., Ningthoujam, R.S., Eds.; Springer: Berlin/Heidelberg, Germany, 2021.
86. Nicolae, S.A.; Au, H.; Modugno, P.; Luo, H.; Szego, A.E.; Qiao, M.; Li, L.; Yin, W.; Heeres, H.J.; Berge, N.; et al. Recent Adv in Hydrothermal Carbonisation: From Tailored Carbon Materials and Biochemicals to Applications and Bioenergy. *Green* **2020**, *22*, 4747–4800. [CrossRef]
87. Ghavami, N.; Özdenkçi, K.; Salierno, G.; Björklund-Sänkiaho, M.; De Blasio, C. Analysis of Operational Issues in Hydroth Liquefaction and Supercritical Water Gasification Processes: A Review. *Biomass Convers. Biorefin.* **2023**, *13*, 12367–12394. [Cros
88. Bhat, V.S.; Supriya, S.; Hegde, G. Review—Biomass Derived Carbon Materials for Electrochemical Sensors. *J. Electrochem* **2020**, *167*, 037526. [CrossRef]
89. Gongxiang, S.; Yankui, W.; Dexin, H.; Hanjian, L.; Abdullahi, S.A.; Jun, X.; Long, J.; Yi, W.; Sheng, S.; Song, H.; et al. The He Rate and Final Temperature Impacts on the Coconut Shell Char Structure Characteristics during Photo-Thermal Pyrolysis. *J. Appl. Pyrolysis* **2022**, *167*, 105695. [CrossRef]
90. Zhang, H.; Zhu, Y.; Liu, Q.; Li, X. Preparation of Porous Carbon Materials from Biomass Pyrolysis Vapors for Hydrogen St *Appl. Energy* **2022**, *306*, 118131. [CrossRef]
91. Pariyar, P.; Kumari, K.; Jain, M.K.; Jadhao, P.S. Evaluation of Change in Biochar Properties Derived from Different Feedstoc Pyrolysis Temperature for Environmental and Agricultural Application. *Sci. Total Environ.* **2020**, *713*, 136433. [CrossRef]
92. Villalgordo-Hernández, D.; Grau-Atienza, A.; García-Marín, A.A.; Ramos-Fernández, E.V.; Narciso, J. Manufacture of C Materials with High Nitrogen Content. *Materials* **2022**, *15*, 2415. [CrossRef]

Cho, S.Y.; Yun, Y.S.; Lee, S.; Jang, D.; Park, K.Y.; Kim, J.K.; Kim, B.H.; Kang, K.; Kaplan, D.L.; Jin, H.J. Carbonization of a Stable β-Sheet-Rich Silk Protein into a Pseudographitic Pyroprotein. *Nat. Commun.* **2015**, *6*, 7145. [CrossRef]

Bouchelta, C.; Medjram, M.S.; Zoubida, M.; Chekkat, F.A.; Ramdane, N.; Bellat, J.P. Effects of Pyrolysis Conditions on the Porous Structure Development of Date Pits Activated Carbon. *J. Anal. Appl. Pyrolysis* **2012**, *94*, 215–222. [CrossRef]

Lua, A.C. A Detailed Study of Pyrolysis Conditions on the Production of Steam-Activated Carbon Derived from Oil-Palm Shell and Its Application in Phenol Adsorption. *Biomass Convers. Biorefin.* **2020**, *10*, 523–533. [CrossRef]

Zhang, S.; Jiang, S.F.; Huang, B.C.; Shen, X.C.; Chen, W.J.; Zhou, T.P.; Cheng, H.Y.; Cheng, B.H.; Wu, C.Z.; Li, W.W.; et al. Sustainable Production of Value-Added Carbon Nanomaterials from Biomass Pyrolysis. *Nat. Sustain.* **2020**, *3*, 753–760. [CrossRef]

Zhang, H.; Yang, K.; Tao, Y.; Yang, Q.; Xu, L.; Liu, C.; Ma, L.; Xiao, R. Biomass Directional Pyrolysis Based on Element Economy to Produce High-Quality Fuels, Chemicals, Carbon Materials–A Review. *Biotechnol. Adv.* **2023**, *69*, 108262. [CrossRef]

Chen, C.; Sun, Z.; Qin, G.; Wang, B.; Liu, M.; Liang, Q.; Li, X.; Pang, R.; Guo, Y.; Li, Y.; et al. Asymmetrically Coordinated Cu Dual-Atom-Sites Enables Selective CO_2 Electroreduction to Ethanol. *Adv. Mater.* **2024**, *2409797*, 1–11. [CrossRef]

Tseng, K.H.; Shiao, Y.F.; Chang, R.F.; Yeh, Y.T. Optimization of Microwave-Based Heating of Cellulosic Biomass Using Taguchi Method. *Materials* **2013**, *6*, 3404–3419. [CrossRef] [PubMed]

Jimenez, G.D.; Monti, T.; Titman, J.J.; Hernandez-Montoya, V.; Kingman, S.W.; Binner, E.R. New Insights into Microwave Pyrolysis of Biomass: Preparation of Carbon-Based Products from Pecan Nutshells and Their Application in Wastewater Treatment. *J. Anal. Appl. Pyrolysis* **2017**, *124*, 113–121. [CrossRef]

Su, Z.; Zhang, M.; Xu, P.; Zhao, Z.; Wang, Z.; Huang, H.; Ouyang, T. Opportunities and Strategies for Multigrade Waste Heat Utilization in Various Industries: A Recent Review. *Energy Convers. Manag.* **2021**, *229*, 113769. [CrossRef]

Mishra, R.R.; Sharma, A.K. Microwave–Material Interaction Phenomena: Heating Mechanisms, Challenges and Opportunities in Material Processing. *Compos. Part A Appl. Sci. Manuf.* **2016**, *81*, 78–97. [CrossRef]

Sha, J.; Li, Y.; Villegas Salvatierra, R.; Wang, T.; Dong, P.; Ji, Y.; Lee, S.K.; Zhang, C.; Zhang, J.; Smith, R.H.; et al. Three-Dimensional Printed Graphene Foams. *ACS Nano* **2017**, *11*, 6860–6867. [CrossRef]

Cheng, L.; Yeung, C.S.; Huang, L.; Ye, G.; Yan, J.; Li, W.; Yiu, C.; Chen, F.-R.; Shen, H.; Tang, B.Z.; et al. Flash Healing of Laser-Induced Graphene. *Nat. Commun.* **2024**, *15*, 2925. [CrossRef] [PubMed]

Zhang, Z.; Zhu, H.; Zhang, W.; Zhang, Z.; Lu, J.; Xu, K.; Liu, Y.; Saetang, V. A Review of Laser-Induced Graphene: From Experimental and Theoretical Fabrication Processes to Emerging Applications. *Carbon* **2023**, *214*, 118356. [CrossRef]

Sayed, M.S.; Aman, D.; Fayed, M.G.; Omran, M.M.; Zaki, T.; Mohamed, S.G. Unravelling the Role of Pore Structure of Biomass-Derived Porous Carbon in Charge Storage Mechanisms for Supercapacitors. *RSC Adv.* **2024**, *14*, 24631–24642. [CrossRef]

Lima, R.M.A.P.; Dos Reis, G.S.; Thyrel, M.; Alcaraz-Espinoza, J.J.; Larsson, S.H.; de Oliveira, H.P. Facile Synthesis of Sustainable Biomass-Derived Porous Biochars as Promising Electrode Materials for High-Performance Supercapacitor Applications. *Nanomaterials* **2022**, *12*, 866. [CrossRef]

Kang, S.; Kim, B.; Lee, S.; Baek, J.; Yoo, J. Tailoring Porosity of Starch-Derived Biocarbon for Enhanced Supercapacitor Performance. *Mater. Technol.* **2024**, *39*, 2338628. [CrossRef]

Szymczykiewicz, E.; Bordun, I.; Maksymych, V.; Klapchuk, M.; Kohut, Z.; Borysiuk, A.; Kulyk, Y.; Ivashchyshyn, F. Charge Storage and Magnetic Properties Nitrogen-Containing Nanoporous Bio-Carbon. *Energies* **2024**, *17*, 903. [CrossRef]

Lee, Y.C.; Jung, S.C. A First-Principles Study on Atomic-Scale Pore Design of Microporous Carbon Electrodes for Lithium-Ion Batteries. *Nanoscale Adv.* **2022**, *4*, 5378–5391. [CrossRef] [PubMed]

Abouelamaiem, D.I.; He, G.; Parkin, I.; Neville, T.P.; Jorge, A.B.; Ji, S.; Wang, R.; Titirici, M.M.; Shearing, P.R.; Brett, D.J.L. Synergistic Relationship between the Three-Dimensional Nanostructure and Electrochemical Performance in Biocarbon Supercapacitor Electrode Materials. *Sustain. Energy Fuels* **2018**, *2*, 772–785. [CrossRef]

Guo, F.; Li, X.; Liu, Y.; Peng, K.; Guo, C.; Rao, Z. Catalytic Cracking of Biomass Pyrolysis Tar over Char-Supported Catalysts. *Energy Convers. Manag.* **2018**, *167*, 81–90. [CrossRef]

Deng, L.; Zhao, Y.; Sun, S.; Feng, D.; Zhang, W. Thermochemical Method for Controlling Pore Structure to Enhance Hydrogen Storage Capacity of Biochar. *Int. J. Hydrogen Energy* **2023**, *48*, 21799–21813. [CrossRef]

Yang, H.; Cui, Y.; Jin, Y.; Lu, X.; Han, T.; Sandström, L.; Jönsson, P.G.; Yang, W. Evaluation of Engineered Biochar-Based Catalysts for Syngas Production in a Biomass Pyrolysis and Catalytic Reforming Process. *Energy Fuels* **2023**, *37*, 5942–5952. [CrossRef]

Wu, S.; Cai, X.; Liao, Z.; He, W.; Shen, J.; Yuan, Y.; Ning, X. Redox Properties of Nano-Sized Biochar Derived from Wheat Straw Biochar. *RSC Adv.* **2022**, *12*, 11039–11046. [CrossRef] [PubMed]

Fan, M.; Li, C.; Shao, Y.; Sun, K.; Zhang, L.; Zhang, S.; Ding, K.; Gholizadeh, M.; Hu, X. Impact of Biochar Catalyst on Pyrolysis of Biomass of the Same Origin. *J. Environ. Chem. Eng.* **2022**, *10*, 108546. [CrossRef]

Jiao, X.; Zheng, K.; Hu, Z.; Zhu, S.; Sun, Y.; Xie, Y. Conversion of Waste Plastics into Value-Added Carbonaceous Fuels under Mild Conditions. *Adv. Mater.* **2021**, *33*, 2005192. [CrossRef] [PubMed]

Zhengyang, E.; Liang, J.; Li, P.; Qiang, S.; Fan, Q. A Review on Photocatalytic Attribution and Process of Pyrolytic Biochar in Environment. *Water Res.* **2024**, *251*, 120994. [CrossRef]

Mian, M.M.; Liu, G. Recent Progress in Biochar-Supported Photocatalysts: Synthesis, Role of Biochar, and Applications. *RSC Adv.* **2018**, *8*, 14237–14248. [CrossRef]

Colmenares, J.C.; Lisowski, P.; Łomot, D. A Novel Biomass-Based Support (Starbon) for TiO_2 Hybrid Photocatalysts: A Versatile Green Tool for Water Purification. *RSC Adv.* **2013**, *3*, 20186–20192. [CrossRef]

121. Zhang, C.; Li, H.; Yang, X.; Tan, X.; Wan, C.; Liu, X. Characterization of Electrodes Modified with Sludge-Derived Biochar a Performance of Electrocatalytic Oxidation of Azo Dyes. *J. Environ. Manag.* **2022**, *324*, 116445. [CrossRef]
122. Saeed, M.; Mohammad, A.; Singh, P.; Lal, B.; Suliman, M.; Alshahrani, M.Y.; Sharma, M. Coconut Waste Valorization to Pr Biochar Catalyst and Its Application in Cellulose-Degrading Enzymes Production via SSF. *Int. J. Biol. Macromol.* **2023**, *240*, 1. [CrossRef]
123. Xiang, Y.; Zhang, H.; Zhang, J.; Xu, Q.; Liu, X.; Yin, D. Sustainable Synthesis of A Novel Zirconium-Coordinated Biochar Ca from Sawdusts for Conversion of Ethyl Levulinate to γ-Valerolactone. *Catal. Lett.* **2024**, *154*, 1931–1944. [CrossRef]
124. Lourenço, M.A.O.; Frade, T.; Bordonhos, M.; Castellino, M.; Pinto, M.L.; Bocchini, S. N-Doped Sponge-like Biochar: A Pron CO_2 Sorbent for CO_2/CH_4 and CO_2/N_2 Gas Separation. *Chem. Eng. J.* **2023**, *470*, 144005. [CrossRef]
125. Simões Dos Reis, G.; Mayandi Subramaniyam, C.; Cárdenas, A.D.; Larsson, S.H.; Thyrel, M.; Lassi, U.; García-Alvarado, F. Synthesis of Sustainable Activated Biochars with Different Pore Structures as Efficient Additive-Carbon-Free Anodes for Li and Sodium-Ion Batteries. *ACS Omega* **2022**, *7*, 42570–42581. [CrossRef] [PubMed]
126. Zhang, Z.; Jiang, S.; Chen, H.; Qi, H.; Chen, Y.; Chen, Y.; Deng, Q.; Wang, S. Efficient Solar-Driven Water Purification Bas Biochar with Multi-Level Pore Bundle Structure for Preparation of Drinking Water. *Foods* **2021**, *10*, 3087. [CrossRef] [Pub
127. Chen, L.; Yang, S.; Zuo, X.; Huang, Y.; Cai, T.; Ding, D. Biochar Modification Significantly Promotes the Activity of Co_3O_4 to Heterogeneous Activation of Peroxymonosulfate. *Chem. Eng. J.* **2018**, *354*, 856–865. [CrossRef]
128. Jin, Z.; Xiao, S.; Dong, H.; Xiao, J.; Tian, R.; Chen, J.; Li, Y.; Li, L. Adsorption and Catalytic Degradation of Organic Contami by Biochar: Overlooked Role of Biochar's Particle Size. *J Hazard. Mater.* **2022**, *422*, 126928. [CrossRef] [PubMed]
129. Díaz-Maroto, C.G.; Mašek, O.; Pizarro, P.; Serrano, D.P.; Moreno, I.; Fermoso, J. Removal of NO at Low Concentrations Polluted Air in Semi-Closed Environments by Activated Biochars from Renewables Feedstocks. *J. Environ. Manag.* **202** 118031. [CrossRef]
130. Cao, Q.; Wang, C.; Tang, D.; Zhang, X.; Wu, P.; Zhang, Y.; Liu, H.; Zheng, Z. Enhanced Elemental Mercury Removal in Fired Flue Gas by Modified Algal Waste-Derived Biochar: Performance and Mechanism. *J. Environ. Manag.* **2023**, *325*, 1 [CrossRef]
131. Wang, B.; Lan, J.; Bo, C.; Gong, B.; Ou, J. Adsorption of Heavy Metal onto Biomass-Derived Activated Carbon: Review. *RS* **2023**, *13*, 4275–4302. [CrossRef]
132. He, L.; Qi, X.; Wei, W.; Zhang, X.; Wang, J.; Gao, Z. Biomass-Activated Carbon-Based Superhydrophobic Sponge with Pho mal Properties for Adsorptive Separation of Waste Oil. *J. Hazard. Mater.* **2024**, *477*, 135222. [CrossRef]
133. He, D.; Xu, J.; Yang, Y.; Zhu, H.; Yu, M.; Li, S.; Xu, S.; Zhou, J.; Wang, X. Preparation of Biomass-Based Gas Separation Meml from Biochar Residue Obtained by Depolymerization of Lignin with ZSM-5 to Promote a Circular Bioeconomy. *Int. Macromol.* **2022**, *214*, 45–53. [CrossRef]
134. Huang, F.; Zhang, S.M.; Wu, R.R.; Zhang, L.; Wang, P.; Xiao, R.B. Magnetic Biochars Have Lower Adsorption but H Separation Effectiveness for Cd^{2+} from Aqueous Solution Compared to Nonmagnetic Biochars. *Environ. Pollut.* **2021**, *275*, 1 [CrossRef] [PubMed]
135. Pipíška, M.; Richveisová, B.M.; Frišták, V.; Horník, M.; Remenárová, L.; Stiller, R.; Soja, G. Sorption Separation of Coba Cadmium by Straw-Derived Biochar: A Radiometric Study. *J. Radioanal. Nucl. Chem.* **2017**, *311*, 85–97. [CrossRef]
136. Nicolaou, E.; Philippou, K.; Anastopoulos, I.; Pashalidis, I. Copper Adsorption by Magnetized Pine-Needle Biochar. *Pr* **2019**, *7*, 903. [CrossRef]
137. Kushwaha, R.; Singh, R.S.; Mohan, D. Comparative Study for Sorption of Arsenic on Peanut Shell Biochar and Modified P Shell Biochar. *Bioresour. Technol.* **2023**, *375*, 128831. [CrossRef]
138. Zhu, Q.; Wu, J.; Wang, L.; Yang, G.; Zhang, X. Adsorption Characteristics of Pb^{2+} onto Wine Lees-Derived Biochar. *Bull. E Contam. Toxicol.* **2016**, *97*, 294–299. [CrossRef]
139. Cai, L.; Zhang, Y.; Zhou, Y.; Zhang, X.; Ji, L.; Song, W.; Zhang, H.; Liu, J. Effective Adsorption of Diesel Oil by Crab-Shell-D Biochar Nanomaterials. *Materials* **2019**, *12*, 236. [CrossRef]
140. Wang, S.; Li, X.; Zhu, Y. Comparison of the Adsorption Capacity and Mechanisms of Mixed Heavy Metals in Wastewa Sheep Manure Biochar and Robinia Pseudoacacia Biochar. *Water Sci. Technol.* **2023**, *87*, 3083–3094. [CrossRef]

MDPI AG
Grosspeteranlage 5
4052 Basel
Switzerland
Tel.: +41 61 683 77 34

Molecules Editorial Office
E-mail: molecules@mdpi.com
www.mdpi.com/journal/molecules

www.ingramcontent.com/pod-product-compliance
Lightning Source LLC
Chambersburg PA
CBHW041406210926
51850CB00013B/3189
9783725864485